W0050936

Shallow Subduction Zones: Seismicity, Mechanics and Seismic Potential Part II

Edited by
Renata Dmowska
Göran Ekström

1994

Birkhäuser Verlag
Basel · Boston · Berlin

Reprint from Pure and Applied Geophysics
(PAGEOPH), Volume 142 (1994), No. 1

The Editors:

Renata Dmowska
Division of Applied Sciences
Harvard University
Cambridge, MA 02138
USA

Göran Ekström
Department of Earth and Planetary Sciences
Harvard University
Cambridge, MA 02138
USA

A CIP catalogue record for this book is available from the Library of Congress,
Washington D.C., USA

Deutsche Bibliothek Cataloging-in-Publication Data

Shallow subduction zones: seismicity. mechanics and seismic
potential / ed. by Renata Dmowska; Göran Ekström. –
Reprint. – Basel ; Boston ; Berlin : Birkhäuser.
NE: Dmowska, Renata [Hrsg.]
Reprint
Pt. 2 (1994)
  Aus: Pure and applied geophysics ; Vol. 142
  ISBN-13: 978-3-7643-2963-1   e-ISBN-13: 978-3-0348-7333-8
  DOI: 10.1007/978-3-0348-7333-8

© 1994 Birkhäuser Verlag, P.O. Box 133, CH-4010 Basel, Switzerland
Printed on acid-free paper produced from chlorine-free pulp

ISBN-13: 978-3-7643-2963-1

9 8 7 6 5 4 3 2 1

Contents

PAGEOPH, Vol. 142, No. 1 (1994)

0033–4553/94/010001–02$1.50 + 0.20/0

Introduction

RENATA DMOWSKA and GÖRAN EKSTRÖM

This is a companion volume to PAGEOPH's vol. 140, N. 2, 1993, and, as its predecessor, it reports recent research on seismicity, mechanics and seismic potential in shallow subduction zones around the world.

The volume opens with four papers concerning Alaska-Aleutians subduction segment, three of them presenting new inversions for two great earthquakes that ruptured that region: the 1957 Aleutian earthquake and 1964 Prince William Sound event.

All available waveform data (body, surface and tsunami waves) are included in the analysis of seismic moment, rupture area and slip distribution of the 9 March 1957 Aleutian earthquake (JOHNSON *et al.*), which is also compared to the 1986 Andreanof Islands that reruptured a segment of the 1957 rupture area.

A new seismic inversion based on *P*-wave data of the March 28, 1964 Prince William Sound earthquake follows (CHRISTENSEN and BECK), the results showing that, except the well-known, very large dominant asperity in the epicentral region, there was a second major, but smaller, asperity in the Kodiak Island area. The historical earthquake data for the region are discussed as well, and a comparison is made with the rupture process and asperity distributions of the 1957 Aleutian, 1965 Rat Islands and 1986 Andreanof Islands earthquakes.

A detailed geodetic inversion of the 1964 Prince William Sound earthquake is presented next (HOLDAHL and SAUBER). Previous seismologic, geologic and geodetic studies of the region were used to constrain the geometry of the fault surface. The inversion provides the most detailed description of the fault geometry and slip to date, consisting of a mosaic of 68 fault planes. The results suggest again a variable slip distribution with a few local maxima, one of them around Kodiak Island (but not exactly in the same place as asperity found by CHRISTENSEN and BECK).

Seismicity trends and potential for large earthquakes in the Alaska-Aleutian region are discussed next (BUFE *et al.*). Analysis of historic earthquake recurrence data and time-to-failure analysis applied to recent decades of instrumental data are used to argue for the high likelihood of a gap-filling thrust earthquake in the Alaska-Aleutian subduction zone within this decade.

The rupture process of large earthquakes in the northern Mexico subduction zone is analyzed next (RUFF and MILLER). The results include, among others, the distribution of spatial concentrations of slip (asperities) for some of the earthquakes. The overall asperity distribution for the northern Mexico segment is found to consist of one clear asperity, in the epicentral region of the 1973 Colima earthquake, and a scattering of diffuse and overlapping regions of high moment release for the remainder of the segment. This character of asperity distribution is then compared to other subduction zones, and implications for future earthquake sizes and sequences are discussed. An important contribution of this work is also the detailed description of a new method of the moment tensor rate function (MTRF) inversion, presented in the Appendix. Although the advantage of the new method is not fully utilized here because of the lack of the mechanism change during the rupture process, the method is a very useful contribution for future studies of earthquake source process.

In the following paper, deviations of slip vector azimuths of interplate thrust earthquakes from expected plate convergence directions for major world trenches are surveyed and interpreted in terms of forearc rheology (MCCAFFREY). This global study shows the variability of forearc deformation, ranging from elastic to viscous. In general, continental forearcs deform less than the oceanic ones. Attempted correlations of the apparent forearc rheology with backarc spreading, convergence rate, slab dip, arc curvature, and downdip length of the thrust contact are poor, however it is found that great subduction zone earthquakes originate in subduction segments which are more elastic (that is deviations of slip in interplate events from directions of plate convergence are small) than others.

The last paper in the present volume (BARRIENTOS) investigates the possible causal relationship between large subduction earthquakes and volcanic eruptions, exemplified by the eruption of the Puyehue-Cordón Caulle volcanic system forty-eight hours after the occurrence of the May 22, 1960 ($M_w = 9.5$) Chile earthquake. It is postulated here that the shallow extensional deformation associated with the occurrence of a large subduction earthquake alters the stress field, allowing the magma to migrate from the deep chambers toward the surface, if the particular volcano is in a mature stage of its eruptive cycle.

PAGEOPH, Vol. 142, No. 1 (1994)

0033–4553/94/010003–26$1.50 + 0.20/0

The 1957 Great Aleutian Earthquake

JEAN M. JOHNSON,[1] YUICHIRO TANIOKA,[1] LARRY J. RUFF,[1] KENJI SATAKE,[1]
HIROO KANAMORI[2] and LYNN R. SYKES[3]

Abstract—The 9 March 1957 Aleutian earthquake has been estimated as the third largest earthquake this century and has the longest aftershock zone of any earthquake ever recorded—1200 km. However, due to a lack of high-quality seismic data, the actual source parameters for this earthquake have been poorly determined. We have examined all the available waveform data to determine the seismic moment, rupture area, and slip distribution. These data include body, surface and tsunami waves. Using body waves, we have estimated the duration of significant moment release as 4 min. From surface wave analysis, we have determined that significant moment release occurred only in the western half of the aftershock zone and that the best estimate for the seismic moment is 50–100×10^{20} Nm. Using the tsunami waveforms, we estimated the source area of the 1957 tsunami by backward propagation. The tsunami source area is smaller than the aftershock zone and is about 850 km long. This does not include the Unalaska Island area in the eastern end of the aftershock zone, making this area a possible seismic gap and a possible site of a future large or great earthquake. We also inverted the tsunami waveforms for the slip distribution. Slip on the 1957 rupture zone was highest in the western half near the epicenter. Little slip occurred in the eastern half. The moment is estimated as 88×10^{20} Nm, or $M_w = 8.6$, making it the seventh largest earthquake during the period 1900 to 1993. We also compare the 1957 earthquake to the 1986 Andreanof Islands earthquake, which occurred within a segment of the 1957 rupture area. The 1986 earthquake represents a rerupturing of the major 1957 asperity.

Key words: Subduction zones, Aleutian Arc, tsunamis, earthquake parameters.

1. Introduction

The Alaska-Aleutian Arc has a history of repeatedly rupturing in great earthquakes. The most recent sequence, beginning in 1938, has ruptured almost the entire arc from southern Alaska to the western Aleutians (Figure 1). Three earthquakes in particular, the 1957 Aleutian, 1964 Alaskan, and 1965 Rat Islands earthquakes are among the largest earthquakes to occur in the 20th century. However, some segments of the arc have apparently not ruptured during this

[1] Department of Geological Sciences, University of Michigan, Ann Arbor, Michigan, U S.A.
[2] Seismological Laboratory, California Institute of Technology, Pasadena, California, U S.A.
[3] Lamont-Doherty Earth Observatory, Department of Geological Sciences, Columbia University, Palisades, New York, U.S.A.

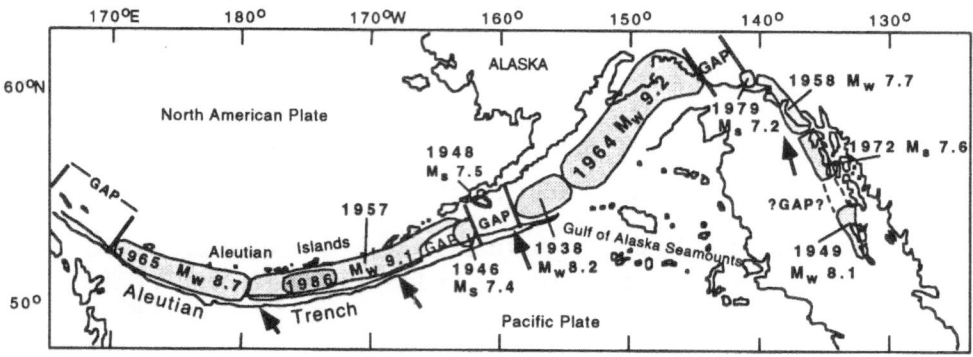

Figure 1
Locations of aftershock zones of major earthquakes and previously identified seismic gaps in Alaska and the Aleutians. Arrows indicate direction of relative convergence. Modified from SYKES *et al.* (1981).

sequence, and these areas are called seismic gaps. These gaps are delineated by the ends of the rupture zones of the adjacent great earthquakes; therefore, it is important to know the bounds of rupture of the great events. Of the three large events, both the 1964 and 1965 earthquakes occurred during the modern instrumental age of seismology, so they have been previously studied and their source parameters are known. However, the 9 March 1957 Aleutian earthquake (epicenter 51.63°N, 175.41°W, at 14:22 GMT, $M_s = 8.1$) has been least understood because it occurred before the introduction of the WWSSN stations, and little seismic data are available. Elementary source parameters such as source area, seismic moment and slip distribution have been inadequately determined.

The source area of an earthquake is often identified as the region containing the aftershocks. The aftershock zone of the 1957 earthquake is the longest of any earthquake ever recorded. It stretches 1200 km along the Aleutian Trench from approximately 163.5°W to 180°W, extending 360 km west and 850 km east of the epicenter. Both SYKES (1971) and KANAMORI (1977) used the 1200 km long aftershock zone of the 1957 earthquake in estimating the seismic moment. But these estimates are more than an order of magnitude different. Sykes' estimate of 30×10^{20} Nm is derived from the length of the aftershock zone and an average slip on the fault of 0.45 m. Kanamori estimated the moment as 585×10^{20} Nm, based on the relationship between source area and moment release, making the 1957 earthquake the third largest this century. However, HOUSE *et al.* (1981) argued that the easternmost end of the aftershock zone near Unalaska Island is anomalous and suggested that this area did not rupture in the 1957 earthquake.

An earthquake occurred on 7 May 1986 in the Andreanof Islands region ($M_s = 7.7$) within the rupture zone of the 1957 earthquake. This earthquake is important for several reasons. First, it was the largest earthquake to occur in this area of the Aleutians since the 1957 earthquake. Second, the epicenter of the 1986 earthquake (51.41°N, 174.83°W) is very close to the epicenter of the 1957 earthquake. Third, this area had previously been thought to have "the lowest seismic potential for the next few decades" (NISHENKO and MCCANN, 1981) due to the occurrence of the 1957 event less than 30 years earlier. In this sense, the 1986 earthquake is a failure of the seismic gap hypothesis, and it is important to understand why. The 1986 earthquake could be related to the 1957 earthquake, filling in areas of low moment release of the 1957 event. Alternatively, the 1986 earthquake could represent a major rerupturing of the arc, marking the end of a complete seismic cycle and the beginning of a new interseismic period. If we are to resolve this question and correctly assess the seismic potential for the central Aleutian Arc, it is even more important to determine the moment release distribution of the 1957 earthquake.

In this paper, we will examine the seismic data, both body waves and surface waves, to determine what information can be derived from the limited data available. We will also examine the tsunami waveform data and how they can be used to give reliable estimates of the earthquake source parameters. This will allow us to determine if the Unalaska Island area is a seismic gap.

2. Seismic Wave Studies

We will first examine the body wave data, which will be used to view the rupture process. If the 1200 km long aftershock area represents the rupture area, typical rupture velocities predict that the source time function could be up to eight minutes long. Then we will examine the surface wave data. For such great earthquakes, long period surface waves provide better estimates of overall seismic moment because of the long wavelength.

2.1. Body Waves

The great difficulty in determining the source parameters for the 1957 earthquake is the dearth of high-quality seismic data. Fortunately, several IGY seismograms are available, but only four long period P waves could be digitized from these. Of these four, Perth, Australia (PER) is at an epicentral distance of 103°; thus, the diffracted P wave provides a four-minute window for the rupture process before the PP arrival. Figure 2 shows the seismograms from PER for the main shock and one aftershock (occurred 9 March 1957, origin time 20:39 G.M.T., $M_s = 7.1$). The duration of significant moment release appears to last only about 4 min, as the P-wave coda has decayed significantly at the time of the PP arrival. It is certainly possible that the P wave continues after 4 min and is masked by the PP arrival, but there is no indication of any arrival within the PP coda that is larger

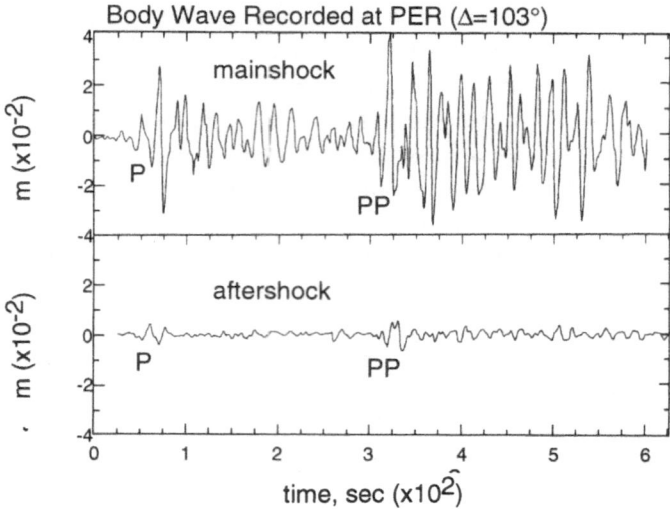

Figure 2
Seismograms of 1957 Aleutian earthquake (upper) and the large aftershock (lower) recorded at Perth, Australia (PER) showing *P* and *PP* arrival.

than the largest pulse that occurs about 1 min after the start of the record. Thus, there is no evidence to indicate significant moment release for longer than 4 min, though there could be minor moment release for up to 8 min. In a qualitative sense, the character of the *P* wave does not suggest an earthquake of a magnitude as large as $M_w = 9.1$, as compared to the body waves of the 1964 Alaska earthquake, $M_w = 9.2$, in which the amplitude continues to grow for more than a minute after the first arrival (RUFF and KANAMORI, 1983).

If we assume that the duration of significant moment release is 4 min, we can approximate the along-strike length of the main rupture. Using rupture velocities of 1.5 km/s and 3.0 km/s as lower and upper limits of rupture velocity, the rupture length from the epicenter is 360 km and 720 km, respectively. Either value is great enough to reach the western end of the aftershock zone. To the east, neither the lower nor the upper rupture length reaches to the eastern end of the aftershock zone. Without knowing anything more about the rupture velocity, we can say that a 4 min source duration is incompatible with a uniform rupture process across the entire aftershock zone.

We deconvolved each body wave record to estimate the single station source time function (RUFF and KANAMORI, 1983), each of which is shown in Figure 3. The PER source time function shows the 4 min duration. For the other three of the four source time functions, the *PP* phase arrives before 4 min. The magnification of the PER instrument is poorly known, but if we assume that the nominal instrument description for PER is correct (magnification of 1000), then the moment determined

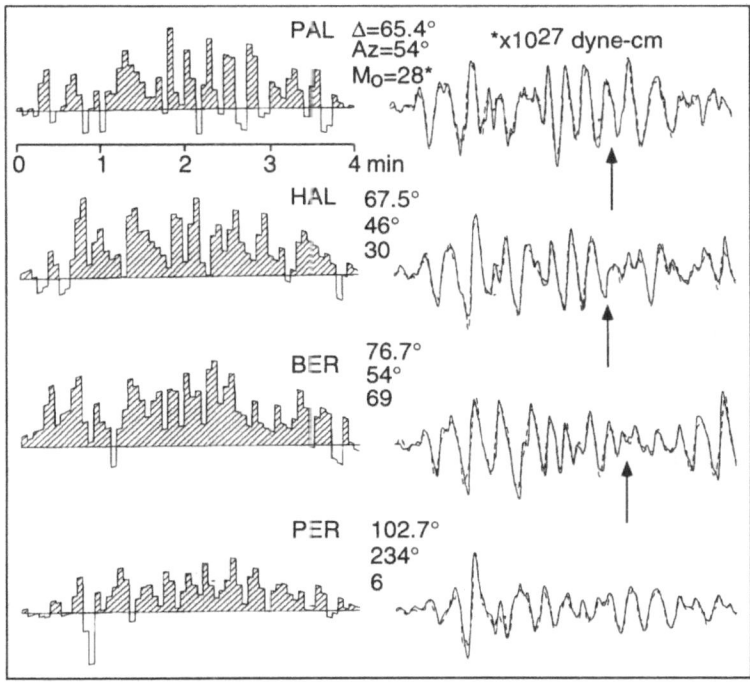

Figure 3
All available *P* wave records (right) and the deconvolved source time functions (left) for the 1957 earthquake. Arrow heads indicate *PP* arrival.

from the PER source time function is only 6×10^{20} Nm. A better moment estimate can be obtained by use of the aftershock *P* wave as an empirical Green's function. This gives the moment ratio of the main shock to the aftershock. The moment of the aftershock is estimated as 1×10^{20} Nm, based on its magnitude. This yields a seismic moment between 20 and 80×10^{20} Nm for the main shock. As IGY instruments have poor long-period response compared to the long source duration (see RUFF and KANAMORI, 1983), the above body wave moment estimates provide a lower bound on seismic moment; thus, seismic moment can be arbitrarily large if the baseline is adjusted to include the large negative pulse at approximately 1 minute. Once again, in a qualitative sense, the 1957 source time function, with a number of small subevents in the first minute, is very different in character to a truly great event such as the 1964 Great Alaskan earthquake, whose source time function is ramping during the first minute (RUFF and KANAMORI, 1983). Interpreting this within the context of the asperity model, we can argue that the plate interface in the central Aleutians is fundamentally different in character than in southern Alaska, the former being made up of many small asperities, the latter having one large asperity.

We can compare the source time function of the 1957 earthquake to the source time function of the 1986 earthquake (Figure 4). The two source time functions are similar during the first minute of rupture. Both build slowly via a number of pulses to the largest pulse of moment release, which can be clearly seen as the largest amplitude arrival in the seismograms. This again suggests that the central Aleutian plate interface is made up of a number of small asperities. Also, the 1957 and 1986 earthquakes are more complex than most subduction zone earthquakes in that each ruptured bilaterally from the epicenter (HWANG and KANAMORI, 1986; BOYD and NÁBÊLEK, 1988; HOUSTON and ENGDAHL, 1989; DAS and KOSTROV, 1990; YOSHIDA, 1992).

2.2. Surface Waves

Several surface waveforms are also available; however, only one is at a nonnodal azimuth—Pietermaritzburg, South Africa (PTM). This record contains on-scale R3 and R4 waveforms which could be used for surface wave analysis. Unfortunately, the PTM instrument response is poorly known. The instrument galvanometer was repaired by the station operator, and the recalibrated instrument response was different from the nominal characteristics recorded on the seismogram. Despite this problem with the PTM record, it can be used to determine an estimate of the 1957 earthquake source parameters.

A. Previous Studies

LANE and BOYD (1990) used a nonlinear inversion (simulated annealing) to study the PTM record. This approach does not require that the instrument response be known, but they could estimate only the rupture length and velocity of the

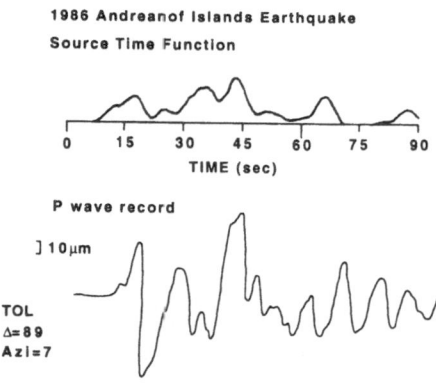

Figure 4
(Upper) Source time function of the 1986 Andreanof Islands earthquake (from HWANG and KANAMORI, 1986). (Lower) Broadband *P*-wave displacement waveform for the 1986 Andreanof Islands earthquake recorded at Toledo, Spain (TOL).

earthquake. They found that rupture extended approximately 600 km to the east and 150 km to the west of the epicenter, with a rupture velocity of 2.9 km/s.

Preliminary work to determine the slip distribution was done by RUFF et al. (1985). They determined the magnification of the PTM instrument by an empirical calibration at periods around 200 s, using the estimated moment of the 1958 Kurile Islands earthquake. In detail, they matched the amplitude of the long period R3 pulse at PTM with a synthetic seismogram calculated for the model of FUKAO and FURUMOTO (1979). The found that, with seismometer and galvanometer periods of 15 and 59 s and damping parameters of 3 and 1 respectively, the apparent magnification of the PTM instrument was 114. Ruff et al. then matched the observed seismogram by inverting for nine point sources, each spatially separated by approximately 110 km, as shown in Figure 5a. The results of a linear inversion for the moment distribution with rupture velocity fixed at 1.6 km/s (Figure 5b) show that the greatest moment release occurred in the western half of the aftershock zone, with no moment release in the eastern half. The seismic moment estimate from these results is 100×10^{20} Nm.

The preliminary results of Ruff et al. are adequate to explain the long period aspects of the observed surface wave record; however, we wish to improve upon these results in several ways. First, Ruff et al. used only fundamental modes in creating the synthetic waveform. We wish to match the complexity of the observed waveform by including all modes down to a shorter period. Second, the previous results were for a homogeneous earth structure. Work of ZHANG and TANIMOTO (1993) has shown that there are significant lateral heterogeneities that can be corrected for in the surface wave velocities (Figure 6). These corrections can produce significant time shifts in our synthetic waveforms. Third, we wish to find the optimum rupture velocity. Ruff et al. fixed the rupture velocity at 1.6 km/s. We also attempted to determine an empirical Green's function for the 1957 earthquake by using the 1986 Andreanof Islands earthquake as a source; however, to our dismay, we found that no usable records are available from anywhere in Africa for the 1986 event, unfortunately closing that avenue of investigation.

B. Surface Wave Inversion

We construct Green's functions for the R3−R4 arrivals at PTM by the following method. We set up eleven subevents, each separated by approximately 100 km, situated along the 1957 aftershock zone. The location and focal mechanism of each subevent is specified (see Figure 7). Details can be found below in the section on tsunami waveform inversion. The start time of each subevent is delayed by the prescribed rupture velocity. Synthetic waveforms are generated for each subevent by adding both fundamental and higher modes down to periods of 45 s, using the earth model 1066A (GILBERT and DZIEWONSKI, 1975). The observed and synthetic waveforms are filtered using a low-pass Butterworth filter such that the amplitudes are 50% at a period of 150 s and 1% at a period of 45 s, with respect to the values

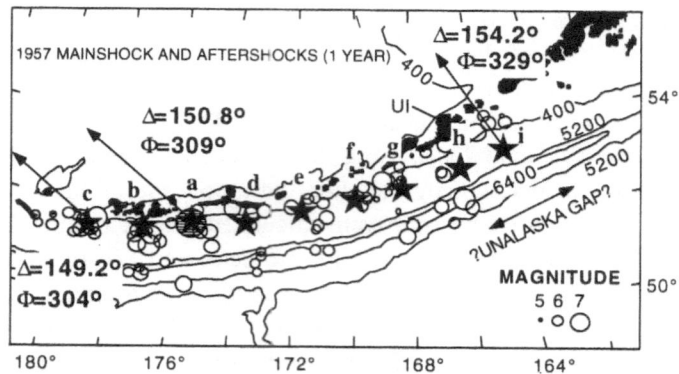

(a)

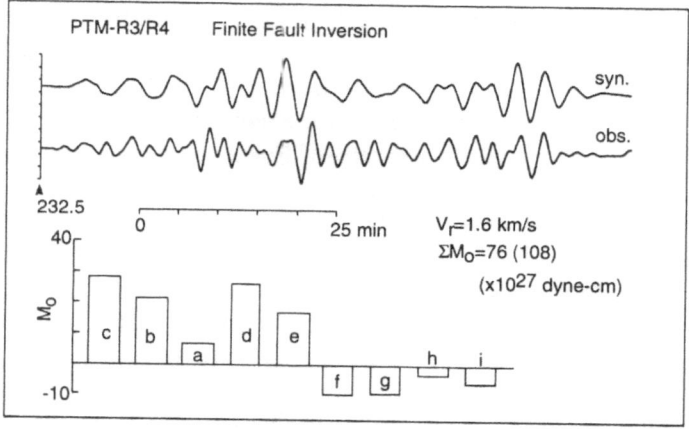

(b)

Figure 5

(a) Locations of point sources used by RUFF *et al.* (1985) overlaid on the aftershock distribution from HOUSE *et al.* (1981). Each point source has dip 20° and slip direction N40°W. The strike changes from 245° in the east to 270° in the west. (b) Moment distribution for the 1957 Aleutian earthquake estimated by Ruff *et al.* from the PTM record. The labeled grid points in the lower figure correspond to those in the upper map. The arrow head gives the start time of the seismograms from the origin time of the earthquake in minutes. Total moment estimate excluding negative values given in parentheses.

at zero frequency. The observed waveform is a linear superposition of the Green's functions, so the moment of each subevent can be determined by solving the linear equation

$$A_{ij} \cdot x_j = b_i \tag{1}$$

where A_{ij} is the computed Green's function at station PTM for unit moment of subevent j, b_i is the observation at PTM, and x_j is the unknown moment of

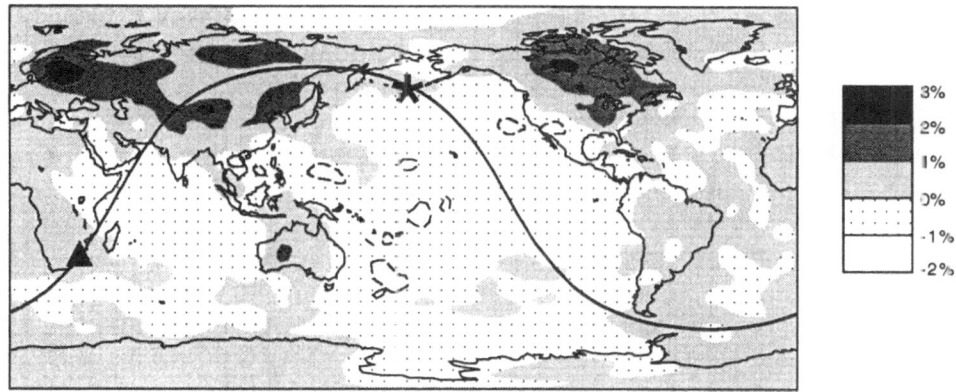

Figure 6
The great circle path connecting the 1957 earthquake (asterisk) and the South African station PTM (triangle), overlaid on the 150 s Rayleigh wave velocity distribution (ZHANG and TANIMOTO, 1993) in percent difference from homogeneous earth model.

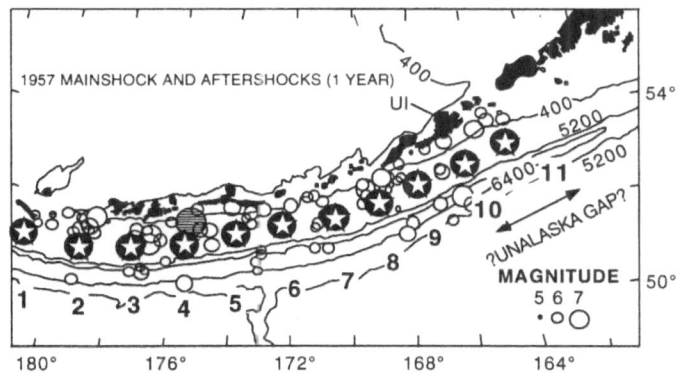

Figure 7
Location of point sources used in this study for inversion of surface wave at PTM overlaid on the aftershock distribution from HOUSE et al. (1981).

subevent j. This equation can be solved by a least-squares method minimizing the misfit between the observed and synthetic waveforms.

We performed 16 inversions for the spatial distribution of moment release using rupture velocities between 1.0 and 4.0 km/s at 0.2 km/s intervals. We also examined the effects of lateral heterogeneities, comparing both the results using the homogeneous earth model 1066A (GILBERT and DZIEWONSKI, 1975) and the laterally heterogeneous earth model of ZHANG and TANIMOTO (1993).

To create Green's functions for a laterally heterogeneous earth, we applied time corrections due to the velocity variations along the great circle path to the fundamental modes at each frequency (WOODHOUSE and DZIEWONSKI, 1984), then summed both the fundamental and higher modes. In detail, we first calculated the average velocity over the great circle path connecting the epicenter and the station. Next we determined the average velocity for the odd and even paths. The difference between the average velocity for the R3 and R4 paths and that of the entire great circle path is very small, with resulting time shifts of 5 seconds maximum. We could therefore ignore velocity variations due to the different velocity structures along the R3 and R4 paths and apply time shifts using the average great circle velocity. The lateral heterogeneity produced a maximum time shift for the great circle path of 40 s from the homogeneous earth model at periods of about 130 s. We did not apply corrections to the higher modes.

Some of the results of 32 inversions for the moment distribution can be found in Figure 8. We also applied a positivity constraint to eliminate negative moment values. These inversion results are similar to the results using standard least squares, though the negative values become zero and the positive values are adjusted slightly. All these results vary greatly with rupture velocity, but several features are similar throughout. As with the previous results of Ruff *et al.*, the greatest moment release occurs in the western half of the aftershock zone, although which point source has the largest moment is a function of the rupture velocity. Also, once again, the results consistently show little or no moment release in the eastern half

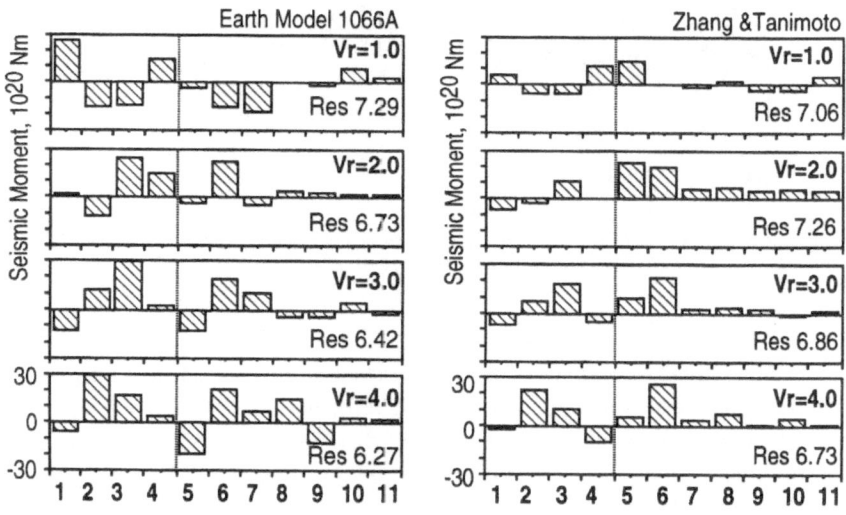

Figure 8
Moment distribution for the 1957 Aleutian earthquake using (left) the homogeneous earth model 1066A (GILBERT and DZIEWONSKI, 1975), and (right) the laterally heterogeneous earth model (ZHANG and TANIMOTO, 1993). V_r is the rupture velocity in km/s, Res is the residual ($\times 10^{-4}$) in m. The dashed line between subevents 4 and 5 marks the location of the epicenter.

of the aftershock zone. This is consistent with the suggestion of HOUSE *et al.* (1981) that the eastern end of the aftershock zone did not rupture in the 1957 event.

Despite the consistency of results, several problems remain in our analysis of the PTM surface wave record. First, we have not been able to determine the optimum rupture velocity. There is very little difference in the fit of the synthetic waveform to the observed as measured by the residuals. In fact, the smallest residuals occur with the most extreme rupture velocities (4.0 km/s). Figure 9 shows the observed and synthetic waveform for a typical rupture velocity of 3.0 km/s. Clearly, the best-fitting model only approximates the amplitude and waveform of the observed.

Another problem that remains concerns the response of the PTM instrument. Though the magnification was determined empirically, work of SCHWARTZ and RUFF (1987) suggests that the moment of the 1958 Kurile Islands earthquake is not known precisely enough to be the sole calibration source of the PTM instrument. Also, there is a possibility that the PTM instrument characteristics changed between the 1957 and 1958 earthquakes. This makes the results from the surface wave analysis suspect, but only as regards the absolute moment release, not the spatial distribution.

We cannot determine the optimum rupture velocity, yet we still wish to ascertain if any features of the moment release distributions we obtained are consistently present, as well as determine their scatter about the mean values for each subevent. To do so, we generate from our results a "global model average" of the slip distribution, applying all the results for the 32 inversions. This "average" slip distribution is shown in Figure 10, along with one standard deviation. This may be

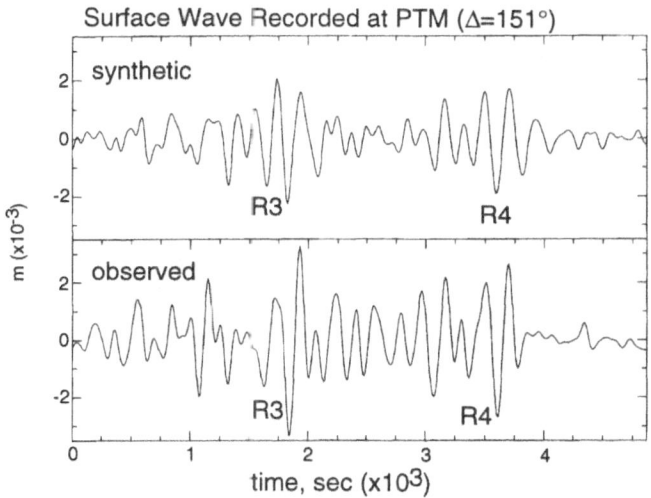

Figure 9
Synthetic (upper) and observed (lower) R3-R4 waves recorded at PTM.

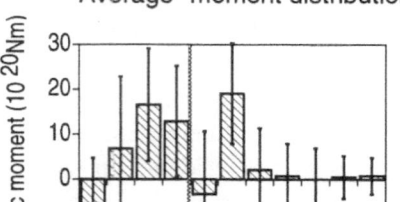

Figure 10
"Global model average" derived from 32 inversions for the slip distribution from the PTM record. The
error bars denote one standard deviation.

the best estimate that can be obtained using the PTM surface wave record. The
errors are large and are significantly nonzero for only two point sources. But again,
the same general trend is readily apparent—large moment release in the western
half of the aftershock zone, low moment release in the eastern half. The moment
estimated from our surface wave analysis is 50.4×10^{20} Nm.

The final consideration must be to ask how reliable is the analysis of a single
surface wave record? Unfortunately, our conclusion is that only gross features of
the slip distribution for the 1957 earthquake can be obtained from the PTM record.
This leaves the actual slip distribution and seismic moment still poorly determined.
Yet it is still vital to determine the unknown parameters if we wish to understand
this earthquake and its relation to the 1986 earthquake.

3. Tsunami Wave Studies

Besides seismic waves, the 1957 earthquake generated a large tsunami, which
was recorded on tide gauges all around the Pacific Ocean. The tsunami waveforms
can be used to determine the source parameters of the 1957 earthquake. ABE (1979)
previously estimated the magnitude of this earthquake from the maximum observed
tsunami heights at tide gauges as $M_t = 9.0$.

3.1. Computation of Tsunami Propagation

A tsunami which is generated by a large earthquake and propagates across the
ocean can be treated as a linear long wave because the wavelength is substantially
larger than the water depth. The wave equation for the small amplitude, linear long

wave is

$$\nabla h^2 = \frac{1}{c^2} \frac{\partial^2 h}{\partial t^2} \qquad (2)$$

where $c = \sqrt{gd}$, h is the height of water displaced from the equilibrium position, g is the acceleration of gravity, and d is water depth. Equivalently, the equation of motion and the equation of continuity are

$$\frac{\partial \mathbf{Q}}{\partial t} = -gd \, \nabla h \qquad (3)$$

$$\frac{\partial h}{\partial t} = -\nabla \cdot \mathbf{Q} \qquad (4)$$

where $\mathbf{Q}$ is the flow rate vector.

Given an initial water height, the equation of motion and equation of continuity can be solved by finite-difference calculations on a staggered grid system. Using highly accurate, digital bathymetry of the Pacific Ocean, the tsunami velocity and, thus, tsunami propagation can be calculated very accurately. A synthetic waveform is computed at the location of the tide gauges where the tsunami was actually observed.

Obviously, the tsunami velocity depends only on the water depth; therefore, the synthetic waveform is very sensitive to the bathymetry. The more accurate the bathymetry, the more accurate the computation. This suggests that a fine grid system be adopted in which to calculate the tsunami propagation. However, very fine grid spacing on the entire northern Pacific Basin would be impractical due to the enormous computational effort. For the majority of the deep Pacific Ocean where the bathymetry changes slowly, the grid space need not be any finer than 5′ (approximately 10 km). However, near coastal areas, the bathymetry changes much more rapidly. Also, islands and harbors where tide gauges are located cannot be adequately represented by a 5′ grid. Therefore, in coastal areas, 1′ (less than 2 km) grid spacing is used. Bathymetry of 1′ accuracy is used for the west coast of North America, the Hawaiian Islands, and around the tide gauges in Alaska.

3.2. Tsunami Source Area

As stated previously, the aftershock zone of the 1957 earthquake is the longest of any recorded earthquake. The eastern end of this aftershock zone has very few aftershocks, and it has been suggested that this area did not rupture in the 1957 main event. An alternative to equating the aftershock area and the rupture area is to determine the source area by tsunami data.

The tsunami source area can be located by a backward computation of the tsunami travel time from the tide gauge to the source. An initial condition is given at the tide gauge location. The tsunami propagation from the tide gauge is calculated for the duration of the observed tsunami travel time to that tide gauge. The location of the leading wavefront, or travel time arc, gives the origin point of the tsunami that reached that tide gauge. When many such travel-time arcs from tide gauges distributed around the earthquake are combined, they bound a region that is the source area of the tsunami. HATORI (1981) used this method to determine the source area of the 1957 earthquake (Figure 11). However, this estimate may be unreliable. Few of the available observations were used. Also, the travel-time arcs associated with both Unalaska Island and Sitka are questionable. The travel time used by Hatori for Sitka does not agree with any published travel time, nor does it agree with the apparent arrival time on the Sitka tide gauge record. Hatori's travel-time arc for Unalaska is not compatible with a backward travel time of 83 minutes from Dutch Harbor, Unalaska where the tide gauge is located. Without these two travel-time arcs, there is no constraint on the eastern end. Most significantly, Hatori drew the inverse refraction diagrams for all the travel times manually from large-scale bathymetric maps of the Pacific Ocean.

Our numerical calculation done on a fine grid of the actual Pacific Ocean bathymetry gives more reliable results. We computed travel time arcs from tide gauges in Alaska, N. America, and Hawaii. Figure 12 shows the travel-time arcs and source area. Our source area is approximately 850 km long, from 180°E to 168°W, and does not include the eastern end of the aftershock zone. Our source area is only slightly smaller than Hatori's original estimate; however, our source area is bounded by a greater number of travel-time arcs. Though there is a large scatter in these travel-time arcs, the best estimate of the source area does not include the Unalaska Island area.

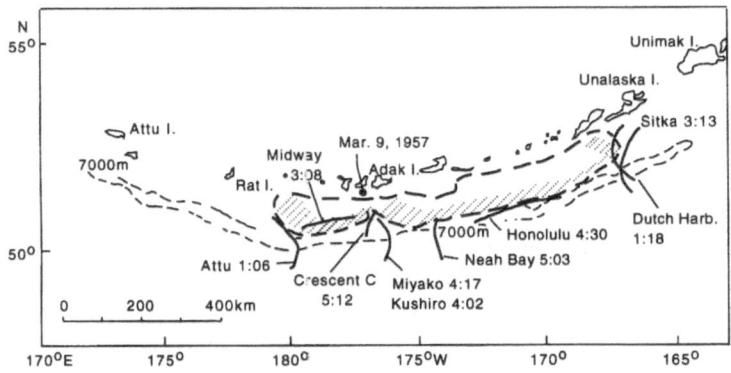

Figure 11

1957 tsunami source area derived by HATORI (1981). Travel time arcs for each tide gauge station are shown with station name and tsunami travel time in hours and minutes. Modified from HATORI (1981).

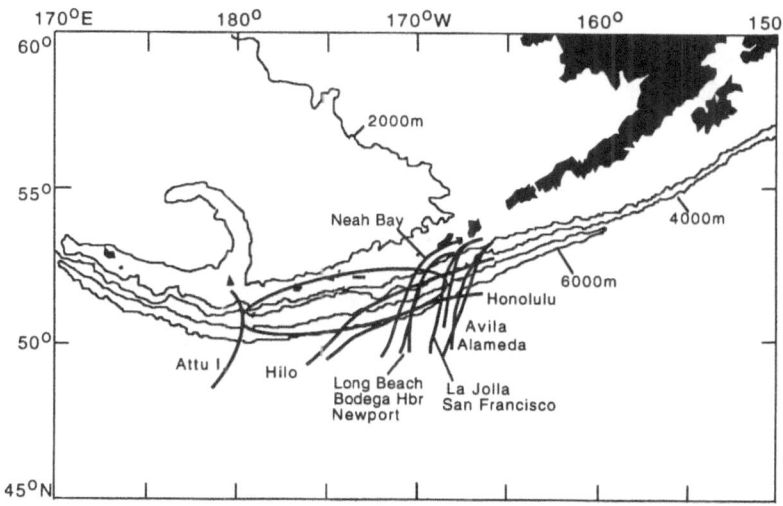

Figure 12
1957 tsunami source area (shaded) derived by backward propagation from tide gauges using finite-difference computation and digital bathymetry. Travel-time arcs for each tide gauge station are shown with station name.

Both the surface wave results and the backward computation of the tsunami support the conclusion of HOUSE *et al.* (1981) that the 1957 earthquake ruptured a smaller section of the arc than can be assumed from the aftershock area.

3.3. Tsunami Waveform Inversion

While the tsunami source area can be estimated from the tsunami travel times by backward propagation, only the extent of the source area can be estimated and no information is provided about the slip distribution. We can obtain the slip distribution, which will give an estimate of the moment, by inverting tsunami waveforms, as has been done previously by SATAKE (1989) for determining the slip distribution of the 1968 Tokachi-Oki and the 1983 Japan Sea earthquakes. Those studies used local and regional tsunami data, while this study is the first to determine slip distribution from far-field tsunami waveforms (JOHNSON and SATAKE, 1993).

A. Method

We divided the aftershock zone of the 1957 earthquake into eleven subfaults. The following fault parameters were the same for each subfault: length 100 km, width 150 km, dip 15°, and depth to the top of the subfault 1 km. (Figure 15 shows the location of the subfaults in relation to the Aleutian Arc.) Each subfault has unit displacement in the direction of Pacific Plate motion relative to N. America. The displacement direction was determined for each subfault individually from the N. America-Pacific Euler pole at 48.7°N, −78.2°E (DEMETS *et al.*, 1990). This means that the slip changes from pure dip-slip in the eastern end of the rupture zone to nearly equal components of dip-slip and strike-slip in the west. The initial condition is specified by the deformation of the ocean floor due to a buried fault as given by a set of equations such as OKADA's (1985). This provides the initial condition in the source area. The tsunami waveform was then computed for each subfault and used as the Green's function. An example of the Green's functions for Attu is shown in Figure 13. The Green's functions are significantly different for each subfault, so the slip distribution can be resolved from them.

Recent studies by EKSTRÖM and ENGDAHL (1989) and MCCAFFREY (1992) show that slip vectors of other earthquakes in the central Aleutians do not conform to the slip direction as predicted by plate motions, but are closer to arc-normal. If this is the case for the 1957 event, the difference between these directions is small, approximately 15°. We tested whether this difference would have an effect on the Green's functions. Our test shows that the waveforms with the more arc-normal slip direction are different from the waveforms from the plate motion slip direction only in small details, and these differences do not affect our results.

Next we inverted the waveforms from 12 tide gauges from Alaska, the Aleutians, Hawaii, and N. America by again solving Equation (1). These tide gauges are

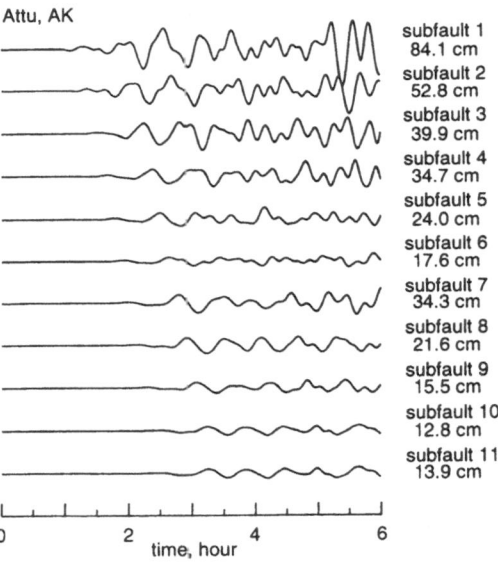

Figure 13
Synthetic tsunami waveforms from each subfault for Attu Island, AK tide gauge. Maximum peak-to-peak amplitudes for each trace are shown.

Attu, Unalaska, and Yakutat, AK; Neah Bay, WA; San Francisco, Alameda, San Pedro, Los Angeles Harbor, Newport Bay, and San Diego, CA; and Hilo, HI. The waveform data at each tide gauge station consist of an average of 110 time points with sampling interval of 1 min and the total number of data points is 1312. Figure 14 displays the observed and computed waveforms from these tide gauges. We performed both a standard least squares inversion and an inversion with a positivity constraint.

B. Slip Distribution

The slip distribution from the solution with a positivity constraint can be seen in Figure 15 and Table 1. It shows that the greatest slip occurred in the western half of the aftershock zone between 174°W and 180°W. The greatest slip occurred on subfault 4 (7 m) and subfault 5 (5 m) between 174° and 177°W. There is very little slip in the eastern half of the aftershock zone, with subfault 8 having the only appreciable slip. There is no slip in the easternmost subfaults (aside from negligible slip on subfault 11) from 164° to 169°W, which corresponds to the results from determination of the tsunami source area. The results for both the standard and constrained solutions are compared in Table 1. The slip distributions in the western half of the rupture zone for both inversions are fairly compatible. In the eastern half, the subfaults that are given negative values in the standard inversion are almost all zero under the positivity constraint. The RMS values for both inversions are almost the same. However, in a qualitative way, the synthetic and observed waveforms match more closely in the constrained solution. In the standard least-squares solution, arrivals are observed in the synthetic waveform prior to the time of the true first arrival in the observed waveform. In the constrained solution, the arrival times of the synthetics match the observed.

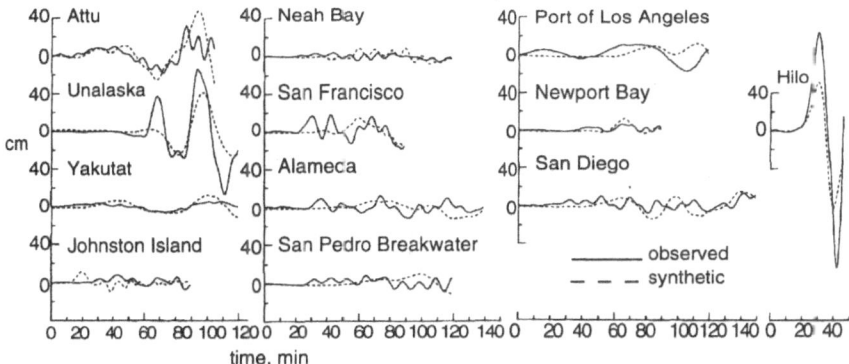

Figure 14
Observed and synthetic waveforms from nonnegative least-squares inversion for eleven subfaults. Start time of each waveform is different.

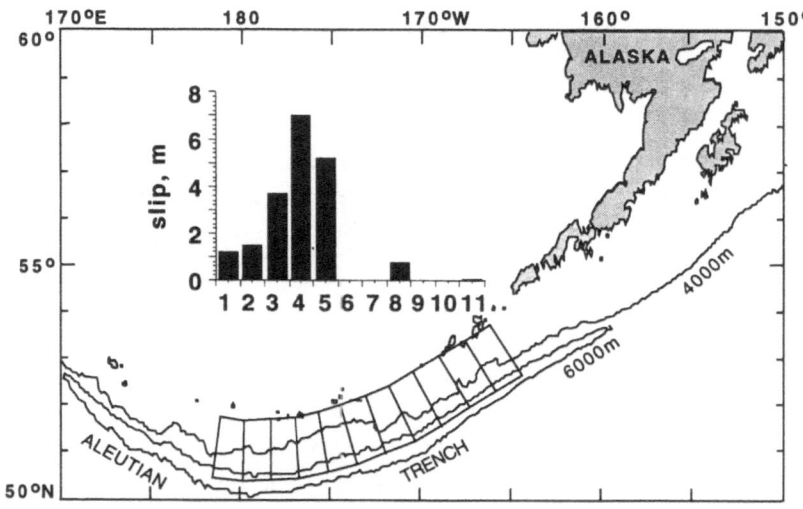

Figure 15
Slip distribution on rupture zone of 1957 earthquake from nonnegative least squares inversion for eleven
subfaults. The numbered segments correspond to the subfault immediately below.

Table 1

Inversion results for 11 subfaults

Subfault #	Nonnegative LS		Standard LS	
	Slip, m	Error, m	Slip, m	Error, m
1	1.1	0.66	1.3	1.93
2	1.5	0.51	1.3	1.17
3	3.7	1.00	3.6	1.31
4	7.0	1.07	6.1	1.15
5	5.2	0.59	4.5	0.74
6	0.0	0.00	−1.4	0.37
7	0.0	0.16	−0.25	0.44
8	0.76	0.32	0.81	0.27
9	0.0	0.35	−0.45	0.47
10	0.0	0.00	−2.5	0.54
11	0.08	0.27	2.5	0.53

RMS error, m	.1002	.0967
Average slip, m	1.77	1.42
M_0, 10^{20} Nm	87.6	70.3

While the computed waveforms for the constrained solution explain the overall features of the observed waveforms, a careful examination of Figure 14 reveals that the first large positive pulses at Unalaska and San Francisco are, among others, poorly matched. Though not shown, the same is true for the standard least-squares inversion results. A larger displacement on subfault 8 can explain the first pulse at Unalaska, and a large displacement on subfault 9 can explain the first pulse at San Francisco. However, large displacements in either of these subfaults are incompatible with the large amplitude wave at Hilo. We hypothesized from an examination of first arrival times at the three tide gauges in question, that a large displacement on a subfault of smaller area and at the down-dip edge of subfaults 8 or 9 might be compatible with all three waveforms. Accordingly, we divided subfaults 8 and 9 into smaller faults. Figure 16 shows the position of the additional subfaults 12 and 13. These subfaults have parameters: length 50 km, width 75 km, dip 15°, and depth to the top of fault 20.4 km. Green's functions were computed for these additional faults, and the inversion was performed again. Figure 17 shows that a displacement of 3.3 m on subfault 12 improves the match of the first pulse on the Unalaska waveform and is still compatible with the Hilo waveform. However, the first pulse on the San Francisco waveform is still poorly matched. The slip distribution results are shown in Figure 16 and Table 2. The solution for 13 subfaults is compatible with our hypothesis of concentration of slip on a smaller subfault in the eastern half of the rupture zone. It is also compatible with the total average slip for the entire rupture zone, as the slip on subfault 12 is approximately four times the slip on

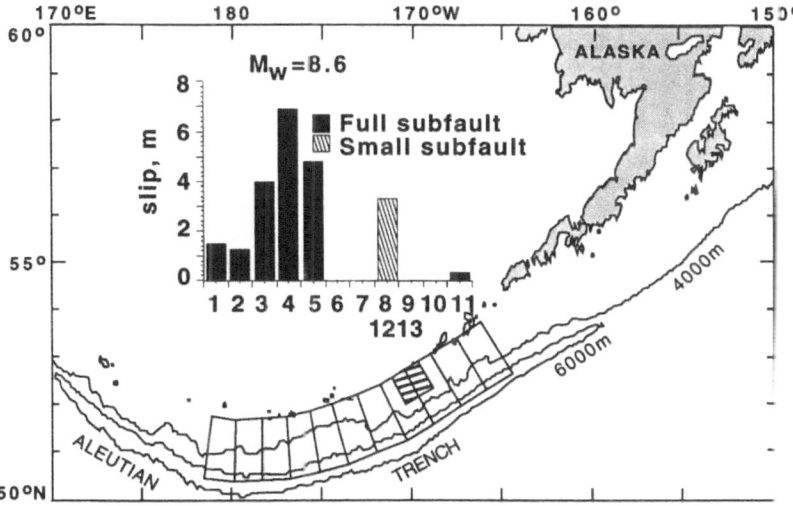

Figure 16
Slip distribution on rupture zone of 1957 earthquake from nonnegative least-squares inversion for thirteen subfaults. The numbered segments correspond to the subfault immediately below. The subfaults corresponding to 12 and 13 are shaded.

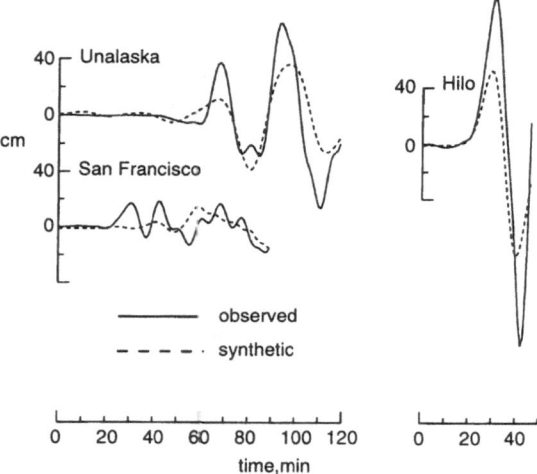

Figure 17

Observed and synthetic waveforms from nonnegative least-squares inversion for thirteen subfaults. Start time of each waveform is different. Not all waveforms used in inversion are shown.

Table 2

Inversion results for 13 subfaults

Subfault #	Nonnegative LS		Standard LS	
	Slip, m	Error, m	Slip, m	Error, m
1	1.5	0.74	1.7	1.77
2	1.3	0.46	0.99	1.30
3	4.0	1.04	4.0	0.53
4	6.9	1.10	6.0	1.01
5	4.8	0.56	3.6	0.65
6	0.0	0.00	−1.4	0.43
7	0.0	0.01	−0.12	0.37
8	0.0	0.13	−0.30	0.75
9	0.0	0.32	−0.55	0.46
10	0.0	0.00	−2.8	1.76
11	0.32	0.27	3.0	1.67
12	3.3	0.26	5.1	0.64
13	0.0	0.03	−1.6	1.49

RMS error, m	.0975	.0925
Average slip, m	1.70	1.36
M_0, 10^{20} Nm	84.2	67.3

subfault 8 from the solution for eleven subfaults. However, the San Francisco record indicates that there may be more slip in the eastern half of the aftershock zone that we cannot resolve.

C. Error Estimates

The formal statistical errors for a standard least-squares inversion cannot always be considered a good estimate of the actual errors (TICHELAAR and RUFF, 1989). Further no formal errors can be estimated for a nonnegative least-squares inversion. Therefore, we applied a resampling technique to determine the errors. We reinverted the tsunami waveforms twelve times, each time dropping one waveform from each of the tide gauge stations from the data. This gives twelve estimates of the slip distribution, and a mean and standard deviation for the slip on each subfault can be determined.

This technique for determining the errors is similar to the jackknifing technique described by TICHELAAR and RUFF (1989). In jackknifing, a fixed number of random data points are deleted to produce a resample that is then inverted for the model parameters. If we treat each waveform as 110 data points out of a total of 1312 data points, then we must consider our resample as a delete-110 jackknife with corresponding errors. These errors are simply the standard errors multiplied by a scale factor. In our case the scale factor is approximately three.

Our original error estimates for the slip distribution using this method were sizable, but still retained the major features discussed. However, since we delete an entire waveform at a time, rather than 110 random data points, the errors determined can be strongly influenced by the presence or absence of certain waveforms. Therefore, we examined each of the jackknife inversions and determined that two waveforms, Hilo and Attu, were necessary to obtain a stable solution. We recomputed the errors using only those jackknives which included both Hilo and Attu. These errors are given in Tables 1 and 2. These error estimates show that the slip distribution is significantly nonzero except for: subfaults 1 and 8 in the inversion for 11 subfaults, and subfault 1 in the inversion for 13 subfaults. These error estimates show that the concentration of slip in the western half of the aftershock zone and the small slip in the eastern half are real.

4. Comparison of Seismic and Tsunami Results

As stated in the introduction, estimates of the seismic moment release of the 1957 earthquake, based on the size of the aftershock zone, vary by as much as an order of magnitude. However, the aftershock zone is an indirect means of deriving the moment. With slip distribution as determined by surface waves and especially tsunami waves, the seismic moment can now be more accurately estimated. The estimate from the constrained inversion of tsunami waveforms is 88×10^{20} Nm.

Table 3

Ten largest earthquakes of 20th century (Modified from KANAMORI, *1977)*

Event	Year	M_w
Chile	1960	9.5
Alaska	1964	9.2
Kamchatka	1952	9.0
Ecuador	1906	8.8
Aleutian	1965	8.7
Assam	1950	8.6*
Aleutian	1957	8.6
Kurile Islands	1963	8.5
Chile	1922	8.5
Banda Sea	1938	8.5

* Estimated moment of 1950 Assam earthquake is 100×10^{20} Nm.

This estimate gives a moment magnitude of $M_w = 8.6$. This is much smaller than the estimate of $M_w = 9.1$ originally assigned by KANAMORI (1977). However, this estimate is in good agreement with the estimate of 100×10^{20} Nm by RUFF *et al.* (1985) and with our estimate of 50.4×10^{20} Nm from surface wave analysis. The slip distribution determined from tsunami inversion also agrees in a qualitative way with the surface wave analysis; namely, that slip was concentrated in the western half of the aftershock zone. The tsunami results show the highest slip in the area of the epicenter, while the surface wave analysis shows that the greatest slip occurred farther away from the epicenter. However, it must be recalled that the surface wave results were inconclusive. The tsunami results were obtained from a larger and more reliable data set. It should be noted, however, that BOYD *et al.* (1992) have speculated from the aftershock sequence that moment release was concentrated in the eastern section of the aftershock zone from 167° to 175°W rather than in the western section. To conclude, our best estimate of the seismic moment is 88×10^{20} Nm, which "demotes" the 1957 Aleutian earthquake to the seventh largest event of this century (Table 3).

5. The 1986 Andreanof Islands Earthquake

We can examine the 7 May 1986 Andreanof Islands earthquake in the light of the new results we have obtained for the 1957 earthquake.

As noted, the epicenters for these two events are nearly coincident. Both earthquakes ruptured bilaterally from the epicenter. Although there is some disagreement between various researchers about the spatial distribution of moment release for the 1986 earthquake (see references cited in section on body waves), it is clear that the aftershock zone in this case coincides with the area of significant moment release. We can compare this to the area of significant moment release for the 1957 earthquake as determined by tsunami waveform inversion. The 1986 earthquake occurred almost entirely within the area of significant moment release of the 1957 earthquake (Figure 18). A small portion of the major rupture of the 1986 earthquake lies to the east of the major rupture area of the 1957 earthquake, but clearly, the 1986 earthquake is not filling in areas of low moment release from the 1957 earthquake, but is a major rerupturing of this segment of the arc.

This leads us to reconsider the seismic potential of the rest of the 1957 rupture area. The entire area may have a low potential for an earthquake of the same magnitude as the 1957 event, but a high potential for earthquakes of the same magnitude as the 1986 event. To look at the historic record, the central Aleutians ruptured in a number of large earthquakes ($M \sim 7.5$) around the turn of the century (SYKES et al., 1981). Though little is known about these earthquakes (i.e., rupture length, moment), they apparently ruptured several segments of the 1957 rupture zone. Taking these turn-of-the-century earthquakes and the better-understood 1957 and 1986 events, we can theorize that the central Aleutians may display a bimodal rupture process. Great earthquakes with large rupture areas may be followed by several smaller earthquakes that break segments of the larger earthquake's rupture area. The 1957–1986 sequence may be similar to the history of subduction earthquakes in Colombia where the 1906 earthquake ruptured a large area and was followed by the 1942, 1958, and 1979 earthquakes (see Figure 19), which successively reruptured the entire 1906 area (KELLEHER, 1972; KANAMORI and McNALLY, 1982). If this is the case, the central Aleutians, particularly the

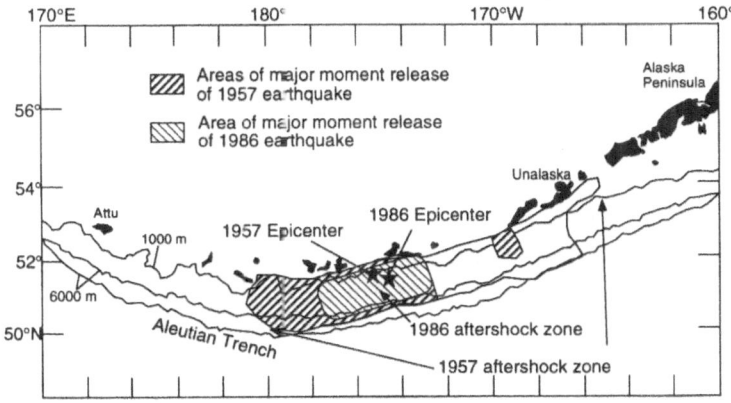

Figure 18
Areas of major moment release of 1957 Aleutian and 1986 Andreanof Islands earthquakes.

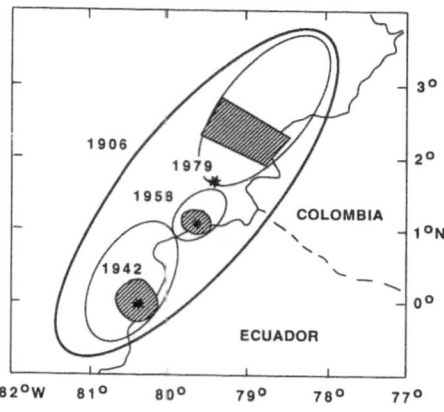

Figure 19
Rupture zones of earthquakes in Colombia-Ecuador subduction zone. Stars denote epicenters. Shaded areas represent asperities. Modified from BECK and RUFF (1987).

western end of the aftershock zone, may have high risk for earthquakes with $M \sim 7-8$ in the future. Our results support this possibility, which is reflected in the latest seismic potential maps of the central Aleutian Islands (NISHENKO, 1991).

Further questions are also raised by our results, which show that low moment release occurred in the eastern half of the 1957 aftershock zone, as well as confirm the existence of the Unalaska Gap. Do these areas have a high seismic potential? Or does the low moment release indicate that convergence in this area has a large aseismic component? Of particular interest is the Unalaska Gap, as no great earthquakes have ever been unambiguously identified as having occurred there. Future research must answer these questions in order to fully assess the seismic potential and seismic and tsunami hazards in the central Aleutians.

Acknowledgments

We wish to thank two anonymous reviewers whose comments helped improve our manuscript. This work was supported by the U.S. Geological Survey (1434–92–G–2187 and 1434–93–G–2320) and the National Science Foundation (EAR–90–19003). This work was partially supported by NSF grant EAR920000N and utilized the Cray-2 system at the National Center for Supercomputing Applications, University of Illinois at Urbana-Champaign. J. M. J. was partially supported by a Shell Graduate Grant.

References

ABE, K. (1979), *Size of Great Earthquakes of 1873–1974 Inferred from Tsunami Data*, J. Geophys. Res. *84*, 1561–1568.

BECK, S., and RUFF, L. (1987), *Rupture Process of the Great 1963 Kurile Islands Earthquake Sequence: Asperity Interaction and Multiple Event Rupture*, J. Geophys. Res. *92*, 14123–14138.

BOYD, T. M., ENGDAHL, E. R., and SPENCE, W., *Analysis of seismicity associated with a complete seismic cycle along the Aleutian arc: 1957–1979*. In *Wadati Conference on Great Subduction Earthquakes*, Sept. 16–19, 1992 (eds. Christensen, D., Wyss, M., Habermann, R. E., and Davies, J.) 43–50 in extended abstracts, 1992.

BOYD, T. M., and NÁBĚLEK, J. L. (1988), *Rupture Process of the Andreanof Islands Earthquake of May 7, 1986*, Bull. Seismol. Soc. Am. *78*, 1653–1673.

DAS, S., and KOSTROV, B. V. (1990), *Inversion for Seismic Slip Rate History and Distribution with Stabilizing Constraints: Application to the 1986 Andreanof Islands Earthquake*, J. Geophys. Res. *95*, 6899–6913.

DEMETS, C., GORDON, R. G., ARGUS, D. F., and STEIN, S. (1990), *Current Plate Motions*, Geophys. J. Int. *101*, 425–478.

EKSTRÖM, G., and ENGDAHL, E. R. (1989), *Earthquake Source Parameters and Stress Distribution in the Adak Island Region of the Central Aleutian Islands, Alaska*, J. Geophys. Res. *94*, 15499–15519.

FUKAO, Y., and FURUMOTO, M. (1979), *Stress Drops, Wave Spectra and Recurrence Intervals of Great Earthquakes—Implications of the Etorofu Earthquake of 1958 November 6*, Geophys. J. R. Astr. Soc. *57*, 23–40.

GILBERT, F., and DZIEWONSKI, A. M. (1975), *An Application of Normal Mode Theory to the Retrieval of Structural Parameters and Source Mechanisms from Seismic Spectra*, Philos. Trans. R. Soc. London, Ser. A *278*, 187–269.

HATORI, T. (1981), *Tsunami Magnitude and Source Area of the Aleutian-Alaska Tsunamis*, Bull. Earthq. Res. Inst., Univ. of Tokyo *56*, 97–110.

HOUSE, L. S., SYKES, L. R., DAVIES, J. N., and JACOB, K. H., *Identification of a possible seismic gap near Unalaska Island, eastern Aleutians, Alaska*. In *Earthquake Prediction—An International Review* (eds. Simpson, D. W., and Richards, P. G.) 81–92 (American Geophysical Union, 1981).

HOUSTON, H., and ENGDAHL, E. R. (1989), *A Comparison of the Spatio-temporal Distribution of Moment Release for the 1986 Andreanof Islands Earthquake with Relocated Seismicity*, Geophys. Res. Lett. *16*, 1421–1424.

HWANG, L. J., and KANAMORI, H. (1986), *Of the May 7, 1986 Andreanof Islands Earthquake Source Parameters*, Geophys. Res. Lett. *13*, 1426–1429.

JOHNSON, J. M., and SATAKE, K. (1993), *Source Parameters of the 1957 Aleutian Earthquake from Tsunami Waveforms*, Geophys. Res. Lett. *20*, 1487–1490.

KANAMORI, H. (1977), *The Energy Release in Great Earthquakes*, J. Geophys. Res. *82*, 2981–2987.

KANAMORI, H., and MCNALLY, K. C. (1982), *Variable Rupture Mode of the Subduction Zone along the Ecuador-Colombia Coast*, Bull. Seismol. Soc. Am. *72*, 1241–1253.

KELLEHER, J. A. (1972), *Rupture Zones of Large South American Earthquakes and Some Predictions*, J. Geophys. Res. *77*, 2087–2103.

LANE, F. D., and BOYD, T. M. (1990), *A Simulated Annealing Approach to the Inversion of Surface Wave Directivities*, EOS *71*, 1468.

MCCAFFREY, R. (1992), *Oblique Plate Convergence, Slip Vectors, and Forearc Deformation*, J. Geophys. Res. *97*, 8905–8915.

NISHENKO, S. P. (1991), *Circum-Pacific Seismic Potential: 1989–1999*, Pure and Appl. Geophys. *135*, 169–259.

NISHENKO, S. P., and MCCANN, W. R., *Seismic potential for the world's major plate boundaries: 1981*. In *Earthquake Prediction—An International Review* (eds. Simpson, D. W., and Richards, P. G.) 20–28 (American Geophysical Union, 1981).

OKADA, Y. (1985), *Surface Deformation due to Shear and Tensile Faults in a Half-space*, Bull. Seismol. Soc. Am. *75*, 1135–1154.

RUFF, L., and KANAMORI, H. (1983), *The Rupture Process and Asperity Distribution of Three Great Earthquakes from Long-period Diffracted P Waves*, Phys. Earth Planet. Inter. *31*, 202–230.

RUFF, L., KANAMORI, H., and SYKES, L. R. (1985), *The 1957 Great Aleutian Earthquake*, EOS *66*, 298.

SATAKE, K. (1989), *Inversion of Tsunami Waveforms for the Estimation of Heterogeneous Fault Motion of Large Submarine Earthquakes: The 1968 Tokachi-oki and the 1983 Japan Sea Earthquakes*, J. Geophys. Res. *94*, 5627–5636.

SCHWARTZ, S. Y., and RUFF, L. J. (1987), *Asperity Distribution and Earthquake Occurrence in the Southern Kuril Islands Arc*, Phys. Earth Planet. Inter. *49*, 54–77.

SYKES, L. (1971), *Aftershock Zones of Great Earthquakes, Seismicity Gaps, and Earthquake Prediction for Alaska and the Aleutians*, J. Geophys. Res. *76*, 8021–8041.

SYKES, L. R., KISSLINGER, J. B., HOUSE, H., DAVIES, J. N., and JACOB, K. H., *Rupture zones and repeat times of great earthquakes along the Alaska-Aleutian Arc, 1784–1980*. In *Earthquake Prediction—An International Review* (eds. Simpson, D. W., and Richards, P. G.) 73–80 (American Geophysical Union, 1981).

TICHELAAR, B. W., and RUFF, L. J. (1989), *How Good are our Best Models? Jackknifing, Bootstrapping, and Earthquake Depth*, EOS *70*, 593, 605–606.

WOODHOUSE, J. H., and DZIEWONSKI, A. M. (1984), *Mapping the Upper Mantle: Three-dimensional Modeling of Earth Structure by Inversion of Seismic Waveforms*, J. Geophys. Res. *89*, 5953–5986.

YOSHIDA, S. (1992), *Waveform Inversion for Rupture Process using a Non-flat Seafloor Model: Application to 1986 Andreanof Islands and 1985 Chile Earthquakes*, Tectonophys. *211*, 45–59.

ZHANG, Y.-S., and TANIMOTO, T. (1993), *High-resolution Global Upper Mantle Structure and Plate Tectonics*, J. Geophys. Res. *98*, 9793–9823.

(Received April 6, 1993, revised September 18, 1993, accepted October 2, 1993)

PAGEOPH, Vol. 142, No. 1 (1994)

0033–4553/94/010029–25$1.50 + 0.20/0
© 1994 Birkhäuser Verlag, Basel

The Rupture Process and Tectonic Implications of the Great 1964 Prince William Sound Earthquake

Douglas H. Christensen[1] and Susan L. Beck[2]

Abstract —We have determined the rupture history of the March 28, 1964, Prince William Sound earthquake ($M_w = 9.2$) from long-period WWSSN *P*-wave seismograms. Source time functions determined from the long-period *P* waves indicate two major pulses of moment release. The first and largest moment pulse has a duration of approximately 100 seconds with a relatively smooth onset which reaches a peak moment release rate at about 75 seconds into the rupture. The second smaller pulse of moment release starts at approximately 160 seconds after the origin time and has a duration of roughly 40 seconds. Because of the large size of this event and thus a deficiency of on-scale, digitizable *P*-wave seismograms, it is impossible to uniquely invert for the location of moment release. However, if we assume a rupture direction based on the aftershock distribution and the results of surface wave directivity studies we are able to locate the spatial distribution of moment along the length of the fault. The first moment pulse most likely initiated near the epicenter at the northeastern down-dip edge of the aftershock area and then spread over the fault surface in a semi-circular fashion until the full width of the fault was activated. The rupture then extended toward the southwest approximately 300 km (Ruff and Kanamori, 1983). The second moment pulse was located in the vicinity of Kodiak Island, starting at ~ 500 km southwest of the epicenter and extending to about 600 km. Although the aftershock area extends southwest past the second moment pulse by at least 100 km, the moment release remained low. We interpret the 1964 Prince William Sound earthquake as a multiple asperity rupture with a very large dominant asperity in the epicentral region and a second major, but smaller, asperity in the Kodiak Island region.

The zone that ruptured in the 1964 earthquake is segmented into two regions corresponding to the two regions of concentrated moment release. Historical earthquake data suggest that these segments behaved independently during previous events. The Kodiak Island region appears to rupture more frequently with previous events occurring in 1900, 1854, 1844, and 1792. In contrast, the Prince William Sound region has much longer recurrence intervals on the order of 400–1000 years.

Key words: Asperities, rupture process, subduction, Alaskan earthquakes, 1964 earthquake, fault segmentation, body waves.

Introduction

The March 28, 1964, Prince William Sound earthquake ($M_w = 9.2$) occurred as a result of underthrusting of the Pacific plate beneath the North American plate along an 800 km long stretch of the Alaska subduction zone as shown in Figure 1.

[1] Geophysical Institute, University of Alaska Fairbanks, Fairbanks, Alaska 99775, U.S.A.
[2] Department of Geosciences, University of Arizona, Tucson, Arizona 85721, U.S.A.

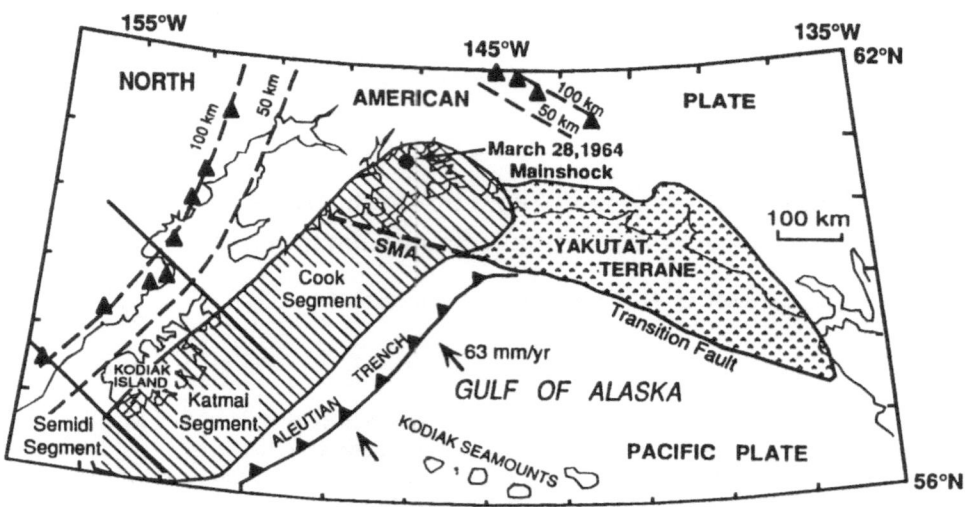

Figure 1
Map of the tectonic setting of the March 28, 1964, Prince William Sound earthquake. The volcanoes (triangles) and 50 and 100 km depth contours associated with the Aleutian and Wrangell Wadati-Benioff zones are shown. The volcanic arc segmentation as defined by KIENLE et al. (1983) and the aftershock area for the 1964 earthquake (hatchured region) are shown. Also shown is the observed surficial boundary of the Yakutat Terrane and the possible extension of the Transition Fault or Slope Magnetic Anomaly (SMA) beneath the accretionary prism under the Prince William Sound region.

The epicenter is located in the Prince William Sound area, approximately 130 km east of Anchorage and 70 km west of Valdez, Alaska (ISC Location, 61.05N, 147.48W; Origin Time, 03:36:13.9). With a seismic moment of 8.2×10^{29} dyne cm (KANAMORI, 1977), the 1964 earthquake is the second largest event in recorded history, surpassed only by the 1960 Chile earthquake ($M_w = 9.5$). A large number of studies have focused on many aspects of this event. Much of this information has been collected in an eight book series published by the NATIONAL ACADEMY OF SCIENCES (1972). Topics range from geology and geophysics, to biology, engineering, and human ecology.

The tectonic setting for this event is somewhat unique. The plate boundary configuration in southern Alaska is complicated by the transition from a convergent (Alaska-Aleutian subduction zone) to a transform (Fairweather, Queen Charlotte faults) plate boundary with a convergence rate of about 62 mm/yr in a northwestward direction (LAHR and PLAFKER, 1980; DEMETS et al., 1990). In addition, the ongoing collision of the Yakutat terrane between these two regions complicates the interaction between the Pacific and North American plates (Figure 1). The plate boundary can also be divided into several segments which behave somewhat independently. The 1964 earthquake zone is truncated on the northeastern end by a geometric boundary between the very shallowly dipping Alaska Wadati-Benioff zone and the more steeply dipping Wrangell Wadati-Benioff zone (PAGE et al., 1989). This margin is not well understood but must represent either a rip or fold of the subducting Pacific plate which may be related to the collision of the Yakutat block. Seismicity drops off sharply in both the overlying plate and the subducting slab in the Wrangell zone relative to the rest of the arc (PAGE et al., 1989). The southwestern end of the 1964 aftershock area abuts the 1938 rupture zone. Historic earthquake occurrences suggest that the boundary between the 1964 and 1938 rupture zones has been an obstruction to earthquake ruptures in previous cycles (SYKES, 1971) and thus it is a plate segment boundary. The 1964 rupture area may also be divided into two segments based on geometry of the subducting plate and historic earthquake occurrences. These segments will be discussed in detail later in this paper. Another interesting feature of the 1964 rupture zone is the drastic change in the trench-volcanic arc gap which ranges from a minimum distance of 325 km at the western end of the rupture to a maximum of 570 km at the eastern end (JACOB et al., 1977; DAVIES and HOUSE, 1979), corresponding to the gradual flattening of the subducting plate (Figure 1).

Some aspects of the rupture process of the 1964 Prince William Sound event have been previously studied. WYSS and BRUNE (1967) analyzed the first 72 seconds of rupture, using short-period seismograms, and found several coherent subevents in the vicinity of the epicenter. In a study of surface waves for this event, KANAMORI (1970) estimated a rupture length of 600 km in the direction of S25°W and a rupture velocity of 3.5 km/sec. FURUMOTO (1965) studied the Rayleigh waves recorded on the strain seismograph at Kipapa, Hawaii and determined a rupture length of 800 km in the S30°W direction and a rupture velocity of 3.0 km/sec. RUFF and KANAMORI (1983) studied body waves from several long-period WWSSN stations to determine the rupture characteristics of this event. They concluded that the rupture as seen through long-period P waveforms was extremely smooth and represented the breaking of at least one large asperity located in the northeastern portion of the aftershock area. KIKUCHI and FUKAO (1987) also analyzed long-period P waveforms and found a large amount of moment release in the first 100 sec of rupture but did not have enough data to clearly delineate later moment release. In this paper we will use a larger data set of long-period P waveforms to extract additional information about the rupture process of this important event.

The March 28, 1964, Prince William Sound Mainshock

The location, aftershock distribution, and focal mechanism of the 1964 earthquake have been studied in detail in several papers (e.g., HARDING and ALGERMISSEN, 1969; STAUDER and BOLLINGER, 1966; KANAMORI, 1970; SHERBURNE et al., 1969). The epicenter along with the one-day aftershock locations from the ISC bulletin are shown in Figure 2. Aftershocks extend from the eastern edge of Prince

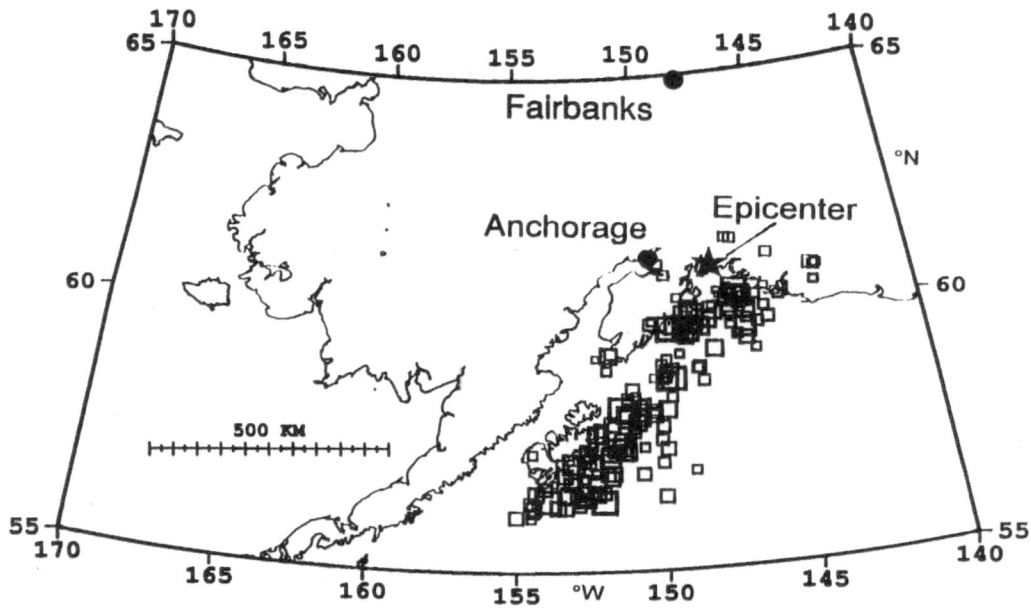

Figure 2
Map of southern Alaska showing the one-day aftershock distribution of the March 28, 1964, Prince
William Sound earthquake. Locations of 178 events with magnitudes larger than 4.0 are taken from the
ISC Bulletin. The size of the symbols are proportional to the magnitude of the earthquakes ranging from
4.0 to 6.2. The epicenter of the mainshock is shown as a star and the two largest cities in Alaska are
shown as solid dots.

William Sound to the southwestern end of Kodiak Island, covering an area of
approximately 200,000 km² (nearly 800 km in length and 250 km in width). The
aftershock area is restricted mostly to the off-shore region between the oceanic
trench and the coastal regions of Prince William Sound, the Kenai Peninsula, and
Kodiak Island.

Data

Obtaining waveform data for an event of this size has always been problematic.
The extremely large amplitudes of the complex *P* waveforms are usually clipped or
off-scale at most of the global seismic network stations. Most of the on-scale
P-wave data are found at diffracted distances on horizontal components. Because
of the large size of this event, diffracted *P* phases are observable out to a distance
of 145°. For the 1964 earthquake there is a secondary problem which is related to
the long duration of the rupture. Given the possible unilateral rupture length of
800 km and assuming a possible rupture velocity of 3.0 km/sec, a rupture process
time of more than 250 seconds is possible. At most distances, secondary phases

(*PP* and *PKP*) could arrive prior to the end of the direct *P*-wave signal, thus complicating the waveforms. There is only a limited epicentral distance range from about 90 to 120° where between 210 and 250 seconds of uncontaminated *P* waveform can be observed prior to the arrival of additional phases.

The large size and long duration of this earthquake severely limit the amount of useful *P*-wave data which are available to be studied. We were able to gather 24 digitizable seismograms from 20 individual seismic stations, eleven of which were at diffracted distances. Of the 24 digitizable records only 13 (from 9 stations) were recorded at distances where more than 210 seconds of uncontaminated *P* waveform could be observed.

P-wave Analysis

The focal mechanism is fairly well determined from *P*-wave first motion studies and the surface wave radiation pattern. *P*-wave first motions from WWSSN stations constrain the steeply dipping nodal plane. HARDING and ALGERMISSEN (1969) preferred a strike of 62° and dip of 82° to the southeast, while STAUDER and BOLLINGER (1966) preferred a strike of 66° and dip of 85° to the southeast. The orientation of the shallow dipping nodal plane, however, is not so well constrained. KANAMORI (1970) used surface waves to help constrain the shallow dipping nodal plane of the focal mechanism. His preferred focal mechanism was strike 62°, dip 70°, and slip 90°. The depth for the surface wave solution of 71 km is deeper than expected from the aftershock distribution and the orientation of the interplate (fault) contact zone. Furthermore, the steeply dipping nodal plane determined from the surface waves does not satisfy the *P*-wave first motions exactly. Given the extreme size of this event it may be possible that the longer period surface waves sample a slightly different source region than shorter period waves leading to slight discrepancies in the focal parameters (KANAMORI, 1970). In this paper we use the values of strike 66°, dip 85°, and slip 90°, in order to satisfy the first motion data and to produce the same sense of motion (pure thrust) as found by the surface wave analysis. The *P* waveforms studied here will be nearly unaffected by small variations in the orientation of the shallow dipping plane.

The depth distribution of this earthquake is difficult to determine from the *P* waveforms of the mainshock. Hence, we have looked at other evidence to estimate the depth of seismic coupling in the Prince William Sound region. TICHELAAR and RUFF (1993) systematically determined the depth of moderate size underthrusting earthquakes located at the down-dip edge of interplate aftershock zones in subduction zone in order to determine the depth of the seismically coupled zone. They were able to analyze only one event (Sept. 4, 1965) located at the down-dip edge of the 1964 Alaska aftershock zone near Kodiak Island and determined a depth of 36–38 km. Although this is only one event, it suggests that the depth of seismic coupling is fairly shallow. Another source of information on the depth of this sequence is the geometry of the interplate contact zone as determined by the results of the Trans-Alaska Crustal Transect (TACT) program (FUIS et al., 1991; FUIS and PLAFKER, 1991). The TACT results show that the interplate contact dips at a shallow angle from the trench to approximately 35 km beneath the northern edge of the Chugach Terrane where the dip then steepens (FUIS and AMBOS, 1986; FUIS and PLAFKER, 1991). The projected depth to the interplate contact beneath the 1964 mainshock epicenter is approximately 25 km. The aftershocks are well trench-ward of the 50 km depth contour of the Wadati-Benioff zone as well as the mainshock epicenter suggesting that much of the rupture was shallow (Figures 1 and 2). We model this earthquake using a distributed source between 0 amd 30 km. However, the results are similar if we assume a distributed source between 0 and 40 km.

First, we examine the single-station deconvolved source time functions. This procedure allows us to explore the range of source time functions which can fit the individual station data, in addition to making comparisons between source time functions at different azimuths and distances. Source time functions are decon-volved from observed P waves assuming a Green's function distributed over a depth range of 0 to 30 km. We tested a range of distributed depths and found that the P waves were not very sensitive to the choice of distributed depth. The seismogram-source time function pairs are plotted in Figure 3 for stations with less than 210 seconds of uncontaminated P waveform and in Figure 4 for stations with over 210 seconds of uncontaminated P waveform. In Figure 3 we observe a single, broad moment pulse similar to that observed by RUFF and KANAMORI (1993) with a duration of about 100 seconds. We obtain an average seismic moment of 3.5×10^{27} dyne cm for the first pulse of moment release using the nondiffracted P waves. As is noted in RUFF and KANAMORI (1983), this pulse is rather smooth and has a very long rise-time and duration as is indicated by the extremely long-period nature of the seismograms for this earthquake. However, the smooth nature of this pulse is in part due to the fact that most stations recorded diffracted waves and thus the higher frequency signals are filtered to some extent. Several of the nondiffracted stations (particularly SCP, GEO, and TRN) show slightly more complicated source time functions with several major pulses of moment release. These complications may be due to the truncation of the rupture process in the up-dip and northeastern directions. We have not been able to separate these multiple pulses in space or time due to the sparse data set, and thus prefer the single large asperity model for this part of the rupture history, acknowledging that some complications, such as those observed by WYSS and BRUNE (1967), exist.

The source time functions in Figure 4 show a more complicated multiple phase character, with at least two major pulses of moment release. The second major moment pulse starts at about 160 seconds after rupture initiation and lasts about 40 seconds. At some stations a smaller third pulse is visible between the two major pulses, however, this pulse is not consistently resolved at all stations and it has a

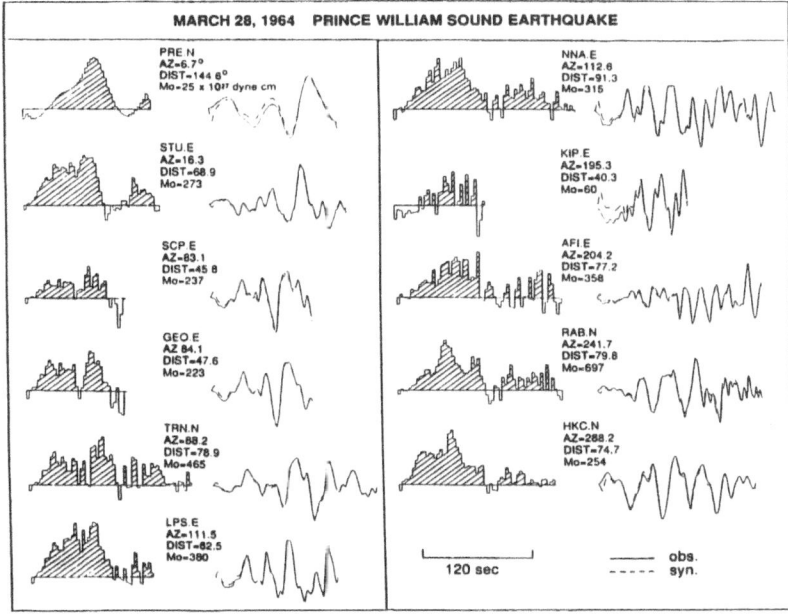

Figure 3

Deconvolved single-station source time function–seismogram pairs for the March 28, 1964, Prince William Sound earthquake. The solid and dashed traces are the observed and synthetic seismograms, respectively (right side of each pair) for the associated source time function (left side of the pair). The station azimuth (AZ), epicentral distance (DIST), component used (Z, N, E), and calculated seismic moment (M_0) in units of 10^{27} dyne cm, are shown for each station. This figure shows only stations for which less than 210 seconds of the P waveform were available.

small amplitude. These plots indicate a total rupture duration of at least 200 seconds. However, it would be impossible to observe a source process time longer than 250 seconds from the P waveforms alone due to the arrival of secondary phases.

The goal of this study is to determine the moment release distribution on the fault surface. Normally this is accomplished by inverting a large number of azimuthally distributed waveforms and/or source time functions to determine the distribution of moment release in one or two dimensions. There are a variety of techniques available which use different assumptions, but in the end these amount to relocating features which have apparent directivity associated with them.

The distribution of available P-wave data is plotted on a lower hemisphere stereographic projection in Figure 5 along with the preferred focal mechanism. The larger symbols represent stations with P-wave windows of over 210 seconds. Examples of seismogram-source time function pairs are shown at several different azimuths. It is apparent from this figure that there is directivity associated with the second major pulse of moment release, grossly indicating that the second pulse is

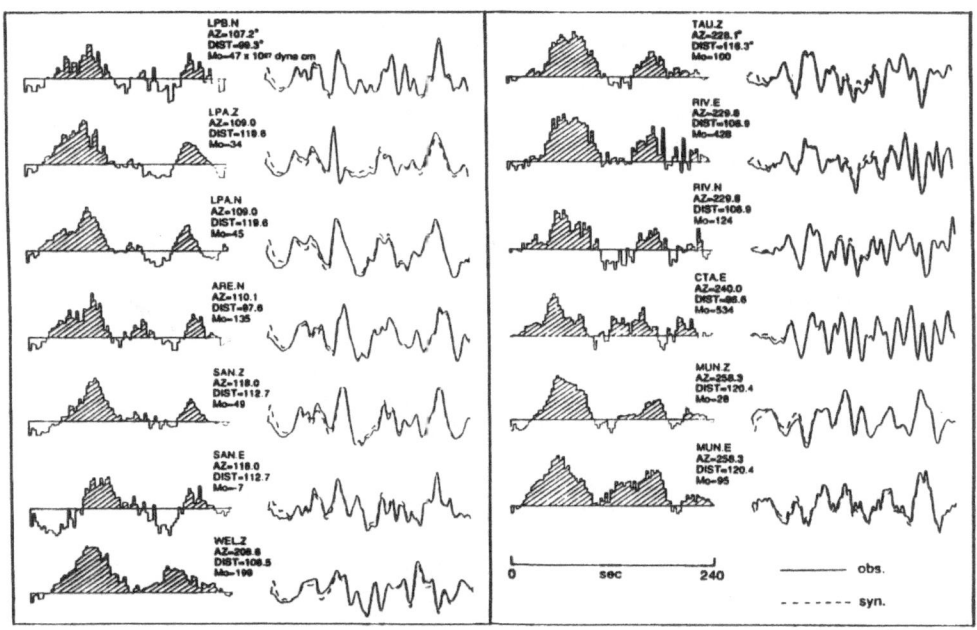

Figure 4

Deconvolved single-station source time function-seismogram pairs for the March 28, 1964, Prince William Sound earthquake. The solid and dashed traces are the observed and synthetic seismograms, respectively (right side of each pair) for the associated source time function (left side of the pair). The station azimuth (AZ), epicentral distance (DIST), component used (Z, N, E), and calculated seismic moment (M_0) in units of 10^{27} dyne cm, are shown for each station. This figure only shows stations for which more than 210 seconds of P waveform were deconvolved.

located southwest of the epicenter. The location of stations which recorded more than 210 seconds of P waveform are concentrated at two azimuths, to the west-southwest ($208°$–$258°$) and east-southeast ($107°$–$118°$), roughly parallel to the strike of the fault. Thus a full scale inversion of the P-wave data is impossible, since the azimuthal distribution is not adequate to uniquely determine a rupture direction.

In order to retrieve more information about this event it is necessary to make some *a priori* assumptions about the rupture history. Since the two azimuths which are represented in the data are nearly in the expected direction of rupture propagation, it is possible to determine the spatial location of the moment (along strike) if we assume the rupture direction. A reasonable assumption is that the rupture started at the epicenter and propagated parallel to the trench in the southwest direction. KANAMORI (1970) and FURUMOTO (1965) determined average rupture directions of S25°W and S30°W and rupture lengths of 600 km and 800 km, respectively, from surface wave data. In this study we use a rupture direction of S30°W (210°) which is consistent with results based on surface wave studies, and the

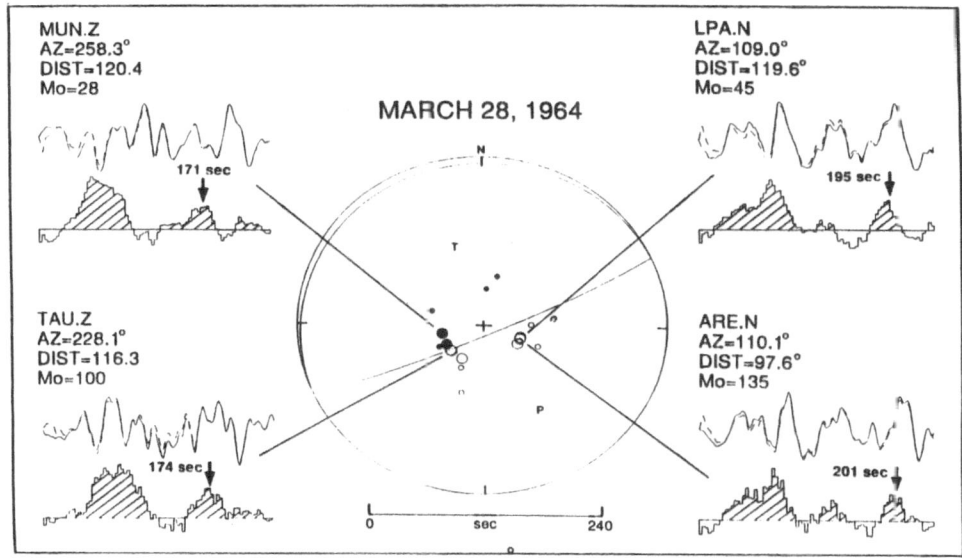

Figure 5

Lower hemisphere stereographic projection showing the focal mechanism and the distribution of waveform data used in this study. Large symbols show stations from which more than 210 seconds of waveform were used and small symbols show stations from which less than 210 seconds of waveform were used. Examples of deconvolved source time functions (bottom) along with observed (solid line) and synthetic (dashed line) seismograms (top) are also shown. A comparison of source time functions at different azimuths shows directivity associated with the second moment pulse.

orientation of the aftershock distribution, which is essentially subparallel with the trench.

Multi-station Inversions

Two inversions schemes have been used to look at the spatial distribution of moment release along the fault. For both inversions we will assume a one-dimensional (ribbon) fault model with a rupture direction of 210°. This is appropriate since the aftershock length of 800 km is much larger than the width of about 200 km. With our data set we cannot resolve details in the dip direction. The rupture velocity can be investigated, but nevertheless cannot be strongly constrained due to the trade-off between location and rupture velocity. KANAMORI (1970) found a rupture velocity of 3.5 km/sec from surface wave studies as did WYSS and BRUNE (1967) using short-period records.

The first simple inversion scheme we used is an iterative pulse stripping technique (KIKUCHI and KANAMORI, 1982). The data consist of the nine independent *P* waves recorded at stations with more than a 210 second window (from Figure 4). The shape and duration of the trapezoid used to construct the unit wavelet for the pulse stripping algorithm was determined from a grid search of the parameter space with the goal of fitting the observed waveforms as well as possible with a minimum number of pulses (40 iterations). A trapezoid with a rise time of 11 seconds and a duration of 32 seconds (i.e., 11–10–11 seconds) was chosen. For this inversion it was necessary to use a slightly altered focal mechanism (strike, 66°; dip, 85°; and slip, 94°) in order for the strike of the shallow dipping nodal plane to match the assumed rupture direction (210°). Results from this pulse stripping method are shown on a space-time plot in Figure 6a along with the fit between the observed and synthetic seismograms in Figure 6b. Figure 6a shows two regions of moment release, one near the epicenter and the second about 525 km southwest of the epicenter. The details of the first pulse near the epicenter should *not* be over-interpreted, since the one-dimensional model does not allow for a realistic two-dimensional rupture radiation which is probably important during the early stages of rupture initiation (RUFF and KANAMORI, 1983). In addition, the extremely long-period waveforms are not ideal for a pulse stripping technique. The large pulse duration used to make the unit wavelet was selected in order to match the shape and dominant period of the seismogram (largest moment pulses). In this way we hoped to constrain the location of the largest two or three major moment pulses. Moment release during the first 80 seconds of rupture is located near the epicentral region, southwest to about 225 km and northeast to 150 km (labeled A in Figure 6a). The location of the second moment pulse is centered about 525 km southwest

MARCH 28, 1964 EARTHQUAKE

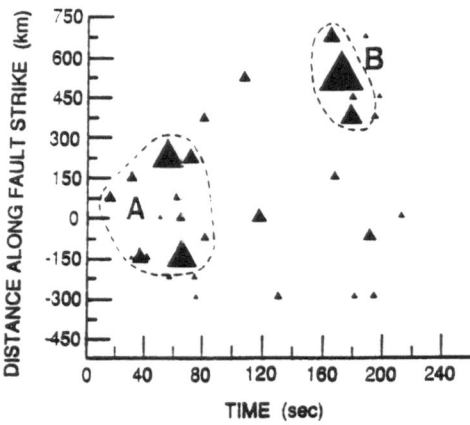

Figure 6(a)

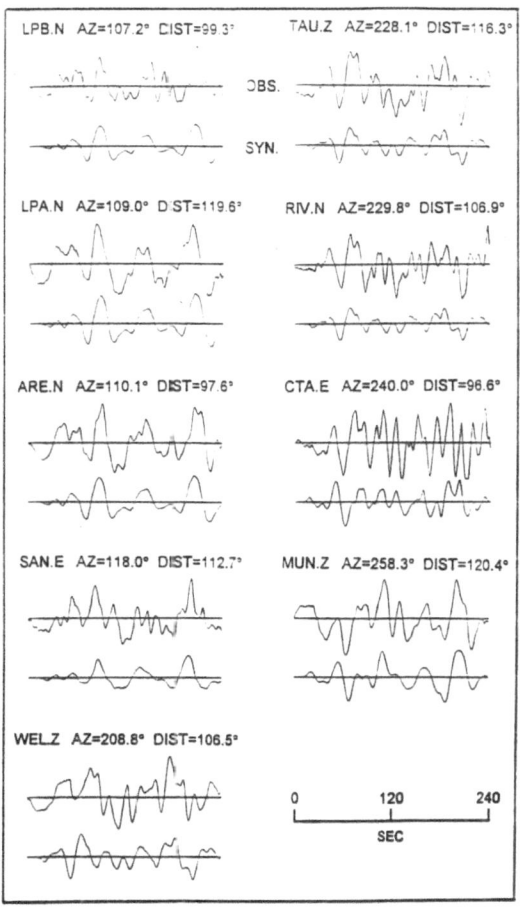

Figure 6(b)

Figure 6

a. Results of the multistation pulse stripping inversion technique of KIKUCHI and KANAMORI (1982). Space-time plot showing the distribution of moment pulses along the strike of a one-dimensional (ribbon) fault model. Each triangle represents the best temporal and spatial location of the subevent for each iteration. Unit wavelets used in the pulse stripping are calculated at a depth of 20 km for a trapezoidal moment pulse with a rise time of 11 seconds and a total duration of 32 seconds (i.e., 11–10–11 sec). The moment (area under the trapezoid) is proportional to the size of the triangle. The epicenter is at 0 km and the spatial dimension extends from −450 km (in the rupture direction of 30 from the epicenter) to +750 km (in the rupture direction of 210 from the epicenter). The fault model was limited to distances between −300 km and 750 km. The temporal scale starts at the origin time and extends to 240 seconds. The inversion is also constrained to disallow pulses with apparent rupture velocities of greater than 5.0 km/sec. The integrated moment density versus time plot is shown at the top of the figure. The two major moment pulses and their associated spatial-temporal locations are labeled A and B. b. Observed (top trace) and synthetic (bottom trace) seismograms for the fault model shown in Figure 6a.

of the epicenter at 175 seconds into the rupture (labeled B in Figure 6a). In this inversion the rupture velocity is not held constant. Given the constraints already mentioned this would suggest an average rupture velocity of about 3 km/sec. Our results suggest a predominant unilateral rupture to the southwest with two major regions of moment release.

The second inversion method we have used to investigate the rupture process of the 1964 earthquake is a tomographic imaging technique devised by RUFF (1987). With this method we invert the previously determined single-station source time functions for coherent features along a one-dimensional fault model. This inversion method uses *a priori* estimates of the rupture azimuth and velocity and iteratively inverts for the moment release distribution that best fits the observed source time functions in a least squares sense. We fix the rupture azimuth to 210° and the rupture velocity to 3 km/sec and invert the source time functions for an initial space-time rupture model (Figure 7a, upper half). We use only the nine source time functions with durations of 240 seconds (Figure 4) in the inversion. The fault model extends from −200 km (200 km in a direction of 30°; opposite the rupture direction) to 700 km in the rupture direction (210°). Synthetic source time functions (dashed lines in Figure 7a) are calculated for the rupture model shown. The match between the observed and synthetic source time functions is measured by the parameter e, the ratio of the error vector length to the data vector length. We iterate to improve the fit between the observed and synthetic source time functions for a final model (Figure 7a, lower half). By comparing the first and tenth iterations we can see that the moment release decreases northeast of the epicenter (negative distances) and increases to the southwest (positive distances). This indicates that the majority of the moment release occurred southwest of the epicenter and can adequately explain the observed source time functions. Additional inversion results in which negative distances were not included in the fault model showed equally good fits to the data. However, we cannot completely rule out any moment release east of the epicenter. Our results show two distinct regions of moment release shown by the solid bars in Figure 7a (labeled A and B). Figure 7b shows the time integrated moment density along the strike of the fault for our preferred model. We obtain a seismic moment of 3×10^{29} dyne cm from the nondiffracted P waves with a duration of more than 210 sec. We know this underestimates the body wave moment because recall that for the data with less than 210 sec of direct P wave we obtained a similar moment for just the first pulse. This is similar to the seismic moment obtained by KIKUCHI and FUKAO (1987) of 1.8×10^{29} dyne cm before they adjusted the moment in the epicentral region due to the semicircular nature of the rupture. The seismic moment release determined from the P waves can only account for about one-third of the total seismic moment identified from the surface waves; therefore we cannot estimate the true displacement at the asperities.

This type of inversion is useful for sparse data sets such as the present one, since it is not possible to uniquely invert for all parameters simultaneously. The two moment pulses are clearly defined, the first located near the epicenter extending to the southwest about 300 km and the second located between 500 km and 600 km southwest of the epicenter (Figure 7a). The location of the first and second moment pulses from this inversion are strongly dependent on the assumed rupture velocity which was constrained at 3 km/sec based on the inversion results from the pulse stripping technique. Figure 8 shows our preferred model for the asperity extent in map view. We include the entire epicentral region extending 300 km to the southwest and 100 km to the east giving rise to an asperity with a length scale of

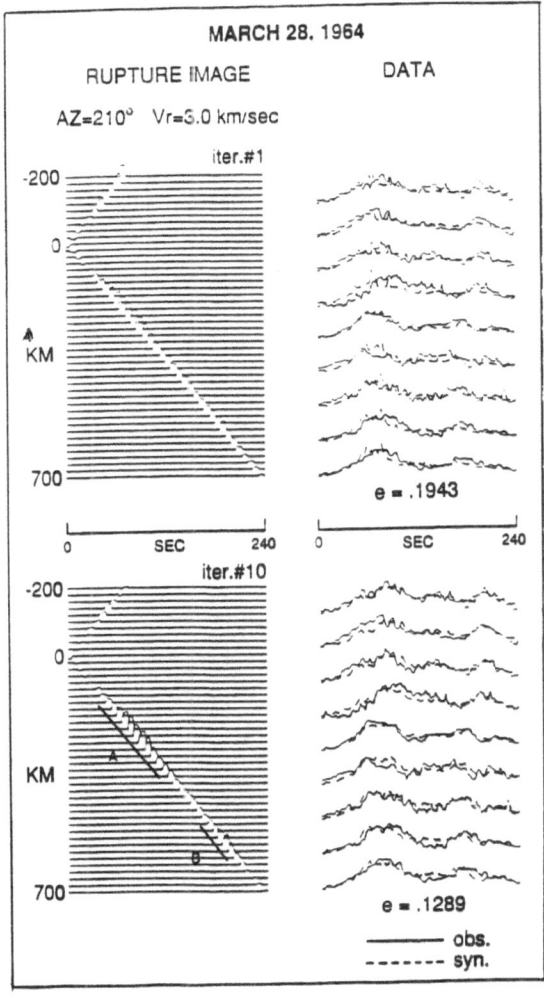

Figure 7(a)

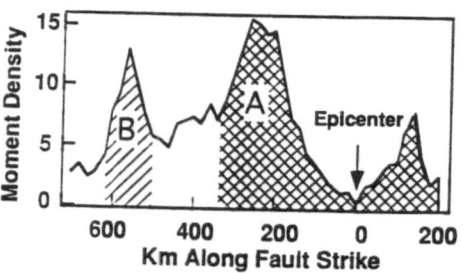

Figure 7(b)

Figure 7

a. Results from a constrained simultaneous inversion of source time functions. The top half of the figure represents the first iteration of the model (left) and the observed and synthetic source time functions (right). The observed source time functions (data) were previously determined from the single-station deconvolutions. The lower frame shows the same after the tenth iteration. The inversion is constrained by assuming a rupture azimuth of 210° (positive distances on the space-time plot) and a rupture velocity of 3.0 km/sec. The reduction of the error (e) is shown for the first and tenth iteration. The rupture process is represented by the space-time model which plots the moment release with time at 20 km increments along the fault. The two major pulses of moment release are highlighted by solid lines and labeled A and B. b. The along strike moment density results from the simultaneous inversion shown in Figure 7a. The moment density has units of 10^{26} dyne cm/km.

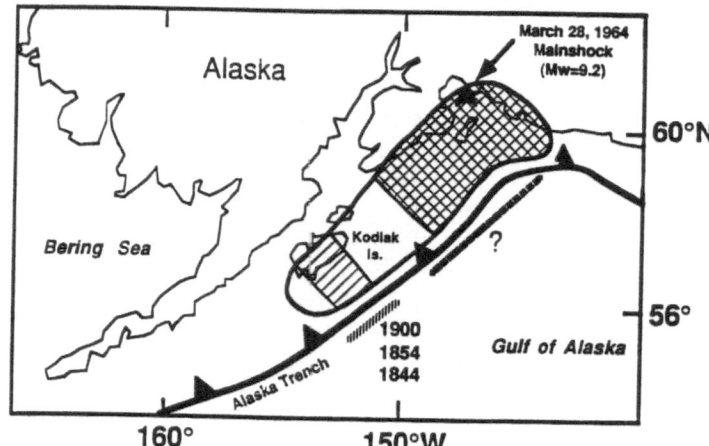

Figure 8

Map view of our interpretation of the rupture process of the March 28, 1964, Prince William Sound earthquake. The two pulses of moment release, representing the major asperities, are shown. The dates of historic earthquakes in the Kodiak Island segment are listed.

~400 km. The large tectonic uplift in the epicentral region is consistent with this asperity distribution (PLAFKER, 1969). The gradual growth of the first moment pulse is believed to be related to the expansion of the semicircular rupture from the epicenter; expanding both oceanward and toward the southwest during the initial stages of rupture (Figure 8). We would interpret this first moment pulse in the same fashion as RUFF and KANAMORI (1983) to represent the rupture of a very large asperity with a length scale of 400 km located near the epicenter. KIKUCHI and FUKAO (1987) also concluded that there was a semicircular expansion of the rupture area during the first 60–80 sec over a region of approximately 200 by 300 km. The second moment pulse represents the breaking of a second major, yet smaller, asperity located in the Kodiak Island region (Figure 8). The 1964 earthquake is thus a multiple asperity rupture.

Asperity Distribution

Two major asperities are associated with the rupture of the 1964 Prince William Sound earthquake. The largest asperity extends throughout the northeastern half of the fault area with a length of approximately 400 km. Our one-dimensional model shows the moment release slowly increasing to a peak at about 75 seconds. We believe that this ramp represents the growth in rupture area which starts from a point (at the hypocenter) and expands in a semicircular fashion until the rupture involves the entire width of the fault zone (RUFF and KANAMORI, 1983). The rupture then continues to extend toward the southwest, for about 300 km where the first moment pulse is truncated. We cannot resolve along-dip variations or comment on the idea that the up-dip portions of faults are less strongly coupled as suggested by BYRNE et al. (1988). The second asperity is located near Kodiak Island, between 500–600 km southwest of the epicenter. Moment release between 300 km and 500 km along the strike of the fault is relatively lower than that released in the asperities.

The maximum coseismic uplift (30 feet) in the Prince William Sound region corresponds to the large epicentral asperity (PLAFKER, 1969). HOLDAHL and SAUBER (1992) inverted these geodetic data for the distribution of displacements on the 1964 fault surface. Their results are remarkably similar to those found in this study. Large displacements in the Prince William Sound and Kenai Peninsula region correspond to the large epicentral asperity. A second region of high displacement was located in the Kodiak Island region corresponding to the second smaller asperity. Displacements in excess of 20 meters were found in the high slip areas surrounded by regions of much lower displacements of 0 to 10 meters. This supplies independent confirmation of the asperity distribution determined in this study.

It has been suggested that temporal variations of stress occur in the subducting slab as well as in the outer-rise as a result of underthrusting earthquake cycles

(CHRISTENSEN and RUFF, 1988; DMOWSKA et al., 1988; ASTIZ et al., 1988; LAY et al., 1989). DMOWSKA and LOVISON (1992) have extended this idea and analyzed the seismicity associated with large asperities in subduction zones. They analyzed the intraplate, intermediate depth and outer-rise events before and after the 1964 mainshock. Seismicity prior to 1964 is hard to evaluate due to the occurrence of most events before the installation of the WWSSN. However, earthquakes (depths > 50 km and m_b > 5.6) occurring up to 25 years after the 1964 mainshock tend to occur down-dip of the epicentral asperity (DMOWSKA and LOVISON, 1992). Similarly, outer-rise seismicity (m_b > 5.0) occurring in a 25-year time period after the mainshock also shows a clustering near the epicentral region as well as a more diffuse cluster trenchward of Kodiak Island near the second smaller asperity. DMOWSKA and LOVISON (1992) conclude that large-scale fault inhomogeneities, such as asperities in large subduction earthquakes, may influence the seismicity in the down-going slab as well as in the outer-rise trenchward of the asperity.

There are several possible contributing factors which may have led to the asperity distribution in the 1964 zone. The physical nature of seismically determined asperities is still in debate, hence, it is important to consider the tectonics and plate boundary segmentation in the region. The segmentation of plate boundaries has been suggested as a contributing factor to asperity distribution. The 1964 earthquake zone can be divided into several segments based on a variety of evidence such as the extent and truncation of historic earthquake aftershock zones, subduction of fracture zones and seamount chains, and rips and folds in the subducting or overriding plate.

The 1964 earthquake zone is truncated at the southwestern end by the 1938 rupture zone. This has apparently been a boundary for previous earthquake cycles (SYKES, 1971) and thus indicates segmentation of the arc. The geometry of the Wadati-Benioff zone changes dramatically at the northeastern edge of the 1964 rupture zone (PAGE et al., 1989) (Figure 9). To the west is the Wadati-Benioff zone associated with the Alaska-Aleutian trench and the associated volcanic arc. To the east is the Wrangell Wadati-Benioff zone with associated Wrangell volcanoes. These two subduction zone segments have very different trends and are apparently not continuous at depth. This margin is not well understood but must represent either a rip or fold of the subducting plate. This change in geometry probably acted as a geometric barrier to the eastern edge of the 1964 earthquake rupture.

KIENLE et al. (1983) suggest a segmentation of the 1964 zone based on changes in the trend or alignment of volcanoes. Three segments are defined, named from east to west the Cook Inlet, Katmai, and Semidi segments. The Katmai and Cook Inlet segments are separated by a pronounced 35° bend in the trend of the volcanic arc northeast of Kodiak Island that can be observed in the 100 km subducting plate contour, while the boundary between the Semidi and Katmai segments is based on a more subtle offset of the volcano alignment (KIENLE et al., 1983) (Figure 1). Changes in the alignment of volcanic chains presumably represent a similar change

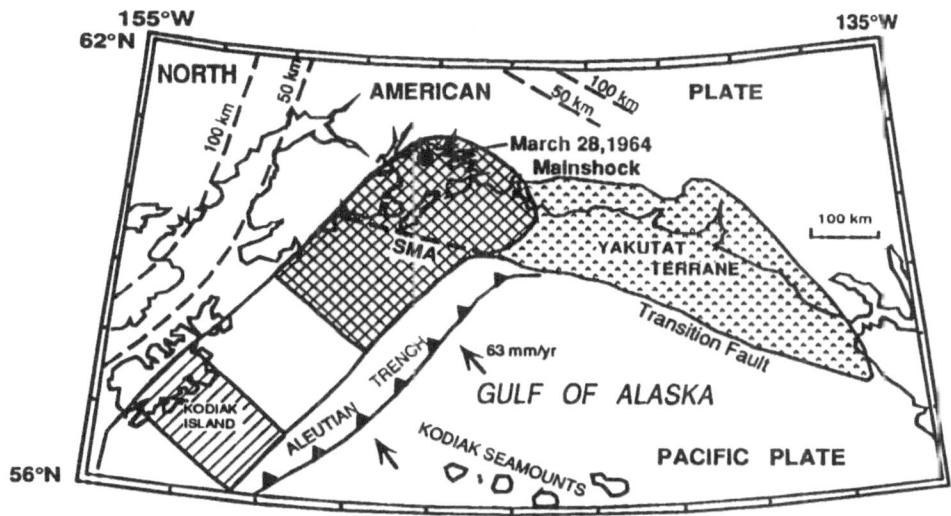

Figure 9

Tectonic map of southern Alaska showing the asperity distribution, epicenter, and aftershock area of the 1964 earthquake. Depth contours (50 and 100 km) of the subducting Pacific plate are shown by the dashed lines. The Alaska Wadati-Benioff zone is located on the left and the Wrangell Wadati-Benioff zone is located on the right. Also shown is the observed surficial boundary of the Yakutat Terrane and the possible extension of the Transition Fault or Slope Magnetic Anomaly (SMA) beneath the accretionary prism and the Prince William Sound region.

in the orientation of the subducting plate and may be caused by a rip, fold, or fault in the plate. The boundary between the Semidi and Katmai segments corresponds to the western edge of the 1964 aftershock area, whereas the Katmai, Cook Inlet segment boundary is located between the epicentral asperity in the Prince William Sound–Kenai Peninsula region and the Kodiak Island asperity (Figures 1 and 9).

There are two fracture zones defined by magnetic lineations on the down-going plate that intersect the Alaska trench between Prince William Sound and Kodiak Island (SCHWAB et al., 1980; NAUGLER and WAGEMAN, 1973; BRUNS, 1983). The first of these is the Slope Magnetic Anomaly (SMA) which is related to the trailing edge of the accreting Yakutat terrane and is discussed below (Figure 9). The second fracture zone does not have a well-defined bathymetry but is identified by the offset in the age of the magnetic anomalies. These fracture zones may also contribute to the segmentation of the 1964 rupture zone as seen by our P-waveform analysis.

The Kodiak seamount chain is being subducted on the down-going Pacific plate in the vicinity of Kodiak Island. The subduction of seamounts has been suggested as a source of asperities (CLOOS, 1992). Often it is difficult to correlate a specific subducted seamount to a specific seismic asperity because the seamount would already have been subducted. In the Kodiak Island region we do not see seamounts actually at the trench directly in front of the Kodiak Island asperity. In addition, the sizes of the seamounts observed are much smaller than the size of the asperity as viewed by long-period P waves. Although it is difficult to completely rule out the role of subducted seamounts we do not consider the seamounts near Kodiak Island as the cause of the asperity in that region.

The 1964 earthquake occurred in a region which is undergoing a fairly dramatic change from a normally dipping subduction zone in the Kodiak Island region (dip 13°) to a very shallowly dipping subduction zone beneath the Kenai Peninsula and Prince William Sound (dip 5°) (JACOB et al., 1977). This shallow dip gives rise to a very wide coupled zone and hence, contributes to the large size of the epicentral asperity. However, a large part of the interface contact zone consists of a large well-developed accretionary prism. BYRNE et al. (1988) suggested that portions of the plate interface associated with accretionary prisms are relatively uncoupled and hence, aseismic. Unfortunately, we cannot isolate the along-dip coseismic moment release in order to comment on the along-dip variations in coupling.

Subduction of sediments has been suggested as a factor contributing to the strength of seismic coupling between plates. As previously mentioned, there is a large accretionary prism associated with the Alaska subduction zone. VON HUENE and SCHOLL (1991) estimate that only 20–30% of the sediments are accreted and 70–80% of the sediments are subducted along the Alaska and eastern Aleutian subduction zone. RUFF (1989) proposed that excess trench sediments may be a contributing factor to the size of earthquakes and hence, the size of asperities.

The collision of the Yakutat terrane along this margin complicates the interaction between the Pacific and North American plates (Figure 9). The Yakutat terrane, which is a newly accreting terrane, is exposed to the east of the 1964 rupture area and may also be subducting beneath Prince William Sound and thus playing an important role in the asperity distribution. The Yakutat terrane as seen today is about 600 km long by 200 km wide and is largely connected to the Pacific plate (PAGE et al., 1989). The south (trailing) edge of the Yakutat block is truncated by the Transition fault which may represent a fossil fracture zone. The Slope Magnetic Anomaly (SMA) which coincides with the Transition fault can be followed under the accretionary prism adjacent to the Prince William Sound area (BRUNS, 1983; SCHWAB et al., 1980). This suggests that the Yakutat terrane may be much more extensive than previously believed and may be subducting beneath the Prince William Sound–Kenai Peninsula area. Based on results from the Trans-Alaska Crustal Transect (TACT) profile, PAGE et al. (1992) suggested that the Yakutat terrane is being subducted in the Prince William Sound region and may be responsible for the strongly coupled nature of this interplate zone. Their results suggest that the eastern section of the 1964 earthquake occurred between the upper plate and the subducting Yakutat terrane. The Yakutat terrane is a composite terrane with both continental and oceanic crustal components (PLAFKER, 1987); hence it may be more buoyant than the Pacific plate. The effect which this may have on the coupling of the region is unclear, except that we might expect that the more buoyant terrane would increase coupling between the Pacific and North American plates and possibly influence the dip of the contact zone between the plates.

In summary, there are at least four factors that may have contributed to the distribution of asperities and the large size of the epicentral asperity of the great 1964 Prince William Sound earthquake: (1) geometric segmentation of the subducting slab related to fracture zones, seamount chains, or rips and tears in the subducting plate, (2) the shallow dip and wide interplate contact zone in the Prince William Sound region, (3) the large amount of trench sediments that are being subducted, and (4) the subduction of the Yakutat terrane which is probably more buoyant than normal oceanic crust.

Previous Earthquakes

The 1964 Alaska earthquake is the largest event to occur along the Alaska-Aleutian subduction zone in recorded history. The historic earthquake record indicates that this portion of the subduction zone is segmented into two regions with different historic earthquake records. The region between 145°W and 152°W (Figure 8), including the 1964 epicentral region, has no major historic earthquakes (since 1780) reported. In contrast, the segment between 152°W and 155°W near Kodiak Island failed previously in 1900 ($M_w = 7.8$), 1854, 1844, 1792 and possibly in 1788 (DAVIES et al., 1981; NISHENKO and JACOBS, 1990). Reports of large and great earthquakes near Kodiak Island during the last 200 years suggest a recurrence interval of approximately 60 years (NISHENKO and JACOBS, 1990).

The geologic data such as uplifted terraces on Middleton Island in the 1964 epicentral segment indicate a geologic recurrence interval of large events (based on radiocarbon dates from 6 terraces) of $1021 + 362$ years (PLAFKER, 1986; NISHENKO and BULAND, 1987). Regional subsidence events as indicated by buried organic soil layers and submerged tree stumps suggest 10 major submergence events during the last 4500 years in the Anchorage and southcentral Alaska area (COMBELLICK and REGER, 1988). The intervals between the submergence events range from 425 to 800 years. Four of these events correspond with the dates for uplifted terraces on Middleton Island. The submergence events suggest a shorter recurrence interval than the uplifted terraces. However, both uplifted terrace and submergence data suggest long recurrence intervals between 400 and 1000 years in the 1964 epicentral region.

These two segments defined by the historic earthquake record correspond to the two segments determined from our study of the P waves of the 1964 Alaska earthquake. Thus the 1964 Alaska earthquake failed in two asperities which correspond to the segmentation of the plate boundary identified in the historic and geologic earthquake record. The very large asperity near the epicenter with a total length of approximately 400 km appears to fail in great earthquakes with a long recurrence interval. The smaller Kodiak Island asperity can fail independently with a much shorter recurrence interval in approximately magnitude 7.5 to 8.0 earthquakes. The 1964 segment of the plate boundary fails differently in successive earthquake cycles. This type of variation in the rupture mode has been identified along many other subduction zones including the Aleutian arc (THATCHER, 1990; BECK and CHRISTENSEN, 1991).

Comparison with the 1957 Aleutian and the 1965 Rat Islands Earthquakes

There have been two other great earthquakes, in addition to the 1964 event, along the Alaska-Aleutian arc in the second half of this century, the 1957 Aleutian ($M_w = 8.6$) and 1965 Rat Islands ($M_w = 8.7$) earthquakes. All three events show multiple pulse ruptures which may be related to segmentation of the rupture zones, and all three segments show variations in the earthquake rupture mode between successive earthquake cycles. However, there are some major differences in the style of these great earthquake ruptures.

The March 9, 1957, earthquake occurred along the central Aleutian arc with an aftershock length of 1000–1200 km. Although the details of the rupture of the 1957 event are poorly known, a source time function determined from a long-period P wave record at HAL suggests a multiple pulse rupture (Figure 10). Investigations of the 1957 earthquake rupture based on numerous kinds of data indicate a heterogeneous rupture, although details vary greatly (HATORI, 1981; HOUSE et al., 1981; SYKES, 1971; BOYD, 1986; WAHR and WYSS, 1980; RUFF et al., 1985; JOHNSON and SATAKE, 1993; JOHNSON et al., present volume). Hence, we would characterize the 1957 earthquake zone as having several small asperities. Part of the 1957 rupture zone subsequently failed again in the 1986 Andreanof Islands earthquake ($M_w = 8.0$) (HOUSTON and ENGDAHL, 1989; BOYD and NABELEK, 1988). A comparison of the 1957 and 1986 waveforms and deconvolved source time functions suggest a similar asperity size and distribution (Figure 10). The historic earthquake record indicates that parts of this segment failed in a series of magnitude 7.5–8.0 events in the early 1900s. The behavior of this segment of the central Aleutian arc suggests a greater fault heterogeneity (i.e., smaller more numerous asperities) than the 1964 Alaska segment.

The February 4, 1965, Rat Islands earthquake occurred along the western part of the Aleutian arc with an aftershock length of 600 km and a source duration of 160 seconds. In a previous study we identified three dominant asperities associated with the rupture of the 1965 event (BECK and CHRISTENSEN, 1991). The scale length of the asperities determined for the 1965 earthquake appears to be similar to the Kodiak Island asperity of the 1964 rupture. The 1965 asperities are much smaller and not nearly as smooth as the dominant epicentral asperity of the 1964 earthquake as recorded by long-period P waves (Figure 10). Similarly the asperities

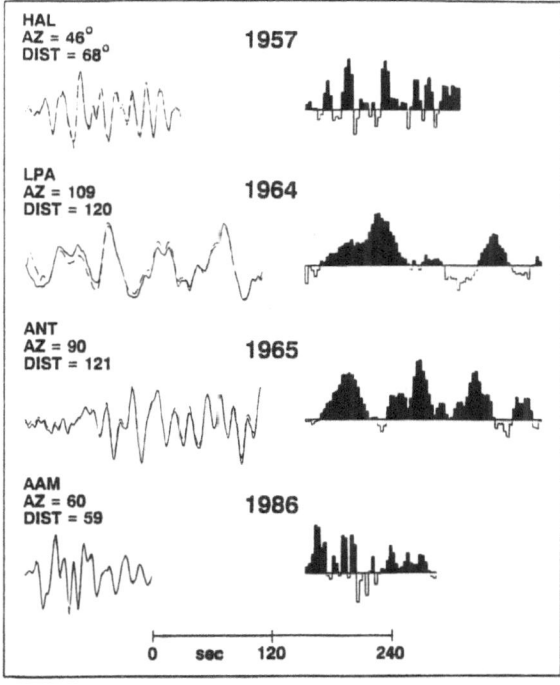

Figure 10
Comparison of typical seismograms and source time functions for four of the largest events in the Alaska-Aleutian arc. The events from top to bottom are the 1957 Aleutian (3/9/57; $M_w = 8.6$), the 1964 Prince William Sound (3/28/64; $M_w = 9.2$), the 1965 Rat Islands (2/4/65; $M_w = 8.7$), and the 1986 Andreanof Islands (5/7/85; $M_w = 8.0$) earthquakes.

associated with the 1957 rupture are slightly smaller and more numerous than those associated with either the 1965 or 1964 ruptures.

One feature that makes the 1964 earthquake different from other great earthquakes is the extremely large size of the epicentral asperity. At this time we know of no other asperity which approaches the size of the 1964 epicentral asperity. The extremely large size of the epicentral asperity set the 1964 Alaska earthquake apart from other great earthquakes.

Conclusions

The March 28, 1964, Prince William Sound earthquake ($M_w = 9.2$) occurred as a result of underthrusting of the Pacific plate beneath the North American plate along an 800 km long stretch of the Alaska subduction zone. This is the second largest recorded earthquake and hence, an extremely important event. Due to its large size there are limited on-scale *P* waves with which to analyze details of the rupture history. However, coupled with results from previous surface wave studies we are able to determine the temporal and spatial moment release for this earthquake. Surface wave directivity studies, the aftershock distribution, and the general strike of the plate boundary all suggest a rupture azimuth of approximately 210°. Constraining the rupture azimuth in our *P*-wave inversions resulted in two pulses of moment release occurring along strike. The first and largest moment pulse has a duration of approximately 100 seconds with a relatively smooth onset which reaches a peak moment release rate at about 75 seconds into the rupture. The second smaller pulse of moment release starts approximately 160 seconds after the origin time and has a duration of approximately 40 seconds. The first moment pulse most likely initiated near the epicenter and then spread over the fault surface in a semicircular fashion, before extending approximately 300 km toward the southwest as suggested by RUFF and KANAMORI (1983). The total spatial extent of this asperity is approximately 400 km. The second moment pulse is located in the vicinity of Kodiak Island, between approximately 500 km and 600 km southwest of the epicenter.

The 1964 Alaska earthquake is the largest event to occur along the Alaska-Aleutian subduction zone in recorded history. Although there remain large uncertainties in the actual displacements and recurrence intervals of historic earthquakes, the historic record indicates that this portion of the subduction zone is segmented into two regions with different historic earthquake records. The 1964 epicentral region (between 145°W and 152°W) has had no major historic earthquakes (since 1780) reported. Geologic data suggested a recurrence interval on the order of 400–1000 years. In contrast, reports of large and great earthquakes near the Kodiak Island segment during the last 200 years suggest a recurrence interval of approximately 60 years (NISHENKO and JACOBS, 1990).

There are at least four factors that may have contributed to the distribution of asperities and the large size of the epicentral asperity of the great 1964 Prince William Sound earthquake: (1) geometric segmentation of the subducting slab related to fracture zones, seamount chains, or rips and tears in the subducting plate, (2) the shallow dip and wide interplate contact zone in the Prince William Sound region, (3) the large amount of trench sediments that are being subducted, and (4) the subduction of the Yakutat terrane which is probably more buoyant than normal oceanic crust.

Acknowledgments

We would like to thank George Zandt, Bob Page, and Hiroo Kanamori for useful discussions and Leslie McCaullough for her help in gathering the data and starting the preliminary study. We also thank R. Dmowska and M. Kikuchi for helpful reviews. This research was supported by NSF grants (EAR89–03660) to D. H. Christensen and (EAR90–17358) to S. L. Beck. Support was also received from IGPP at Lawrence Livermore National Laboratory. This is SASO contribution number 19.

References

ASTIZ, L., LAY, T., and KANAMORI, H. (1988), *Large Intermediate Depth Earthquakes and the Subduction Process*, Phys. Earth Planet. Inter. *53*, 80–166.

BECK, S. L., and CHRISTENSEN, D. H. (1991), *Rupture Process of the February 4, 1965, Rat Islands Earthquake*, J. Geophys. Res. *96*, 2205–2221.

BOYD, T. M. (1986), *The Andreanof Islands Earthquake: Previous Rupture History* (abstract), EOS *67*, 1082.

BOYD, T. M., and NABELEK, J. L. (1988), *Rupture Process of the Andreanof Islands Earthquake of May 7, 1986*, Bull. Seismol. Soc. Am. *78*, 1653–1673.

BRUNS, T. R. (1983), *Model for the Origin of the Yakutat Block, an Accreting Terrane in the Northern Gulf of Alaska*, Geology *11*, 718–721.

BYRNE, D. E., DAVIS, D. M., and SYKES, L. R. (1988), *Loci and Maximum Size of Thrust Earthquakes and the Mechanics of the Shallow Region of Subduction Zones*, Tectonics *7*, 833–857.

CHRISTENSEN, D. H., and RUFF, L. J. (1988), *Seismic Coupling and Outer-rise Earthquakes*, J. Geophys. Res. *93*, 13,421–13,444.

CLOOS, M. (1992), *Thrust-type Subduction-zone Earthquakes and Seamount Asperities: A Physical Model for Seismic Rupture*, Geology *20*, 601–604.

COMBELLICK, R. A., and REGER, R. D. (1988), *Evaluation of Holocene Subsidence Events of Cook Inlet Estuarine Flats near Anchorage Alaska, as a Basis for Assessing Seismic Hazards in Southcentral Alaska*, U.S. Geol. Sur. Open File Rep. *88-434*, 463–464.

DAVIES, J. N., and HOUSE, L. (1979), *Aleutian Subduction Zone Seismicity, Volcano-trench Separation, and their Relation to Great Thrust-type Earthquakes*, J. Geophys. Res. *84*, 4583–4591.

DAVIES, J. N., SYKES, L., HOUSE, L., and JACOB, K. (1981), *Shumagin Seismic Gap, Alaskan Peninsula: History of Great Earthquakes, Tectonic Setting, and Evidence for High Seismic Potential*, J. Geophys. Res. *86*, 3821–3855.

DEMETS, C., GORDON, R. G., ARGUS, D. F., and STEIN, S. (1990), *Current Plate Motions*, Geophys. J. Inter. *101*, 425–478.

DMOWSKA, R., RICE, J. R., LOVISON, L. C., and JOSELL, D. (1988), *Stress Transfer and Seismic Phenomena in Coupled Subduction Zones during the Earthquake Cycle*, J. Geophys. Res. *93*, 7869–7884.

DMOWSKA, R., and LOVISON, L. C. (1992), *Influence of Asperities along Subduction Interfaces on the Stressing and Seismicity of Adjacent Areas*, Tectonophysics *211*, 23–43.

FUIS, G. S., and AMBOS, E. L., *Deep structure of the Contact Fault and Prince William Terrane: Preliminary results of the 1985 TACT seismic-refraction survey*. In *The U.S. Geological Survey in Alaska—Accomplishments during 1985* (Bartsch-Winkler, S. ed.) (U.S. Geol. Surv. Circ. *978*, 1986) pp. 41–45.

FUIS, G. S., AMBOS, E. L., MOONEY, W. D., CHRISTENSEN, N. I., and GEIST, E. (1991), *Crustal Structure of Accreted Terranes in Southern Alaska, Chugach Mountains and Copper River Basin, from Seismic Refraction Results*, J. Geophys. Res. *96*, 4187–4227.

FUIS, G. S., and PLAFKER, G. (1991), *Evolution of Deep Structure along the Trans-Alaska Crustal Transect, Chugach Mountains and Copper River Basin, Southern Alaska*, J. Geophys. Res. *96*, 4229–4253.

FURUMOTO, A. S. (1965), *Analysis of Rayleigh Wave. Part II in Source Mechanism Study of the Alaska Earthquake and Tsunami of March 27, 1964*, Report HIG–65–17, Honolulu: University of Hawaii, Institute of Geophysics, December, 31–42.

HARDING, S. T., and ALGERMISSEN, S. T. (1969), *Focal Mechanism of the Prince William Sound, Alaska Earthquake of March 28, 1964*, Bull. Seismol. Soc. Am. *59*, 799–811.

HATORI, T. (1981), *Tsunami Magnitude and Source Area of Aleutian-Alaska Tsunamis*, Bull. Earthquake Res. Inst. *56*, 97–110.

HOLDAHL, S. R., and SAUBER, J. M. (1992), *Coseismic Slip and Post-seismic Deformation Associated with the 1964 Prince William Sound Earthquake*, Wadati Conference on Great Subduction Earthquakes, Sept. 16–19, 1992, pp. 83–89.

HOUSE, L. S., SYKES, L. R., DAVIES, J. N., and JACOB, K. H., *Identification of a possible seismic gap near Unalaska Island, eastern Aleutians, Alaska*. In *Earthquake Prediction — An International Review*, Maurice Ewing Ser., vol. 4 (Simpson, D. W., and Richards, P. G., eds.) (American Geophysical Union, Washington, D.C. 1981) pp. 81–92.

HOUSTON, H., and ENGDAHL, E. R. (1989), *A Comparison of the Spatio-temporal Distribution of Moment Release for the 1986 Andreanof Islands Earthquake with Relocated Seismicity*, Geophys. Res. Lett. *16*, 1421–1424.

JACOB, K. H., NAKAMURA, K., and DAVIES, J. N., *Trench-volcano gap along the Alaska-Aleutian arc: Facts, and speculations on the role of terrigenous sediments for subduction*. In *Island Arcs, Deep Sea Trenches, and Back-arc Basins*, Maurice Ewing Series, vol. 1 (Talwani, M., and Pitman III, W. C., eds.) (Am. Geophys. Union, Washington, D.C. 1977) pp. 243–258.

JOHNSON, J. M., and SATAKE, K. (1993), *Source Parameters of the 1957 Aleutian Earthquake from Tsunami Waveforms*, Geophys. Res. Lett. *20*, 1487–1490.

JOHNSON, J. M., TANIOKA, Y., RUFF, L. J., SATAKE, K., KANAMORI, H., and SYKES, L. R. (1994), *The 1957 Great Aleutian Earthquake*, Pure and Appl. Geophys, present volume.

KANAMORI, H. (1970), *The Alaska Earthquake of 1964: Radiation of Long-period Surface Waves and Source Mechanism*, J. Geophys. Res. *75*, 5029–5040.

KANAMORI, H. (1977), *The Energy Release in Great Earthquakes*, J. Geophys. Res. *82*, 2981–2987.

KIENLE, J., SWANSON, S. E., and PULPAN, H., *Magmatism and subduction in the eastern Aleutian arc*. In *Arc Volcanism: Physics and Tectonics* (Shimozuru, D., and Yokoyama, I., eds.) (Terra Scientific Publishing Company, Tokyo 1983) pp. 191–224.

KIKUCHI, M., and KANAMORI, H. (1982), *Inversion of Complex Body Waves*, Bull. Seismol. Soc. Am. *72*, 491–506.

KIKUCHI, M., and FUKAO, Y. (1987), *Inversion of Long-period P-waves from Great Earthquakes along Subduction Zones*, Tectonophysics *144*, 231–247.

LAHR, J. C., and PLAFKER, G. (1980), *Holocene Pacific-North American Plate Interaction in Southern Alaska: Implications for the Yakataga Seismic Gap*, Geology *8*, 483–486.

LAY, T., ASTIZ, L., KANAMORI, H., and CHRISTENSEN, D. H. (1989), *Temporal Variation of Large Interplate Earthquakes in Coupled Subduction Zones*, Phys. Earth Planet. Inter. *54*, 258–312.

NATIONAL ACADEMY OF SCIENCES, *The Great Alaska Earthquake of 1964*, 8 Volumes (Washington, D.C. 1972).

NAUGLER, F. P., and WAGEMAN, J. M. (1973), *Gulf of Alaska: Magnetic Anomalies, Fracture Zones, and Plate Interactions*, Geol. Soc. Am. Bull. *84*, 1575–1584.

NISHENKO, S. P., and BULAND, R. (1987), *A Generic Recurrence Interval Distribution for Earthquake Forecasting*, Bull. Seismol. Soc. Am. *77*, 1382–1399.

NISHENKO, S. P., and JACOB, K. H. (1990), *Seismic Potential of the Queen Charlotte-Alaska-Aleutian Seismic Zone*, J. Geophys. Res. *95*, 2511–2532.

PAGE, R. A., STEPHENS, C. D., and LAHR, J. C. (1989), *Seismicity of the Wrangell and Aleutian Wadati-Benioff Zones and the North American Plate along the Trans-Alaska Crustal Transect, Chugach Mountains and Copper River Basin, Southern Alaska*, J. Geophys. Res. *94*, 16059–16082.

PAGE, R. A., LAHR, J. C., STEPHENS, C. D., FOGLEMAN, K. A., BROCHER, T. M., and FISHER, M. A. (1992), *Seismicity and Stress Orientation in the Alaska Subduction Zone after the Great 1964 Earthquake and Speculation on the Origin of a Giant Asperity*, Wadati Conference on Great Subduction Earthquakes, Sept. 16–19, 1992, 31–32.

PLAFKER, G. (1969), *Tectonics of the March 27, 1964 Alaska Earthquake*, U.S. Geol. Surv. Prof. Pap. *543-I*, 74 pp.

PLAFKER, G. (1986), *Geologic Studies Related to Earthquake Potential and Recurrence in the "Yakataga Seismic Gap,"* U.S. Geol. Surv. Open File Rep. *86–92*, 135–143.

PLAFKER, G., *Regional geology and petroleum potential of the northern Gulf of Alaska continental margin*. In *Geology and Resource Potential of the Continental Margin of Western North America and Adjacent Ocean Basins*, Earth Sci. Ser., vol. 6 (Scholl, D. W., Grantz, A., and Vedder, J. G., eds.) (American Association of Petroleum Geologists, Tulsa, Okla. 1987) pp. 229–268.

RUFF, L., and KANAMORI, H. (1983), *The Rupture Process and Asperity Distribution of Great Earthquakes from Long-period Diffracted P Waves*, Phys. Earth Planet. Inter. *31*, 202–230.

RUFF, L., *Tomographic imaging of seismic sources*. In *Seismic Tomography* (Nolet, G., ed.) (D. Riedel Publishing Company, Dordrecht, Holland 1987) pp. 339–366.

RUFF, L. (1989), *Do Trench Sediments Affect Great Earthquake Occurrence in Subduction Zones?*, Pure and Appl. Geophys. *129*, 263–282.

RUFF, L., KANAMORI, H., and SYKES, L. (1985), *The 1957 Great Aleutian Earthquake* (abstract), EOS *66*, 298.

SCHWAB, W. C., BRUNS, T. R., and VON HUENE, R. (1980), *Maps Showing Structural Interpretation of Magnetic Lineaments in the Northern Gulf of Alaska, Scale 1:1,500,000*, U.S. Geol. Surv. Misc. Field Stud. Map, MF–1245.

SHERBURNE, R. W., ALGERMISSEN, S. T., and HARDING, S. T., *The hypocenter, origin time, and magnitude of the Prince William Sound earthquake of March 28, 1964*. In *The Prince William Sound, Alaska, Earthquake of 1964 and Aftershocks*, vol. 2 (Leipold, L. E., ed.) (U.S. Department of Commerce, Environmental Science Services Administration, Washington, D.C. 1969) pp. 49–69.

STAUDER, W., and BOLLINGER, G. A. (1966), *The Focal Mechanism of the Alaska Earthquake of March 28, 1964, and of its Aftershock Sequence*, J. Geophys. Res. *71*, 5283–5296.

SYKES, L. R. (1971), *Aftershock Zones of Great Earthquakes, Seismicity Gaps, and Earthquake Prediction for Alaska and the Aleutians*, J. Geophys. Res. *76*, 8021–8041.

THATCHER, W. (1990), *Order and Diversity in the Modes of Circum-Pacific Earthquake Recurrence*, J. Geophys. Res. *95*, 2609–2623.

TICHELAAR, B. W., and RUFF, L. J. (1993), *Depth of Seismic Coupling along Subduction Zones*, J. Geophys. Res. *98*, 2017–2037.

VON HUENE, R., and SCHOLL, D. W. (1991), *Observations at Convergent Margins Concerning Sediment Subduction, Subduction Erosion, and the Growth of Continental Crust*, Reviews of Geophys. *29*, 279–316.

WAHR, J., and WYSS, M. (1980), *Interpretation of Postseismic Deformation with a Viscoelastic Relaxation Model*, J. Geophys. Res. *85*, 6471–6477.

WYSS, M., and BRUNE, J. N. (1967), *The Alaska Earthquake of 28 March 1964: A Complex Multiple Rupture*, Bull. Seismol. Soc. Am. *57*, 1017–1023.

(Received September 20, 1993, revised/accepted January 27, 1994)

PAGEOPH, Vol. 142, No. 1 (1994)

0033–4553/94/010055–28$1.50 + 0.20/0
© 1994 Birkhäuser Verlag, Basel

Coseismic Slip in the 1964 Prince William Sound Earthquake: A New Geodetic Inversion

Sandford R. Holdahl[1] and Jeanne Sauber[2]

Abstract —The 1964 Prince William Sound earthquake (March 28, 1964; $M_w = 9.2$) caused crustal deformation over an area of approximately 140,000 km² in south central Alaska. In this study geodetic and geologic measurements of this surface deformation were inverted for the slip distribution on the 1964 rupture surface. Previous seismologic, geologic, and geodetic studies of this region were used to constrain the geometry of the fault surface. In the Kodiak Island region, 28 rectangular planes (50 by 50 km each) oriented ~218°N, with a dip varying from 8° nearest the Aleutian trench to 9° below Kodiak Island, define the rupture surface. In the Prince William Sound region 39 planes with variable dimensions (~40 by 50 km near the trench, ~64 by 50 km inland) and orientation (218°N in the west and 270°N in the east) were used to approximate the complex faulting. Prior information was introduced to constrain offshore dip-slip values, the strike-slip component, and slip variation between adjacent planes. Our results suggest a variable dip-slip component with local slip maximums occurring near Montague Island (up to ~30 m), further to the east near Kayak Island (up to ~14 m), and trenchward of the northeast segment of Kodiak Island (up to ~17 m). A single fault plane dipping 30°NW, corresponding to the Patton Bay fault, with a slip value of ~8 m modeled the localized but large uplift on Montague Island. The moment calculated on the basis of our geodetically derived slip model of 5.0×10^{29} dyne cm is 30% less than the seismic moment of 7.5×10^{29} dyne cm calculated from long-period surface waves (Kanamori, 1970) but is close to the seismic moment of 5.9×10^{29} dyne cm obtained by Kikuchi and Fukao (1987).

Key words: Alaska earthquake, Prince William Sound, modeling, Kodiak, coseismic slip, geodetic, Kenai Peninsula.

Introduction

The 1964 Prince William Sound earthquake (March 28, 1964, $M_w = 9.2$) occurred in the eastern Aleutians of Alaska where the oceanic Pacific plate is being subducted under continental North American plate (Figure 1). The predicted rate of relative plate motion between the Pacific and North American plates is approximately 6 cm/yr at N22°W (DeMets *et al.*, 1987). Following the occurrence of the

---

[1] National Geodetic Survey, C&GS, National Ocean Service, NOAA, Silver Spring, MD 20910-3282, U.S.A.
[2] Geodynamics Branch, Lab. for Terrestrial Physics, NASA/Goddard Space Flight Center, Greenbelt, MD 20771, U.S.A.

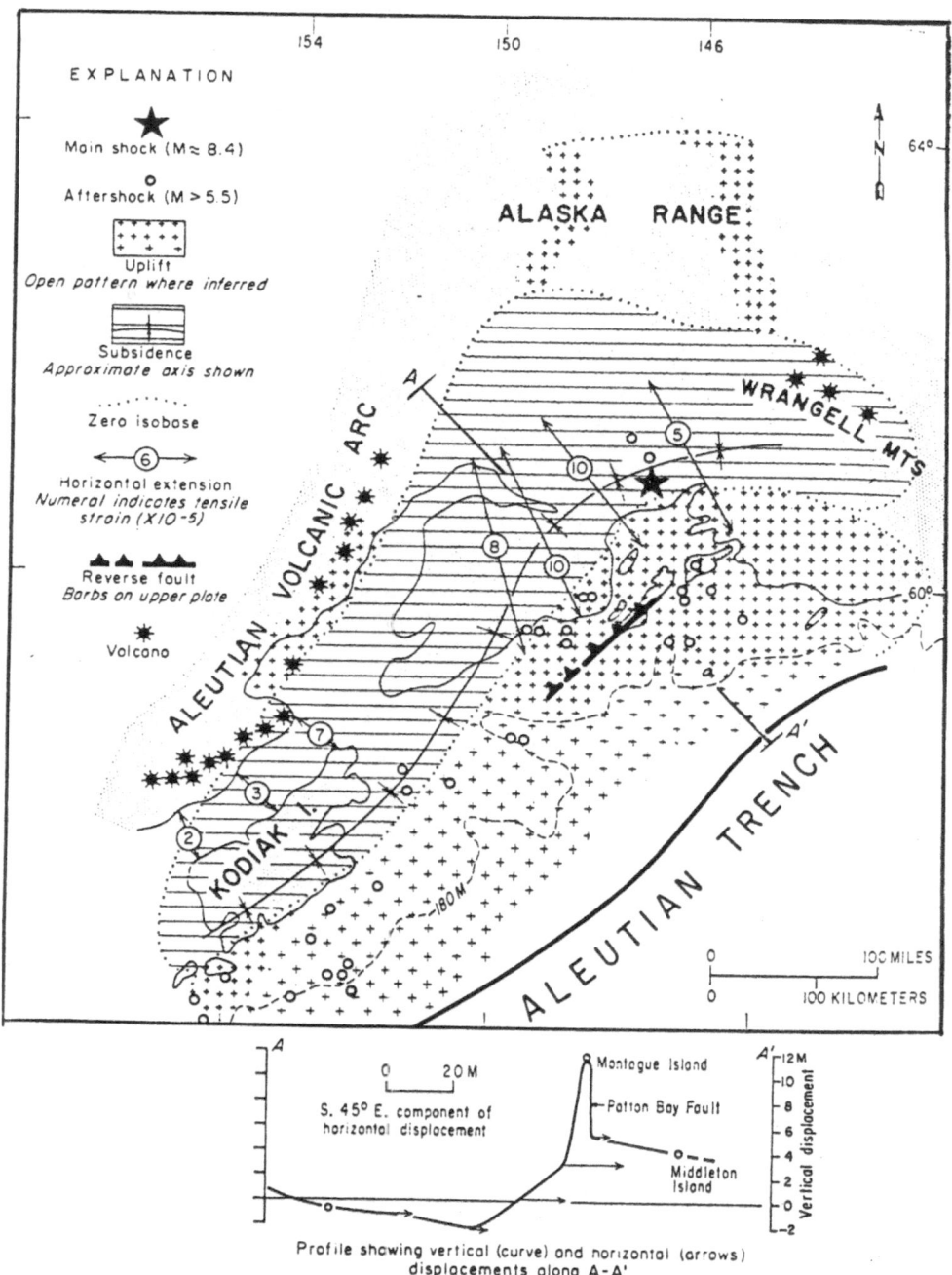

Figure 1

Tectonic displacements and seismicity associated with the 1964 Alaskan earthquake relative to the
Aleutian trench and volcanic arc (from PLAFKER, 1972). Tectonic data after PLAFKER (1969) and
unpublished USCGS data; epicenters from PAGE (1968).

1964 earthquake, numerous studies have added to our understanding of this great earthquake [a summary of the studies made shortly after the earthquake is given by WOOD (1966) and PLAFKER (1969, 1972)]. The extent of the zone that ruptured in the main shock has been inferred by aftershock studies (STAUDER and BOLLINGER, 1966; ALGERMISSEN et al., 1969) and modeling of geologic and geodetic data (PLAFKER, 1969, 1972; HASTIE and SAVAGE, 1970; PRESCOTT and LISOWSKI, 1977; MIYASHITA and MATSU'URA, 1978). The aftershock region was centered near Montague Island and extended outward approximately 300 km to the east and 500 km to the southwest. Body- and surface-wave modeling of the main shock (WYSS and BRUNE, 1967; KANAMORI, 1970; RUFF and KANAMORI, 1983; CHRIS-TENSEN and BECK, 1993), and the pattern of aftershocks suggest the Alaskan earthquake ruptured in a series of subevents. More recent seismicity studies (DAVIES and HOUSE, 1979; PAGE et al., 1989) and reflection/refraction profiles (BROCHER et al., 1993; MOORE et al., 1989) support the interpretation that the rupture plane corresponds to a shallow, dipping megathrust that accommodates subduction of the Pacific plate.

Geologic and geodetic data were used to delineate a zone of uplift which extended west and southwest from the vicinity of Cape Yakataga to the southern-most tip of Kodiak Island (Figure 1). The maximum uplift of 11 m was located at the southwest end of Montague Island. To the northwest of the uplift zone, there was a parallel trough of subsidence that included most of Kodiak Island, the Kenai Peninsula, and continued east to the Wrangell Mountains (Figure 1). The maxi-mum subsidence was 2.5 m, centered 30 km northeast of Whittier.

In this study we improve upon earlier geodetically derived coseismic slip models (HASTIE and SAVAGE, 1970; MIYASHITA and MATSU'URA, 1978) by inverting simultaneously vertical and horizontal geodetic, tide gauge, and geologic data for a more detailed distribution of coseismic slip by using the fault geometry inferred from other geophysical information. We examine the hypothesis that there were regions of high slip separated by regions of more moderate slip.

Data from the Coseismic Time Interval

Figure 2 shows the locations of leveling, triangulation, tide gauge and coastline height change measurements used in our study. Table 1 summarizes the uncertain-ties of the measurement types. The leveling data were input as observed differences of elevation (from observations made both before and after the earthquake), and the remaining vertical data types (tides, barnacles, etc.) were input as point observations of coseismic elevation change. A total of 216 leveled height differences and 140 coseismic elevation changes were included. One hundred and forty-six horizontal coseismic displacements computed by SNAY et al. (1987) from triangula-tion data were also used. Specific details on the data are given in the Appendix. Some of the geodetic and geologic measurements were made up to 14 months after

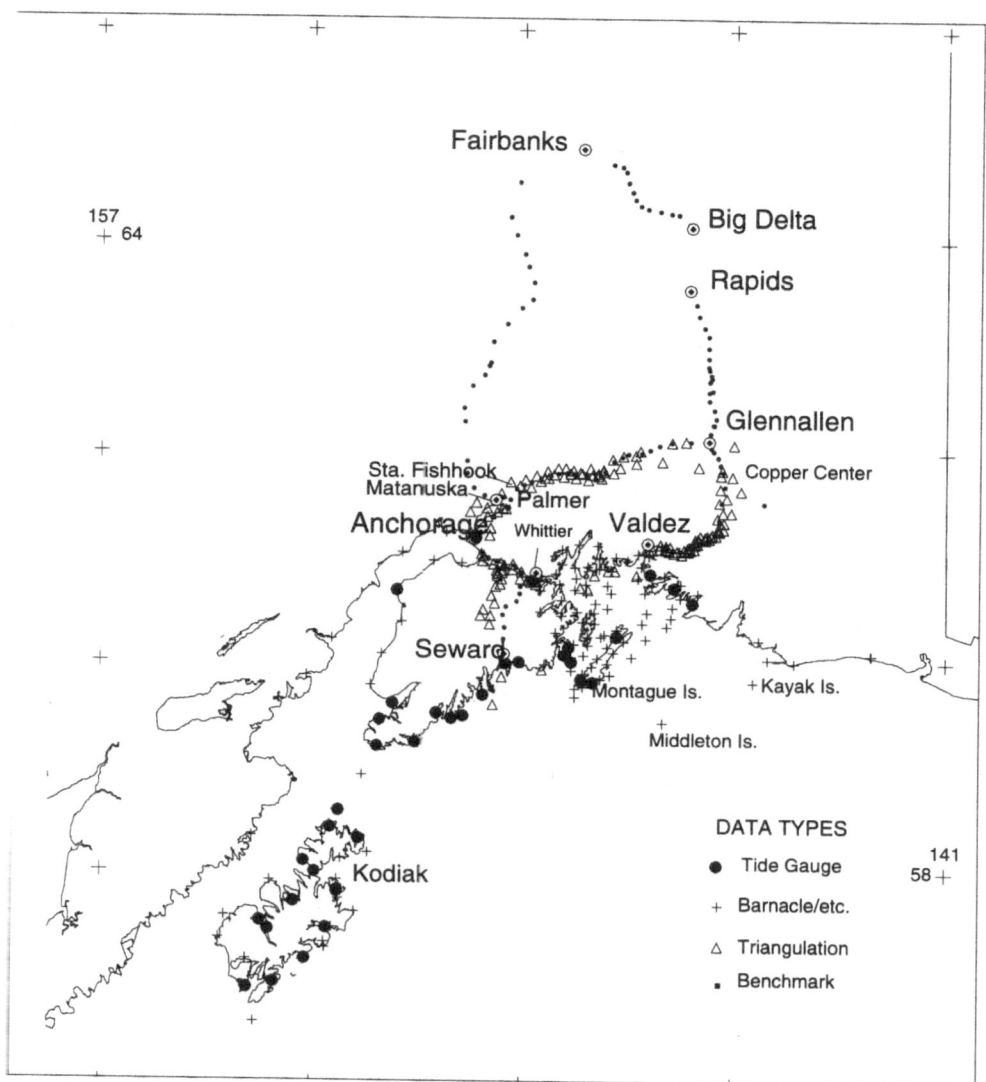

Figure 2
Distribution of tide gauges, barnacle data (this includes all coast line height changes listed in Table 1),
triangulation, and leveling benchmarks used in this study in the region of the 1964 earthquake. The open
circles with a centered dot indicate cities. The triangulation station Fishhook is given for reference.

the main shock. Depending on the temporal window of the observations, the
measured coseismic deformation may include deformation (1) associated with the
aftershocks, and (2) aseismic processes, such as afterslip (creep) on a down-dip
extension of the fault plane and/or viscous relaxation within the lower crust and
upper mantle.

Table 1

Summary of measurement uncertainties

Observation Type	Standard Deviation (σ)
Coastal height change	
barnacles	15–90 cm
vegetation	15–30 cm
storm beach	1 m
marker	15–30 cm
Change in tidal datum plane	3–10 cm
Pre- and post-leveling	1.0 mm × (distance km)$^{1/2}$
Pre- and post-bathymetry	3 m
Horizontal displacements	1.5 m in each component

Coseismic Dislocation Model

Initially, we iterated through trial solutions to derive the approximate geometry and magnitude of coseismic displacements that were most consistent with the data. The slip surface was parameterized by five rectangular planes and a range of fault slip parameters was explored. Four planes corresponded to the megathrust surface and one plane simulated the Patton Bay fault. A single fault parameter was changed at each iteration of the dislocation modeling (MANISINHA and SMYLIE, 1971). We assumed a homogeneous half-space. Additional detail is given in HOLDAHL and SAUBER (1992).

The beginning location and orientation of the fault planes was guided by the distribution of aftershocks and by the 4-plane model of MIYASHITA and MATSU-'URA (1978). We chose the dip of the fault plane to approximate profiles of the plate interface given by MOORE *et al.* (1989) and BROCHER *et al.* (1994) and recent seismicity studies (DAVIES and HOUSE, 1979; PAGE *et al.*, 1989). The direction of horizontal motion had been calculated previously by PARKIN (1969) and by SNAY *et al.* (1987). Therefore, it was possible to approximately orient the fault planes such that the strike-slip component would be minimal if most of the coseismic slip occurred on the megathrust. The focal mechanisms of aftershocks in the Kodiak region were primarily thrust-type events (STAUDER and BOLLINGER, 1966). In Prince William Sound, aftershock mechanisms were more varied and reflect the complex geometry of the region.

In the second phase of model development we modified the geometrical configuration for the fault surface and then used a least-squares adjustment procedure (HOLDAHL, 1992) to solve for the magnitude of fault slip. To resolve a more detailed distribution of slip, the rupture surface was subdivided into a mosaic of smaller planes (~ 50 by 50 km) that could remain connected or be fanned out to fit regional geometry. This approach has been used by HARRIS and SEGALL (1987) to discern regions of high coseismic slip in the 1966 Parkfield earthquake and by BARRIENTOS and WARD (1990) to estimate the slip distribution for the 1960 Chile earthquake.

Figure 3 shows the mosaic of 68 fault planes used in our detailed model. The specific fault parameters for each subfault are given in Table 2. The rectangular submosaic near Kodiak Island had 4 rows and a total of 28 planes. Each of the fault planes in this Kodiak region was contiguous to its neighboring planes, had a

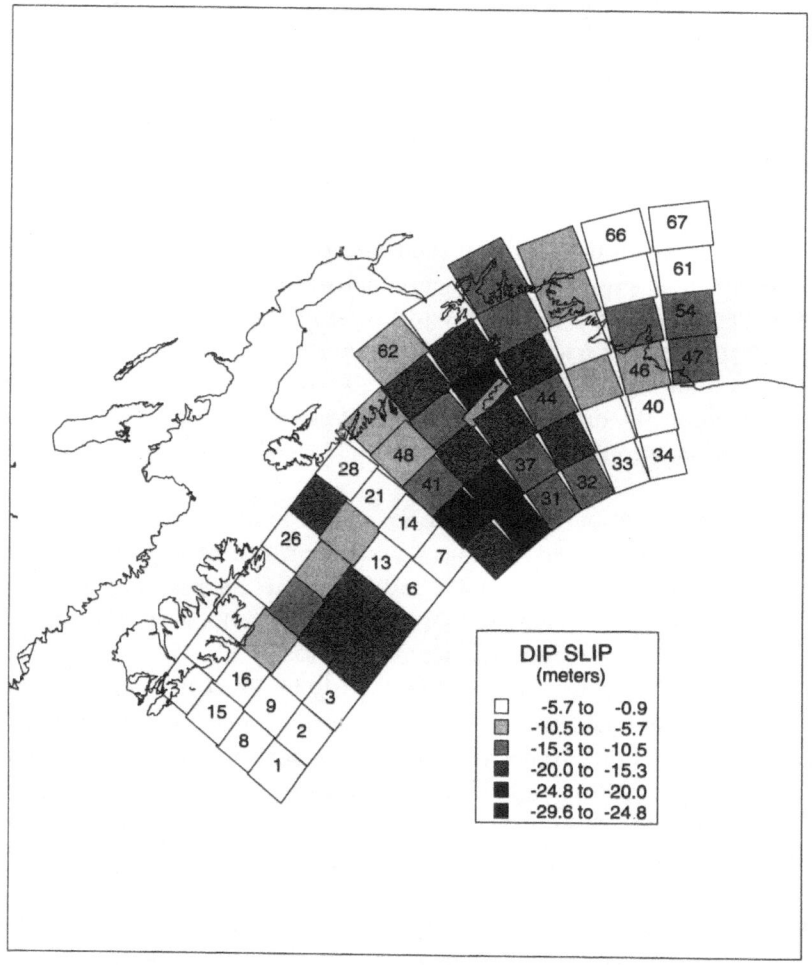

Figure 3

Coseismic dip slip on the mosaic of fault planes for the model resulting from use of all geodetic data types, smoothing, and tsunami slip constraints near the leading edge of the main thrust zone (Model 1, Table 3). The numbered planes indicate the numbering convention used in Tables 2 and 3.

Table 2

Fault plane parameters

Fault No.	Strike (°N, cw)	Half Length (m)	Center Lat. (deg.)	Center Long. (deg.)	Dip (deg.)	Distance, along dip Upper (m)	Distance, along dip Lower (m)
1	218	25000	55.6	209.6	8.0	79038	129038
2	218	24999	56.0	210.1	8.0	79038	129038
3	218	24999	56.4	210.6	8.0	79038	129038
4	218	24999	56.7	211.1	8.0	79038	129038
5	218	25000	57.1	211.7	8.0	79038	129038
6	218	24999	57.4	212.2	8.0	79038	129038
7	218	25000	57.8	212.7	8.0	79038	129038
8	218	24931	55.6	209.6	8.0	129038	179038
9	218	24930	56.0	210.1	8.0	129038	179038
10	218	24928	56.4	210.6	8.0	129038	179038
11	218	24927	56.7	211.2	8.0	129038	179038
12	218	24928	57.1	211.7	8.0	129038	179038
13	218	24926	57.4	212.2	8.0	129038	179038
14	218	24924	57.8	212.8	8.0	129038	179038
15	217	24862	55.7	209.5	8.5	168577	218577
16	217	24862	56.1	210.1	8.5	168577	218577
17	217	24859	56.4	210.6	8.5	168577	218577
18	217	24857	56.8	211.1	8.5	168577	218577
19	217	24856	57.1	211.6	8.5	168577	218577
20	217	24853	57.5	212.2	8.5	168577	218577
21	217	24852	57.8	212.7	8.5	168577	218577
22	217	24796	55.8	209.5	9.0	206525	255525
23	217	24793	56.1	210.0	9.0	206525	255525
24	217	24789	56.5	210.5	9.0	206525	255525
25	217	24787	56.8	211.0	9.0	206525	255525
26	217	24784	57.2	211.6	9.0	206525	255525
27	217	24781	57.5	212.1	9.0	206525	256525
28	217	24779	57.9	212.6	9.0	206525	256525
29	225	20000	58.0	212.9	8.0	28741	78741
30	230	20000	58.3	213.4	8.0	28741	78741
31	236	19999	58.5	213.9	8.0	28741	78741
32	242	19999	58.6	214.5	8.0	28741	78741
33	249	19999	58.8	215.1	8.0	28741	78741
34	256	19999	58.9	215.7	8.0	28741	78741
35	224	22409	58.0	213.0	8.0	78741	128741
36	229	22384	58.3	213.4	8.0	78741	128741
37	235	22352	58.5	213.9	8.0	78741	128741
38	241	22321	58.6	214.5	8.0	78741	128741
39	249	22287	58.8	215.1	8.0	78741	128741
40	256	22261	58.9	215.7	8.0	78741	128741
41	224	24820	58.1	213.0	8.0	128741	178741
42	229	24770	58.3	213.5	8.0	128741	178741
43	235	24706	58.5	214.0	8.0	128741	178741
44	241	24643	58.6	214.5	8.0	128741	178741
45	248	24575	58.8	215.1	8.0	128741	178741
46	255	24521	58.9	215.7	8.0	128741	178741
47	263	24486	58.9	216.3	8.0	128741	178741

Table 2 (*Contd*)

Fault No.	Strike (°N, cw)	Half Length (m)	Center Lat. (deg.)	Center Long. (deg.)	Dip (deg.)	Distance, along dip Upper (m)	Distance, along dip Lower (m)
48	224	27232	58.1	212.9	8.5	168297	218297
49	229	27155	58.4	213.4	8.5	168297	218297
50	235	27059	58.6	214.0	8.5	168297	218297
51	241	26964	58.7	214.5	8.5	168297	218297
52	248	26863	58.9	215.1	8.5	168297	218297
53	255	26781	59.0	215.7	8.5	168297	218297
54	263	26729	59.0	216.3	8.5	168297	218297
55	224	29644	58.2	212.9	9.0	206261	256261
56	229	29543	58.4	213.4	9.0	206261	256261
57	235	29415	58.7	214.0	9.0	206261	256261
58	241	29286	58.8	214.5	9.0	206261	256261
59	248	29150	59.0	215.1	9.0	206261	256261
60	255	29040	59.1	215.7	9.0	206261	256261
61	262	28970	59.1	216.3	9.0	206261	256261
62	228	31930	58.9	212.8	12.0	192813	242813
63	234	31770	59.1	213.4	12.0	192813	242813
64	240	31609	59.3	214.1	12.0	192813	260000
65	248	31438	59.5	214.8	12.0	192813	242813
66	255	31300	59.6	215.5	12.0	192813	242813
67	262	31211	59.7	216.2	12.0	192813	242813
68	219	29999	59.8	212.2	60.0	100	30000

strike of 218° and a dip varying from 8° nearest the Aleutian Trench to 9° below Kodiak Island. To the east, the 39 planes spanning Prince William Sound (PWS) were fanned out to bend with the geography and to make the columns approximately parallel to the local direction of horizontal slip. The length of the PWS fault planes widened toward the mainland to minimize space between laterally adjacent planes. The dimensions of the planes vary from ~40 by 50 km near the trench to ~64 by 50 km on the deepest segment. Dip for the PWS planes ranged from 8° nearest the Aleutian Trench to 12° near the perimeter of Prince William Sound. The strike varied from 218° at the west end to 270° at the east end. The PWS fault planes are continuous along dip but the columns separate slightly with greater depth (Figure 3). For the single fault plane used to model the large uplift (Figures 1 and 3) corresponding to the Patton Bay fault on Montague Island, a dip of 60° was assigned. The dimension of the plane was ~72 km along strike by 30 km in the direction of dip. Unfortunately the location of the southwest terminus of the Patton Bay fault is unknown (Figure 1). We placed the fault center coincident with the southwest end of Montague Island.

Prior Information

Since the spatial variation of fault slip often exceeds the resolution obtainable using the available geodetic and geologic data, prior information was introduced using a method similar to that described by SNAY (1989). In this approach quasi-observations are used to represent the prior information. A strike-slip value of 0.0 ± 1.0 m was introduced for all fault planes (Table 3) with the exception of the Patton Bay fault which had a prespecified strike-slip of 0.0 ± 0.1 m and a dip slip

Table 3

Estimated dip-slip (m) and strike-slip (m) motion in 1964 earthquake for a model with prior tsunami information offshore (Model 1) and a uniform offshore slip model (Model 2). The geometry of Models 1 and 2 are identical. For the strike-slip component a positive value indicates left-lateral motion.

	MODEL 1					MODEL 2	
FAULT	S-Slip	±(σ)	D-Slip	±(σ)	Prior D-Slip	D-Slip	Prior D-Slip
1	0.1	2.2	−1.9	2.4	−2.0	−9.8	10.0
2	0.1	2.2	−1.8	2.4	−2.0	−9.8	10.0
3	0.2	2.2	−2.0	2.4	−2.0	−9.8	10.0
4	0.1	2.2	−17.7	2.4	−18.0	−9.8	10.0
5	0.2	2.2	−17.7	2.4	−18.0	−9.8	10.0
6	0.1	2.2	−2.9	2.4	−3.0	−9.7	10.0
7	0.1	2.2	−3.0	2.4	−3.0	−9.6	10.0
8	0.1	2.2	−1.5	2.4	−2.0	−6.9	10.0
9	0.3	2.2	−1.9	2.4	−2.0	−9.5	10.0
10	0.2	2.2	−2.2	2.4	−2.0	−9.8	10.0
11	0.1	2.2	−17.8	2.4	−18.0	−9.8	10.0
12	0.2	2.2	−17.7	2.4	−18.0	−9.6	10.0
13	0.2	2.2	−2.9	2.4	−3.0	−9.6	10.0
14	0.2	2.2	−2.5	2.4	−3.0	−9.5	10.0
15	0.5	2.1	−1.2	2.3		−4.1	
16	1.0	2.1	−4.7	4.3		−3.7	
17	−0.3	2.1	−9.6	5.0		−8.2	
18	−0.1	2.1	−12.5	7.3		−9.5	
19	−0.0	2.2	−6.8	10.8		−7.9	
20	0.3	2.2	−7.9	3.8		−9.2	
21	0.4	2.2	−5.4	2.7		−8.3	
22	1.5	1.4	−3.0	3.2		−4.1	
23	1.4	1.7	−3.6	3.1		−3.6	
24	−0.4	1.9	−2.5	2.8		−2.6	
25	0.1	1.9	−5.2	4.6		−4.1	
26	−0.1	2.1	−3.3	6.7		−5.8	
27	0.1	2.1	−17.3	5.4		−11.3	
28	0.1	1.9	−3.9	2.8		−6.5	
29	0.2	2.2	−19.9	2.4	−20.0	−9.9	10.0
30	0.2	2.2	−20.1	2.4	−20.0	−10.3	10.0
31	−0.1	2.2	−14.1	2.4	−14.0	−10.6	10.0
32	−0.1	2.2	−13.2	2.3	−14.0	−9.1	10.0
33	−0.1	2.2	−2.4	2.4	−2.3	−9.9	10.0
34	−0.0	2.2	−2.3	2.4	−2.3	−9.9	10.0

Table 3 (*Contd*)

FAULT	MODEL 1					MODEL 2	
	S-Slip	±(σ)	D-Slip	±(σ)	Prior D-Slip	D-Slip	Prior D-Slip
35	0.4	2.2	−23.5	2.4	−24.0	−9.7	10.0
36	−0.1	2.1	−22.7	2.4	−24.0	−15.5	10.0
37	−1.2	2.1	−14.3	2.3	−18.0	−26.4	10.0
38	0.2	2.1	−18.1	1.7	−18.0	−12.7	10.0
39	1.0	2.1	−4.2	2.4	−6.3	−6.9	10.0
40	2.5	1.9	−5.3	2.2	−6.3	−8.2	10.0
41	−0.2	2.2	−12.6	1.3		−9.6	
42	−2.0	2.1	−16.6	1.2		−15.5	
43	−2.1	1.9	−18.2	1.1		−26.4	
44	1.3	1.9	−14.8	0.9		−12.7	
45	0.5	2.1	−8.1	1.1		−6.9	
46	0.7	1.9	−9.3	1.3		−8.2	
47	−0.9	1.8	−14.1	1.8		−13.9	
48	−0.9	2.1	−8.1	3.4		−13.4	
49	−4.7	1.8	−13.5	3.1		−15.5	
50	−2.4	1.5	−29.6	1.3		−26.6	
51	0.4	1.4	−16.3	1.7		−12.8	
52	−0.3	1.4	−5.6	1.9		−6.9	
53	3.8	1.8	−10.7	3.8		−14.9	
54	0.7	2.1	−14.3	9.8		−10.6	
55	−3.2	1.5	−6.7	1.2		−9.3	
56	−5.9	1.3	−15.4	1.3		−17.5	
57	−4.3	1.6	−16.5	1.0		−17.9	
58	1.0	1.5	−14.7	1.4		−11.0	
59	0.0	1.1	−9.9	1.3		−10.9	
60	−0.0	1.5	−3.9	2.1		−2.3	
61	1.4	2.1	−4.4	3.7		−0.9	
62	−0.5	1.3	−9.5	1.1		−12.9	
63	1.1	1.1	−5.0	0.7		−6.3	
64	−0.4	1.0	−10.8	0.9		−4.1	
65	5.4	1.5	−5.8	0.9		−5.8	
66	1.6	1.4	−0.9	1.1		−0.6	
67	1.6	1.5	−5.3	2.1		−4.3	
68	0.4	0.8	−8.6	0.4	−7.0	−8.4	

of 7.0 ± 0.1 m (after PLAFKER, 1972). Dip-slip magnitude was provided as prior information for the rows nearest the Aleutian Trench. In Model 1, dip slips near the leading edge of the main thrust zone were obtained from preliminary modeling of tsunami data (JOHNSON and SATAKE, 1993). The geometry used by JOHNSON and SATAKE was the same as Model 1 but with some simplifications. These values were input with a standard deviation of 1 m. Slip differences between adjacent fault planes were controlled by using quasi-observations, one for each plane, which essentially required the strike or dip slip on that plane to be the average of adjacent fault planes. A range of weights was tried for these constraints so as to get adequate smoothing without unduly obscuring slip contrast. The weight assigned to the strike-slip and dip-slip averages assumed a standard deviation of 2.0 m for strike slip and 6.0 m for dip slip. These constraints were weighted more heavily to bridge large data gaps found seaward of Montague and Kodiak Islands. The bridging weights for average dip slip were based on standard deviations of 1.0 m for the Kodiak region, and 0.2 m for the PWS zone.

Model Results

Figure 3 shows the distribution of dip slip for Model 1, Table 3. Figure 4 shows contours of the corresponding coseismic uplift and subsidence. The adjusted slip distribution (Figure 3) shows three zones of higher slip, corresponding to (1) a large region seaward of Prince William Sound and including parts of the Kenai Peninsula, (2) off Kodiak Island, and (3) a small, localized region near Kayak Island. The region of highest slip (30 m) occurs near Montague Island. The strike-slip component (Model 1, Table 3) in the Kodiak region is small (<1.6 m) and statistically insignificant. In the PWS mosaic, a few of the planes resolve a significant right-lateral strike-slip component of up to 6 m (planes 49, 50, 55, 56, 57, Table 3) and on the northeastern planes a left-lateral component of up to 5 m.

A major concern arising from our use of prior information is whether the results given in Figure 3 are unique. As a test of the influence of the prior dip-slip values derived from tsunami data, 10 m of uniform slip was substituted for the offshore fault planes (first two rows) as an alternative model (Model 2, Table 3). The results of using the uniform dip-slip model are given in Table 3 (Model 2). Since there is very little data to conflict with the prior information near the trench (Figure 2) the estimated dip slip clearly reflects the prior information (1–14, 29–46) and Models 1 and 2 are clearly different offshore. Closer to the coast, the dip-slip values in the two models corresponding to the fault planes (15–28) show little statistical difference; whereas the dip-slip values for planes 48–67 show some significant variation but the general dip-slip features are similar. Since the preliminary tsunami results reflect the best existing estimate of offshore coseismic slip, results from Model 1 (Table 3) are preferred.

Evaluation of Model

Table 4 summarizes the statistical fit of the model to the various types of measurements and constraints that were used as input. A comparison of the predicted and observed vertical offset along routes leveled before and after the earthquake are given in Figures 5, 6 and 7. The overall fit of the predicted barnacle height changes to the observed values is given in Figure 8. The barnacle plot illustrates the generally good fit of Model 1, although the predicted values average slightly higher (7 cm) than the observed values. A specific profile for the Kodiak

PWS EARTHQUAKE OF 1964.
COSEISMIC HEIGHT CHANGE (CM)

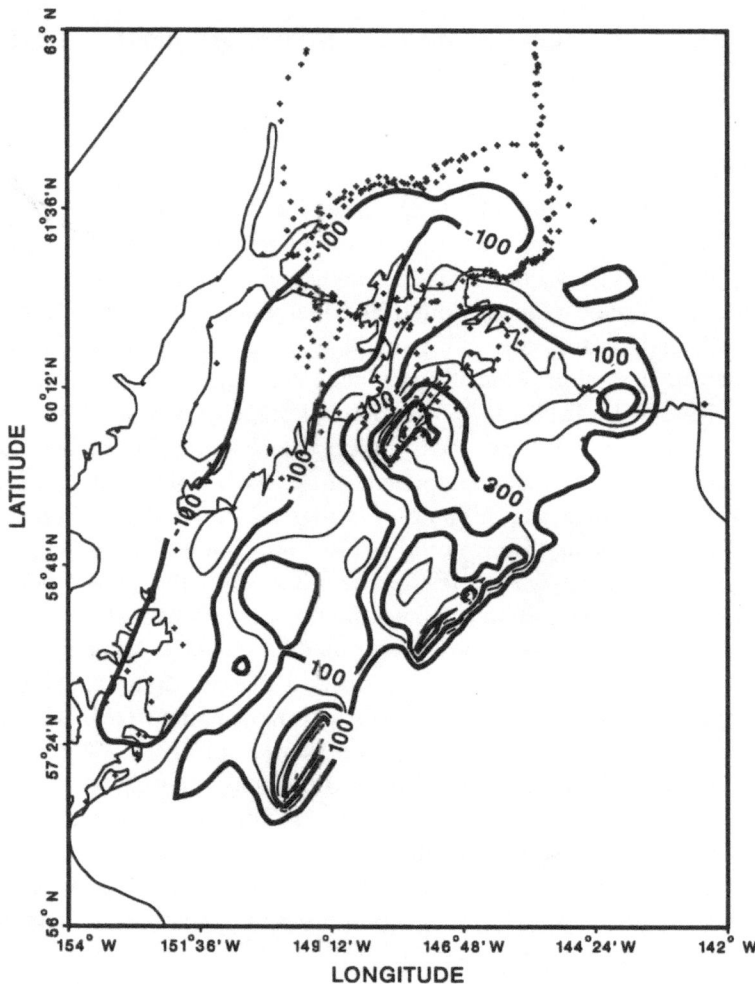

Figure 4
Contours of the vertical static displacement field predicted by the model illustrated in Figure 3. Contours occur every 100 cm. Input data are given by small crosses.

Island region is given in Figure 9. A comparison of the predicted uplift along a profile from Anchorage to Middleton Island (Figure 10) was compared to the observed data. Once again the fit to the data is good; the large, localized uplift on the eastern side of Montague Island is underpredicted, however, by 1.5 m. The model was used also to calculate horizontal displacements and these are compared

Table 4

Unweighted RMS for input data types

Observation Type	RMS (m)
Leveled height difference	0.03
Height changes from barnacles and tide gauges	0.56
Horizontal displacements:	
East component	1.1
North component	1.3

Prior information/Constraint	RMS (m)
Strike slip	1.6
Dip slip	1.0
Average strike slip	1.2
Average dip slip	4.5

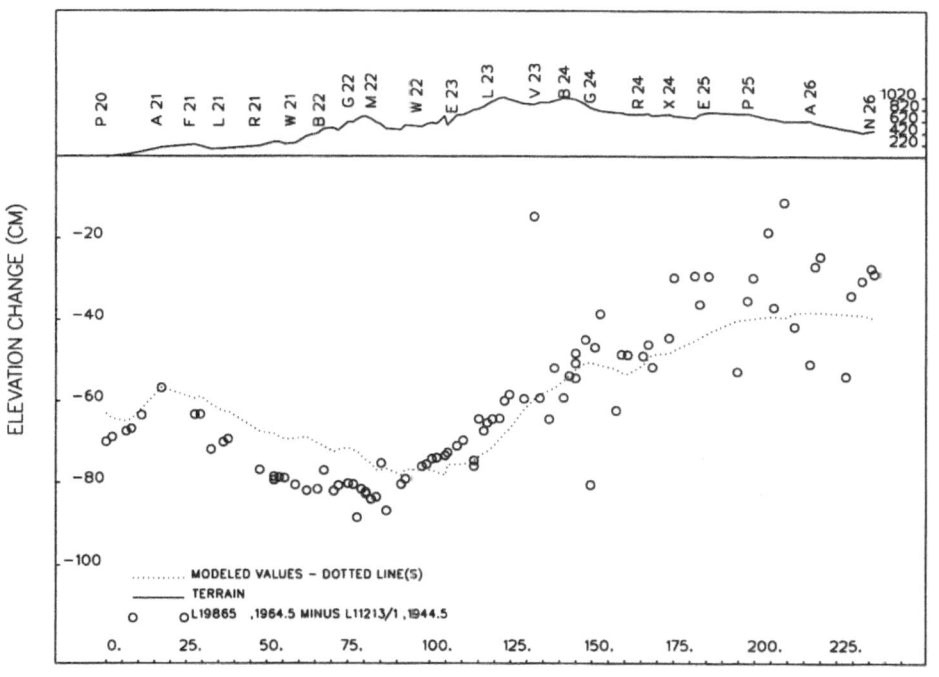

Figure 5

Profile of observed and predicted vertical displacement during the coseismic time interval: Palmer to Glennallen.

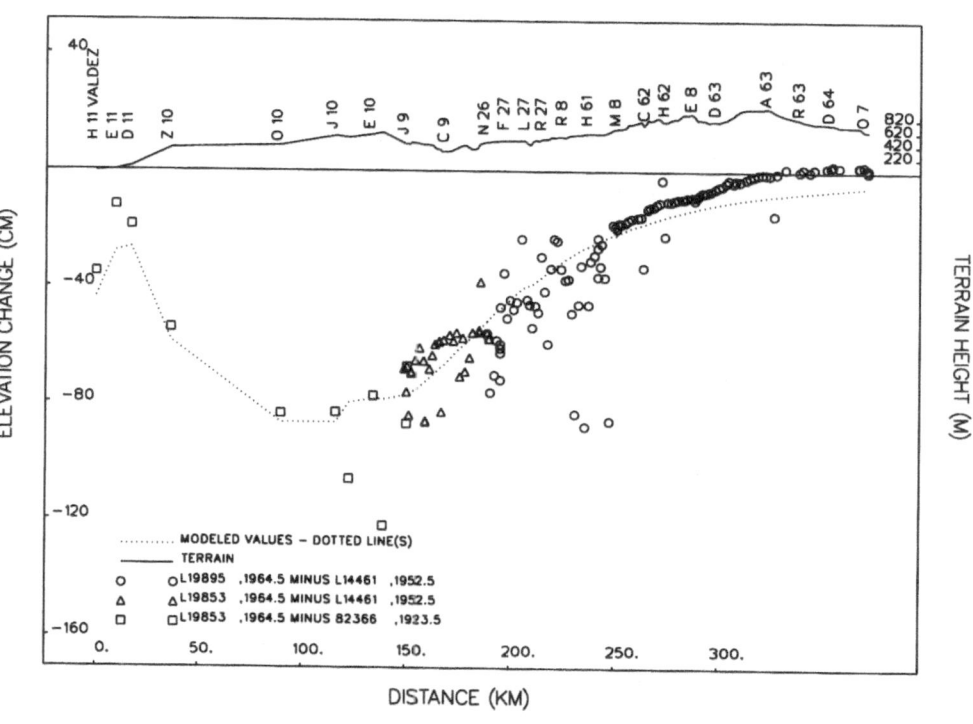

Figure 6
Profile of observed and predicted vertical displacement during the coseismic time interval: Valdez via
Copper Center and Glennallen to Rapids, Alaska.

with the displacements previously calculated by SNAY *et al.* (1987) refined by the scale, rotation, and translation parameters derived by our model (Figure 11). The rms for the two horizontal components (~ 1.2 m, Table 4) are similar to the estimated standard deviation of the observed horizontal displacement (1.5 m, Table 1). The rms values for the constraints on average slip (1.2 m for strike slip and 4.5 m for dip slip, Table 4) are a general indication of the average contrast required between slip values on adjacent planes.

To evaluate the robustness of our strike-slip results (Model 1, Table 3) we tried using a weaker strike-slip constraint of 0 ± 4 m. The results are similar to those in Table 3; i.e., no systematic strike-slip motion was resolved in the Kodiak Island region and in the PWS mosaic both right-lateral (up to 9 m) and left-lateral (up to 7 m) were resolved for similar regions. Because Model 1 fits the horizontal displacements well, it is difficult to make a case for more strike slip. When the strike-slip constraints are relaxed, it appears that misfit of the model caused by imperfect geometry may be observed in the strike-slip component.

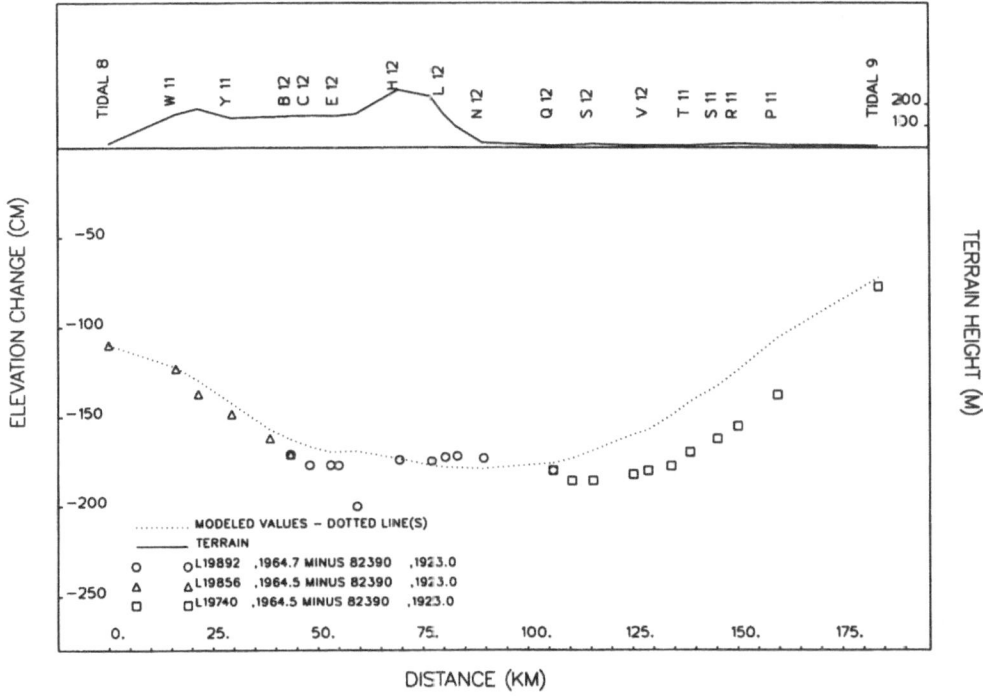

Figure 7
Profile of observed and predicted vertical displacement during the coseismic time interval: Seward via
Portage to Anchorage.

Discussion

Comparison of Coseismic Slip Model to Previous Geodetic and Seismologic Results

From earlier modeling of the deformation data average coseismic slip values have been estimated for large fault planes. For example, HASTIE and SAVAGE (1970) estimated an average dip slip of ~ 12 m on a plane dipping 3.7° and ~ 10 m of left-lateral slip over a ~ 200 by 600 km region. The average dip-slip value of our model was approximately 10 m. The results given in Table 3 suggest no systematic strike-slip component. As discussed earlier, our original motivation for constraining the strike-slip component was based on aftershock focal mechanism studies which suggest little strike-slip motion in the Kodiak region and a variable component in the PWS region. Additionally the orientation of coseismic strain inferred from triangulation data suggests coseismic strain perpendicular to the coastline (Figure 1). The large, left-lateral component in the one-plane model of HASTIE and SAVAGE (1970) may have been required to compensate for the mismatch with the complex geometry.

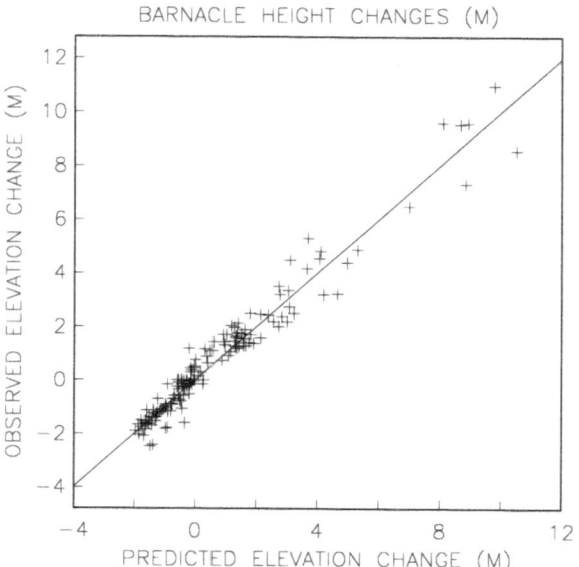

Figure 8
Observed versus predicted barnacle height changes.

Additional geometrical complexity was introduced in the four-plane coseismic slip model of MIYASHITA and MATSU'URA (1978). Although they started with initial estimates of fault dip, orientation and slip magnitude, all these parameters were solved for in their rigorous inversion study. Over a region extending northeast-ward from Kodiak Island to the southwest side of Prince William Sound including Montague Island, MIYASHITA and MATSU'URA (1978) estimated an average dip slip of ∼19 m on a plane dipping 25°NW. In the PWS region they estimated ∼10 m of dip slip on a 7°NW dipping rupture plane and in the Kayak Island region ∼8 m of dip slip on a 20°NE plane. Since the publication of MIYASHITA and MATSU'URA (1978) detailed seismicity studies (DAVIES and HOUSE, 1979; PAGE *et al.*, 1989) and reflection/refraction profiles (BROCHER *et al.*, 1994) have been made that further constrain the geometry of the main thrust zone. Therefore, after initially exploring alternative fault configurations we assumed a fixed rupture surface geometry and solved for slip magnitude only. The results given in Figure 3 would be roughly similar (within 2–3 m) to MIYASHITA and MATSU'URA if averaged over the same area. Our results using smaller (spatial) grid elements suggest large variations that were of course masked by using the larger fault planes of MIYASHITA and MATSU'URA. For example in the largest of their planes used, our slip estimates ranged from 1 to 30 m.

A large number of the aftershock epicenters (March 28, 1964 through Dec. 31, 1965, Figure 12) are located in the region of highest slip. Unfortunately there are

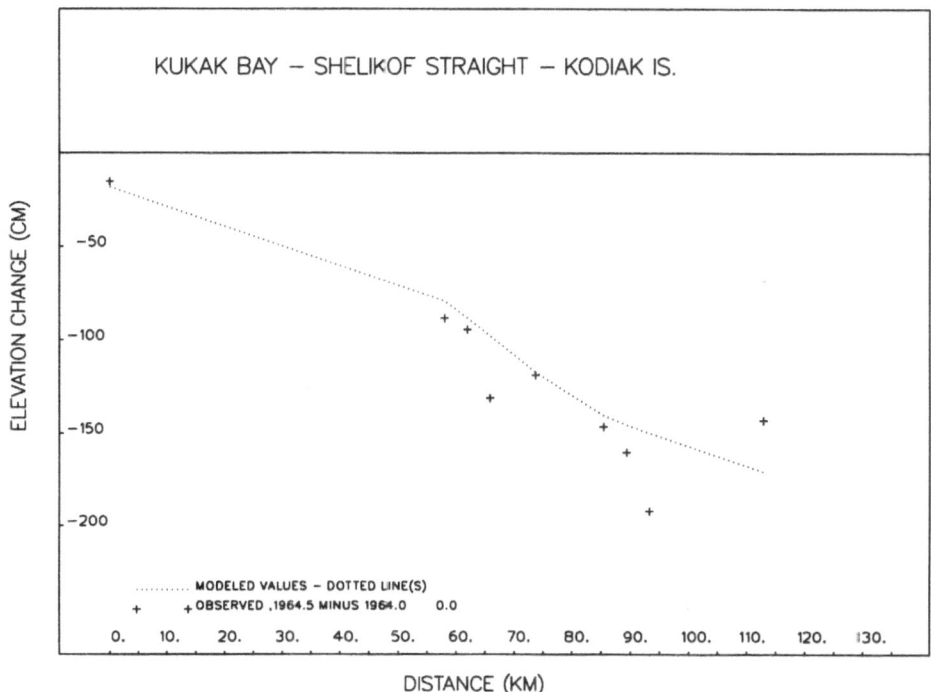

Figure 9

Observed (plus signs) and predicted (dotted line) height change versus distance across the Kodiak Island region.

large uncertainties in the location of these aftershocks. Beyond a gross visual characterization no systematic correlation seems to exist between the level of aftershock activity and the magnitude of coseismic slip. For example, our results indicate an average of 18.2 m of slip on planes 29–32, 35–38 but the seismicity level is quite low. In contrast, the region of highest slip near Montague Island had numerous large aftershocks.

From seismological modeling, moment release as a function of time has been determined and this has been used to infer spatial patterns in moment release in the 1964 earthquake. Based on an examination of the P-wave portion of the seismograms, WYSS and BRUNE (1967) suggested multiple distinct events superimposed on a continuous low level of energy release. Most of the proposed subevents (Figure 3 of WYSS and BRUNE, 1967) were in the Prince William Sound region. RUFF and KANAMORI (1983) suggested that the primary rupture occurred over a 200 km region in a southeast direction from the epicenter (Figure 1). In addition to the primary moment release in the Prince William Sound region, KIKUCHI and FUKAO (1987) suggest additional subevents in the Kodiak Island region. In a more

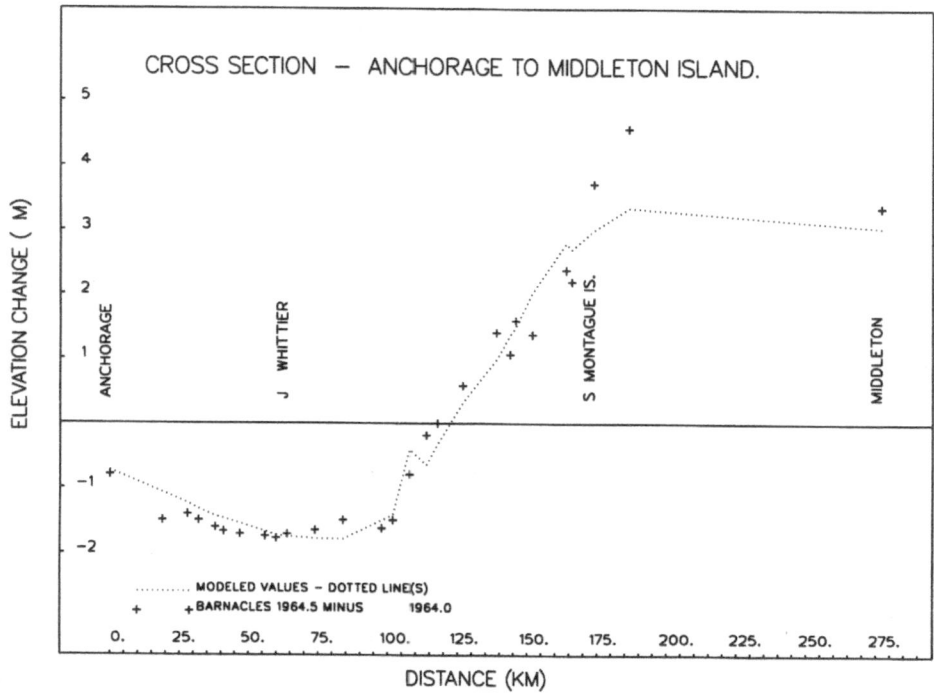

Figure 10
Predicted versus observed vertical uplift profile from Anchorage to Middleton Island.

recent study, CHRISTENSEN and BECK (1993) determined the rupture history of the 1964 event from modeling of the long-period P-wave seismograms. Their results suggest that the primary moment release occurred seaward of Prince William Sound also. They located a second (smaller) moment-release pulse in the vicinity of Kodiak Island. Similarly, the results given in Figure 3 and Table 3 indicate that most of the coseimic slip occurred seaward of and within the Prince William Sound region. A smaller region of localized slip off Kodiak Island, indicated in Figure 3, has been inferred by this study. However, the region identified by CHRISTENSEN and BECK (1993) is a little farther to the southeast. The local slip high (~ 9–14 m, on planes 46, 47, 53, 54) near Kayak Island has not been suggested from seismological studies; however large, localized uplift is supported by coastline height changes (PLAFKER, 1969). The moment calculated on the basis of our geodetically derived slip model of 5.0×10^{29} dyne cm is 30% less than the seismic moment of 7.5×10^{29} dyne cm calculated from long-period surface waves (KANAMORI, 1970; RUFF and KANAMORI, 1983) but is close to the seismic moment of 5.9×10^{29} dyne cm obtained by KIKUCHI and FUKAO (1987). A shear modulus of 3×10^{10} Pa was assumed.

PWS EARTHQUAKE OF 1964.
OBSERVED-vs-COMPUTED HORIZONTAL DISPLACEMENTS.

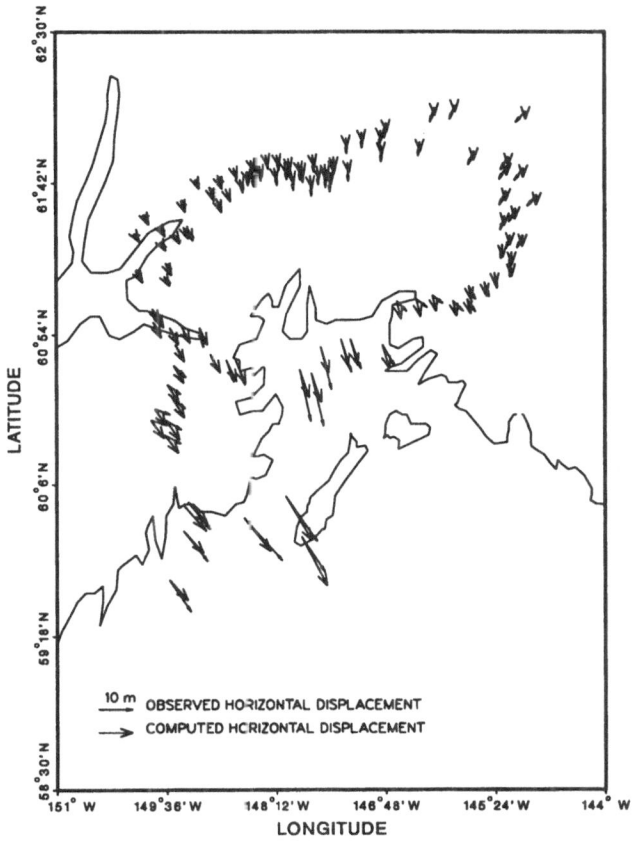

Figure 11

Observed and predicted horizontal coseismic displacements for representative stations in the Prince William Sound region. Some stations located very close together were removed to avoid additional crowding.

Alternative Rupture Surface Models

Geometrically, the modeled rupture surface is consistent with the dip of the inferred interface between the North American and Pacific plates and the general spatial extent of aftershocks. Based on their interpretation of a reflection/refraction profile BROCHER *et al.* (1994) have suggested, however, that 1964 coseismic slip in the Prince William Sound (PWS) region may have occurred on a shallow (dip ~3°) prominent reflector which corresponds to the interface between the Yakutat terrane (which might be sutured to the top of the Pacific plate) and the overriding North American plate. Such a model was tested by altering the PWS subfaults of Model

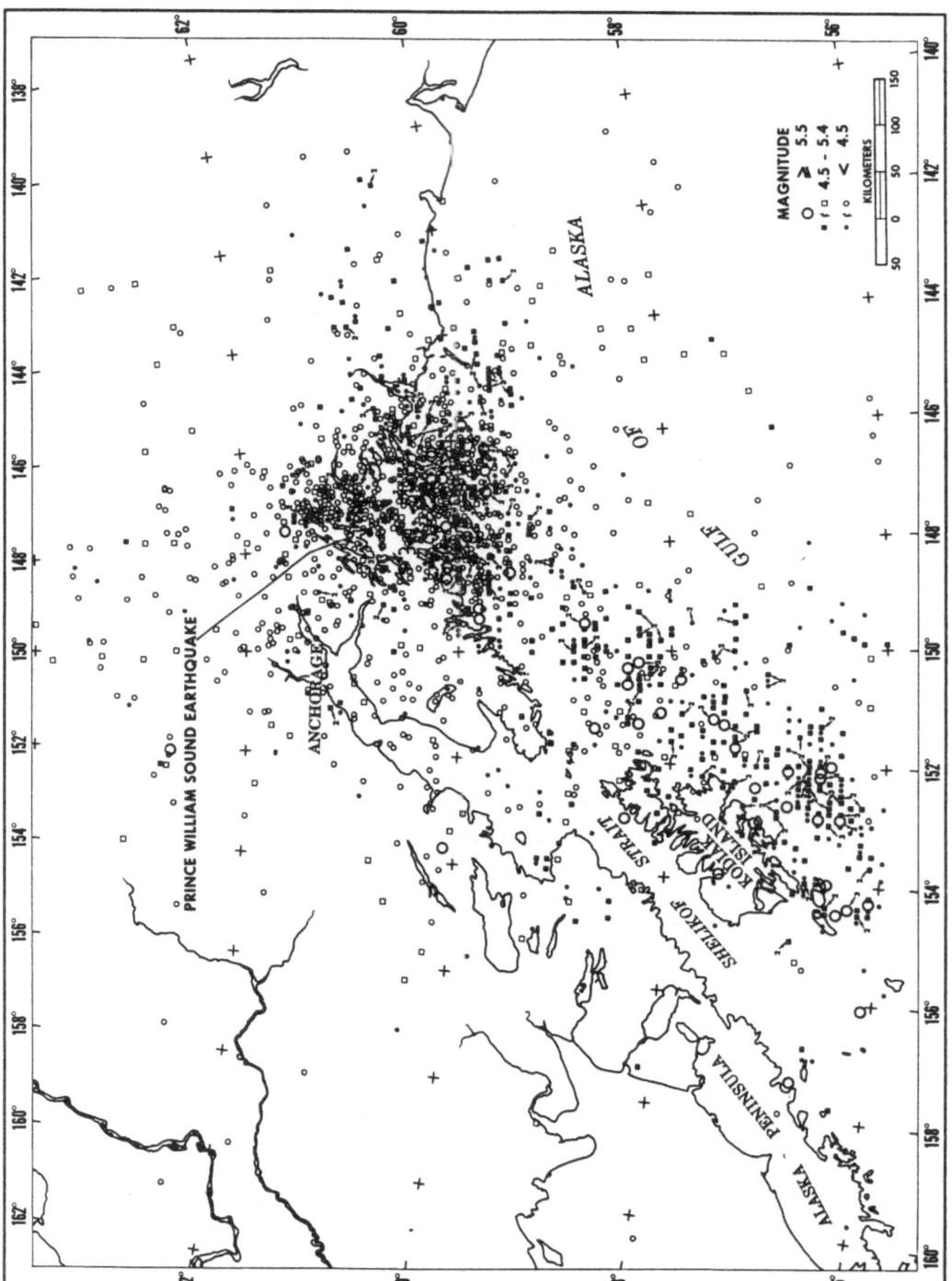

Figure 12

All epicenters located between March 28, 1964 and December 31, 1965 (from ALGERMISSEN *et al.*, 1969). The small open circles are aftershocks located with data from a temporary station network. All other events were located using teleseismic data.

1 such that the dip on subfaults near the center of PWS changed from $8-9°$ to $2.5°$ at a down-dip distance of 150 km. Our Model 1 had a better fit to the data, as indicated by a 30% lower variance of unit weight and lower rms values for most individual data types. However, these statistical differences do not rule out such an alternative model, as we may not have defined the mosaic geometry in the best way to depict the interface between the Yakutat terrane and the North American interface.

Comparison to Slip Distribution and Static Stress Drop in the Great 1960 Chilean Earthquake

The 1964 Alaskan and the 1960 Chilean ($M_w = 9.5$) earthquakes are similar in that they both ruptured shallow main thrust zones and showed similar horizontal strain and uplift/subsidence patterns (PLAFKER, 1972). The Chilean event ruptured a zone approximately 900 km long and 150 km wide (BARRIENTOS and WARD, 1990). Using sea-level change, elevation differences and horizontal strain, BARRIENTOS and WARD (1990) used a variable slip planar model to invert for slip distribution. They found several subregions or asperities which ranged in size from 50 to 200 km and had slips of up to 41 m. The introduction of tsunami results from JOHNSON and SATAKE (1993) enabled us to better constrain the offshore slip values for the 1964 earthquake.

The mean static stress drop ($\Delta\tau = \mu\Delta u/\sqrt{A}$, where μ is the rigidity, Δu the coseismic slip, and A is the area) for the 1960 event was 0.85 MPa (8.5 bars) with local static stress drops of up to 68.5 MPa (685 bars). For the 1964 Alaska earthquake we calculated a mean static stress drop of approximately 0.72 MPa (7.2 bars) and a local static stress drop of up to 17.1 MPa (171 bars). BARRIENTOS and WARD used 600 sources to represent slip in the 1960 Chilean event and were thus able to resolve smaller scale asperities of higher stress drop.

Coseismic Slip and Recurrence Intervals

As summarized by NISHENKO and JACOB (1990) the Prince William Sound and Kodiak Island segments of the Pacific/North American plate boundary have different earthquake histories. For example from radiocarbon dates of terraces on Middleton Island (Figure 2) PLAFKER and RUBIN (1978), inferred interevent times of 500 to 1300 years. On the other hand, reports of large earthquakes on Kodiak Island over the last 200 years suggest an average repeat time of ~ 60 years (NISHENKO and JACOB, 1990), significantly different than the Prince William Sound region. Based on a survey of great circum-Pacific earthquakes, THATCHER (1990) suggested that zones of concentrated moment release slip comparable amounts in each cycle and have high shear strength. Furthermore, the recurrence time of great earthquakes is more likely to be controlled by maximum rather than average slip values. We examined this assertion by calculating a crude estimate of recurrence based on the maximum slip magnitude and then compared this to geological estimates of former earthquake occurrence in the Prince William Sound region. For the 3 columns of fault planes with the highest slip (36–56, 37–57, 38–58) the average dip slip was about 23 m, with a range between 16 and 30 m. If we make the simple assumption that for the region of highest slip most of the relative plate motion of ~6 cm/yr occurs during great earthquakes then we can use either 30 m or 23 m (averaged over 9 planes) to estimate a recharge time of ~500 or 380 years, similar to the lower bound of the geologically inferred recurrence time. For weaker regions (which would correspond to our regions of lower coseismic slip such as the Kodiak Island region) THATCHER (1990) suggests that they slip by differing amounts in each event. This is consistent with the earthquake history from Kodiak Island which suggests that in addition to the 1964 great earthquake, large earthquakes ($M_s = 7–8$) have occurred near Kodiak Island in the last 200 years.

Implications for Post-1964 Geodetic Studies

The results given in Figure 3 suggest large variations in coseismic slip that have implications for post-1964 geodetic studies. In particular, postseismic, time-dependent processes such as creep on the down-dip extension of the rupture plane and distributed viscous relaxation in the lower crust and upper mantle are partially controlled by the magnitude and distribution of coseismic slip. In earlier work we explored these alternative postseismic mechanisms (HOLDAHL and SAUBER, 1992) by forward modeling of post-1964 leveling (BROWN et al., 1977) and VLBI (MA et al., 1990) data. Unfortunately, the post-1964 surface deformation data do not provide sufficient information to distinguish between these alternative processes. However, since the coseismic slip magnitude is larger in the PWS region than the Kodiak region, post-1964 geodetic measurements from the PWS are more likely to show time-dependent behavior. The ongoing geodetic measurements being made across southeastern Kodiak (LISOWSKI et al., 1993), across northwestern Kodiak Island (J. Sauber and others), across the Kenai Peninsula (COHEN et al., 1994) and from Middleton Island to Anchorage (J. Savage and others) will provide new constraints on the rate and distribution of interseismic strain within the region that ruptured in the 1964 earthquake.

Summary

In this study we have resolved spatial variations in coseismic slip during the great 1964 earthquake with the use of geodetic and geologic data. To do this with the available data required introduction of some prior information. We examined the robustness of our particular solution by systematically varying these prior constraints. Although the specific dip-slip values have large uncertainties and should not be over-interpreted we suggest that the general features of our results are robust. The results suggest a broad region ($\sim$ 150 by 200 km) with an average slip of about 23 m near Montague Island. Two other regions near Kodiak and Kayak Islands have local slip maximums that are smaller with lower average slip values. These results can be used in future studies which examine the relative importance of postseismic processes such as viscous relaxation in the lower crust and upper mantle and time-dependent creep. Additionally since there are a large number of ongoing geodetic studies in this region, the results can be used to test if there are variations in the rate and distribution of strain accumulation associated with regions of different coseismic slip patterns. The results reported in this paper represent a step toward a more detailed characterization of the 1964 coseismic slip distribution that can be used as a starting point for a three-dimensional model of the complete seismic cycle.

Appendix: Details of Data Used

Leveling

Elevation changes at inland locations were estimated from first-order spirit leveling measurements made by reoccupation in the summers of 1964 and 1965 of benchmarks established in 1923, 1944, and 1952 (see SMALL and WHARTON, 1969). By itself, leveling is a relative height measurement. However, the levelings in Alaska were connected to tide gauges where coseismic height changes had been measured directly relative to sea level. A 1965 level survey connected Anchorage and Fairbanks, a segment not affected strongly by the earthquake. The 1964–65 and earlier leveling segments were assigned a standard deviation, m, which is given by: $m = aD^{1/2}$ where $a = 1$ (standard deviation of leveling is 1 mm per 1 km), and D is the distance in km (BOMFORD, 1962). It was possible to form misclosures of these levelings against revised tidal planes at Homer, Seward, Valdez, and Anchorage; and against a benchmark considered stable at Fairbanks. The largest misclosure in the 1964 leveling was 17 cm, in a circuit connecting the Seward and Homer tide gauges by way of Moose Pass and Soldatna. Only the Seward to Moose Pass segment was included in our study, because no pre-1964 leveling existed for the remainder of the route. The remaining circuits of 1964 leveling had misclosures of -3.2, -77.8, 67.2 and -128.5 mm. The latter misclosure corresponds to the large circuit connecting Anchorage, Fairbanks, and Glennallen which uses an assumption of stability at Fairbanks as well as the 1964 revised tidal datum plane at Anchorage. A 1965 releveling between Matanuska and Fairbanks can be used to close this circuit; the misclosure is -268.0 mm. This misclosure has questionable interpretation value because the 1965 leveling contains up to 14 months of possible post-earthquake vertical motion.

The primary uncertainty in the computation of coseismic vertical motion is due to monument instability between the time of the earthquake and the previous survey. The level line segment from Glennallen to Rapids is one of the few indicators of the possible magnitude of interseismic vertical motions prior to 1964. SMALL and WHARTON (1969) noted a 129 cm subsidence feature (1923 to 1944) near Sourdough, which continued to subside another 37.7 cm in the period from 1944–1952. They also noted numerous local instabilities in the range of 5–10 cm. The distribution and number of pre-1964 relevelings were insufficient to permit conclusions as to whether such interseismic height change may have been widespread in south central Alaska prior to 1964. In the present study, the problem of individual benchmark instability was minimized by segmenting the leveling data using only connecting benchmarks which showed vertical agreement of a few centimeters with most monuments within several kilometers; that is, marks which were clearly part of the local trend of vertical deformation.

Tide Gauge Data

Tidal observations were made several months after the March 27, 1964 earthquake to determine the amount of height change and to revise tidal datum planes (SMALL and WHARTON, 1969). This involved reoccupation of 11 tide gauge sites (Figure 2). Five new control stations were established at the same time. In the summer of 1965, 26 sites were reoccupied, including six which had been reoccupied in 1964. The duration of observations varied at each location but was typically two months. The tidal observations were reduced to mean values by simultaneous comparison with a control station at Sitka. Sitka was the nearest control tide station which was believed to be unaffected by the earthquake. The coseismic height change at these sites is the difference between the pre- and post-1964 mean sea-level values. Post-1964 height change was computed for the six stations that were occupied in both the 1964 and 1965 summer measurement campaigns. One of the six, Valdez, subsided an additional 79 cm during the 1964–65 period; Whittier subsided an additional 9 cm; and the other 4 stations showed insignificant change. The remaining 20 stations where tidal measurements were made in 1965 may also reflect postseismic vertical movement between March, 1964 and the summer of 1965. The tide gauge results entered the adjustment as observations of coseismic height change, each having a standard deviation of 3–10 cm, depending mainly on the length of series in both the pre- and post-1964 occupations.

Geologic Data

Early models of coseismic deformation were constructed on the basis of measurements related to changes in heights of barnacles, other types of sea growth, or coastal markers. All of the input data used in our study are described by

PLAFKER (1969), and include (1) changes in upper growth limits of coastal organisms relative to sea level, (2) changes in lower growth limits of terrestrial vegetation, (3) changes in heights of pre- and post-earthquake storm beaches, (4) changes in the position of shoreline markers by local residents, (5) offshore changes in height deduced from repeated bathymetric surveys. The irregular shape of Alaska's mainland and offshore island coastlines permitted a nice distribution of these measurements, especially important because greatest deformation occurred near, or just seaward from the mainland coastline. Not all of the data reported by PLAFKER (1969) were used in our study. Selection of the points was done to retain a good spatial distribution and to use data points with low standard deviations. Locations with a high height change gradient retained a heavier density of data points in order to more clearly resolve deformation features.

Triangulation

Horizontal displacements were derived by SNAY et al. (1987) from a comparison of pre- and post-1964 triangulation surveys described by PARKIN (1969). The original 1941–42 primary arc extended from Anchorage to Valdez via Palmer and Glennallen and was a first-order survey. A second-order arc extended across northern Prince William Sound, connecting Valdez to Perry Island and was accomplished in 1947–48. The remainder of the pre-1964 surveys covered southern Prince William Sound and the Kenai Peninsula and was of third-order quality made primarily for chart control. Because the displacements associated with the 1964 earthquake were large in the Prince William Sound area, the third-order surveys were sufficiently precise to warrant their inclusion in our study. The scale of the pre-1964 network was furnished by 15 taped baselines, and orientation came from 5 Laplace azimuths. The post-earthquake surveys made in the summers of 1964 and 1965 were to first-order specifications. Orientation was supplied by 5 Laplace azimuths; and scale was from 8 Geodiameter lengths, and 146 Tellurometer lengths. The pre- and post-earthquake networks were adjusted independently; both adjustments assumed station Fishhook (near Palmer) was horizontally fixed. The resulting displacement vectors computed by PARKIN (1969) generally pointed to the southeast, and ranged in length up to 21 meters. SNAY et al. (1987) readjusted this data set after using an improved reduction procedure and computed horizontal displacements which ranged up to 14 m. In the present study, the displacements computed by SNAY were used as input, with the exception of the vectors west of longitude 151°W. The latter vectors were considered to be poorly determined and unreliable due to weak attachment to the network surrounding Prince William Sound. An uncertainty of 1.5 m was assigned to each component (North or East) of the retained displacement vectors.

Because of the low-dip angle of the slip surface for the 1964 event, the horizontal displacements derived from retriangulation should strengthen derivation of a coseismic model. However, because the scale and orientation of the pre- and post-1964 triangulation surveys were weak, and because the assumption of stability at station Fishhook is also somewhat uncertain, four additional unknown parameters were introduced corresponding to scale, orientation, and two translations for Fishhook. This avoids the assumption of stability at Fishhook, and allows the displacements to collectively rotate and change scale to find optimal consistency with the other types of data.

Acknowledgments

We would like to thank R. Snay for valuable discussions on the use of prior information and his assistance with using the triangulation results, J. Johnson for giving us her preliminary tsunami results in advance of publication, D. Christensen for sending us a preprint of his paper on seismological modeling of the 1964 earthquake, and R. Dmowska and an anonymous reader for their reviews of the paper. This research was partially supported by the NASA grant 465–921–13–10.

REFERENCES

ALGERMISSEN, S. T., RINEHART, W. A., SHERBURNE, R. W., and DILLINGER, W. Jr. (1969), *Preshocks and Aftershocks of the Prince William Sound Earthquake of March 28, 1964*, Coast and Geodetic Survey Publication 10–3, vol. II, parts B and C.

BARRIENTOS, S. E., and WARD, S. N. (1990), *The 1960 Chile Earthquake: Inversion for Slip Distribution from Surface Deformation*, Geophys. J. Int. *103*, 589–598.

BOMFORD, G., *Geodesy*, 2nd Ed. (Oxford Press, 1962).

BROCHER, T. M., FUIS, G. S., FISHER, M. A., PLAFKER, G., MOSES, M. J., and TABER, J. J. (1994), *Mapping the Megathrust beneath the Northern Gulf of Alaska Using Wide-angle Seismic Data*, J. Geophys. Res.

BROWN, L. D., REILINGER, R. E., HOLDAHL, S. R., and BALAZS, E. I. (1977), *Postseismic Crustal Uplift near Anchorage, Alaska*, J. Geophys. Res. *82*, 3369–3378.

CHRISTENSEN, D., and BECK, S. (1994), *The 1964 Prince William Sound Earthquake: Rupture Process and Plate Segementation*, Pure and Appl. Geophys. *142*, 29–53.

COHEN, S. C., HOLDAHL, S., CAPRETTE, D., HILLA, S., SAFFORD, R. W., and SCHULTZ, D. (1994). *Observations and Models of Subduction Zone Tectonics, Kenai Peninsula, Alaska* (abstract), International Assoc. of Seism. and Phys. Earth Interior, 27th General Assembly, Wellington, New Zealand.

DAVIES, J. N., and HOUSE, L. (1979), *Aleutian Subduction Zone Seismicity, Volcano-trench Separation, and their Relation to Great Thrust-type Earthquakes*, J. Geophys. Res. *84*, 4583–4591.

DEMETS, C., GORDON, R. G., STEIN, S., and ARGUS, D. F. (1987), *A Revised Estimate of Pacific-North American Motion and Implications for Western North America Plate Boundary Zone Tectonics*, Geophys. Res. Lett. *14*, 911–914.

HARRIS, R. A., and SEGALL, P. (1987), *1987: Detection of a Locked Zone at Depth on the Parkfield, California, Segment of the San Andreas Fault*, J. Geophys. Res. *92*, 7945–7962 (1987).

HASTIE, L. M., and SAVAGE, J. C. (1970), *A Dislocation Model for the 1964 Alaska Earthquake*, Bull. Seismol. Soc. Am. *60* (4), 1389–1392.

HOLDAHL, S. R. (1992), *A Dynamic Vertical Reference System*, Surveying and Land Information Systems *52* (2), 92–103.

HOLDAHL, S., and SAUBER, J. (1992), *Coseismic Slip and Post-seismic Deformation Associated with the 1964 Prince William Sound Earthquake* (extended abstract), Wadati Conference on Great Subduction Earthquakes, Sept. 16–19, 83–89, Fairbanks, Alaska.

JOHNSON, J., and SATAKE, K. (1993), *Slip Distribution of the 1964 Alaska Earthquake from Inversion of Tsunami Waveforms* (abstract), EOS, Trans. Am. Geophys. Un. Supplement *74*, 95, Oct. 1993.

KANAMORI, H. (1970), *The Alaska Earthquake of 1964: Radiation of Long-period Surface Waves and Source Mechanism*, J. Geophys. Res. *75*, 5011–5027.

KIKUCHI, M., and FUKAO, Y. (1987), *Inversion of Long-period P Waves from Great Earthquakes along Subduction Zones*, Tectonophysics *144*, 231–247.

LISOWSKI, M., SAVAGE, J. C., SVARC, J. L., and PRESCOTT, W. H. (1993), *Deformation across the Alaska-Aleutian Subduction Zone near Kodiak, Alaska* (abstract), EOS, Trans. Am. Geophys. Un., Supplement, 191, Oct. 1993.

MA, C., SAUBER, J., BELL, L., CLARK, T., GORDON, D., HIMWICH, W., and RYAN, J. (1990), *Measurement of Horizontal Motions in Alaska Using Very Long Baseline Interferometry*, J. Geophys. Res. *95*, 21991–22011.

MANSINHA, L., and SMYLIE, D. E. (1971), *The Displacements Fields of Inclined Faults*, Bull. Seismol. Soc. Am. *66*, 204–206.

MIYASHITA, K., and MATSU'URA, M. (1978), *Inversion Analysis of Static Displacement Data Associated with the Alaska Earthquake of 1964*, J. Phys. Earth. *26*, 333–349.

MOORE, J. C., DIEBOLD, J., FISHER, M., DAVIES, J., SAMPLE, J., VON HUENE, R., BROCHER, T., PAGE, B., STONE, P., TALWANI, M., and EWING, J. (1989), *Geological and Geophysical Context of the EDGE Deep Seismic Reflection Line: Gulf of Alaska* (abstract), EOS Trans. AGU *70*, 1338.

NISHENKO, S. P., and JACOB, K. H. (1990), *Seismic Potential of the Queen Charlotte-Alaska-Aleutian Seismic Zone*, J. Geophys. Res. *95*, 2511–2532.

PAGE, R. (1968), *Aftershocks and Microaftershocks of the Great Alaska Earthquake of 1964*, Seismol. Soc. Am. Bull. *58*, 1131.

PAGE, R. A., STEPHENS, C. D., and LAHR, J. C. (1989), *Seismicity of the Wrangell and Aleutian Wadati-Benioff Zones and the North American Plate along the Trans-Alaska Crustal Transect, Chugach Mountains and Copper River Basin, Southern Alaska*, J. Geophys. Res. *94*, 16059–16082.

PARKIN, E. J. (1969), *Horizontal Crustal Movements Determined from Surveys after the Alaskan Earthquake of 1964, the Prince William Sound, Alaska, Earthquake of 1964 and Aftershocks*, U.S. Dept. of Comm. Coast and Geodetic Survey, vol. III.

PLAFKER, G. (1969), *Tectonics of the March 27, 1964 Alaska Earthquake*, U.S. Geol. Surv. Prof. Pap. *543-I*, 1–74.

PLAFKER, G. (1972), *Alaskan Earthquake of 1964 and Chilean Earthquake of 1960: Implications for Arc Tectonics*, J. Geophys. Res. *77*, 901–925.

PLAFKER, G., and RUBIN, M. (1978), *Uplift History and Earthquake Recurrence as Deduced from Marine Terraces on Middleton Island, Alaska*, U.S. Geol. Surv. Open File Rep. *78-943*, 687–721.

PRESCOTT, W., and LISOWSKI, M. (1977), *Deformation at Middleton Island, Alaska, during the Decade after the Alaska Earthquake of 1964*, Bull. Seismol. Soc. Am. *67*, 579–586.

RUFF, L., and KANAMORI, H. (1983), *The Rupture Process and Asperity Distribution of Three Great Earthquakes from Long-period Diffracted P Waves*, Phys. Earth Planet Inter. *31*, 202–230.

SAVAGE, J. C., and PLAFKER, G. (1991), *1990: Tide-gauge Measurements of Uplift along the South Coast of Alaska*, J. Geophys. Res. *96*, 4325–4335.

SMALL, J. B., and WHARTON, L. C. (1969), *Vertical Displacements Determined by Surveys after the Alaska Earthquake of March 1964, the Prince William Sound Earthquake of 1964 and Aftershocks*, U.S. Dept. of Comm., Coast and Geodetic Survey, vol. III.

SNAY, R. A. (1989), *Enhancing the Geodetic Resolution of Fault Slip by Introducing Prior Information*, Manuscripta Geodaetica *14*, 391–403.

SNAY, R. A., CLINE, M. W., and TIMMERMAN, E. L. (1987), *Project REDEAM: Models for Historical Horizontal Deformation*, NOAA Technical Report NOS 125 NGS 42.

STAUDER, W., and BOLLINGER, G. A. (1966), *The Focal Mechanism of the Alaska Earthquake of March 28, 1964, and of its Aftershock Sequence.* J. Geophys. Res. *71*, 5283–5296.

THATCHER, W. (1990), *Order and Diversity in the Modes of Circum-Pacific Earthquake Recurrence*, J. Geophys. Res. *95*, 2609–2623.

WOOD, F. J. (editor) (1966), *The Prince William Sound, Alaska, Earthquake of 1964 and Aftershocks*, U.S. Dept. of Comm., Coast and Geodetic Survey, 3 vols.

WYSS, M., and BRUNE, J. N. (1967), *The Alaska Earthquake of 28 March 1964: A Complex Multiple Rupture*, Bull. Seismol. Soc. Am. *57*, 1017–1023.

(Received November 8, 1993, revised/accepted February 1, 1994)

PAGEOPH, Vol. 142, No. 1 (1994)

0033–4553/94/010083–17$1.50 + 0.20/0

Seismicity Trends and Potential for Large Earthquakes in the Alaska-Aleutian Region

CHARLES G. BUFE,[1] STUART P. NISHENKO,[1] and DAVID J. VARNES[1]

Abstract —The high likelihood of a gap-filling thrust earthquake in the Alaska subduction zone within this decade is indicated by two independent methods: analysis of historic earthquake recurrence data and time-to-failure analysis applied to recent decades of instrumental data. Recent (May 1993) earthquake activity in the Shumagin Islands gap is consistent with previous projections of increases in seismic release, indicating that this segment, along with the Alaska Peninsula segment, is approaching failure. Based on this pattern of accelerating seismic release, we project the occurrence of one or more $M \geq 7.3$ earthquakes in the Shumagin-Alaska Peninsula region during 1994–1996. Different segments of the Alaska-Aleutian seismic zone behave differently in the decade or two preceding great earthquakes, some showing acceleration of seismic release (type "A" zones), while others show deceleration (type "D" zones). The largest Alaska-Aleutian earthquakes—in 1957, 1964, and 1965—originated in zones that exhibit type D behavior. Type A zones currently showing accelerating release are the Shumagin, Alaska Peninsula, Delarof, and Kommandorski segments. Time-to-failure analysis suggests that the large earthquakes could occur in these latter zones within the next few years.

Key words: Alaska-Aleutian seismic zone, Shumagin seismic gap, accelerating moment release, time-to-failure.

Introduction

Interplate thrust earthquake activity in the Shumagin Islands region of the Alaska-Aleutian seismic zone on 13 May 1993 (M_w 7.1) (TANIOKA *et al.*, 1993; BEAVAN *et al.*, 1993) has renewed anticipation for the occurrence of an even larger "gap-filling" earthquake in the near future.

The region known as the Shumagin seismic gap was first identified as a possible site for a future major earthquake by KELLEHER (1970). A pronounced east to west progression of great earthquakes along the Queen Charlotte-Alaska seismic zone (1949 Queen Charlotte Islands, M_w 8.1; 1958 Lituya Bay, M_w 8.2; 1964 Prince William Sound, M_w 9.2) led KELLEHER (1970) to extrapolate that the region at 56 N, 158 W (offshore of the Alaska Peninsula) would be the nucleation point for the next event in this progression of plate boundary earthquakes. At that time the

---

[1] U.S. Geological Survey, Denver, CO, U.S.A.

Shumagin Islands region was thought to have ruptured as part of the 1938 M_w 8.2 Alaska Peninsula event (SYKES, 1971). KELLEHER *et al.* (1973) identified the Alaska Peninsula region as fulfilling both initial and supplementary criteria for the likely location of future earthquakes (more than 30 years elapsed since the last major event, and next in line in a progression of earthquake activity, respectively). Subsequent analysis by DAVIES *et al.* (1981) however, indicated that earthquakes in the Shumagin Islands region, previously identified as aftershocks were actually deeper intraplate earthquakes not directly related to the 1938 earthquake. New analysis and relocation of early 20th century seismicity along the Alaska-Aleutian arc by BOYD and LERNER-LAM (1988) and ESTABROOK and BOYD (1989) indicate that the Shumagin gap may have ruptured as three distinct segments (1899 (M_s 7.2), 1917 (M_s 7.4), and 1948 (M_s 7.5)) during this century.

JACOB (1984), and NISHENKO and JACOB (1990) assessed the long-term (i.e., decade scale) seismic potential of the Alaska-Aleutian seismic zone, based on estimates of average recurrence intervals for large and great earthquakes in each individual segment and the amount of time elapsed since the last gap-rupturing earthquake. NISHENKO and JACOB (1990) identified a number of areas with high (i.e., >0.60) conditional probabilities for the recurrence of either large or great earthquakes (depending on the segment) during the interval 1988–2008. These areas include the Yakataga, Shumagin Islands, Fox Islands, Delarof Islands, and Near Islands segments of the Alaska-Aleutian arc.

BUFE *et al.* (1990, 1992) and JAUMÉ and ESTABROOK (1992) have identified the Shumagin Islands/Alaska Peninsula segments as being near the end of a seismic cycle for shallow interplate thrust earthquakes, based on an increase in the regional rate of seismic release since 1985 and the occurrence of a compressional outer-rise earthquake seaward of the Alaska Peninsula in 1990 (M_w 5.3). However, the latter evidence is not definitive, as a more recent (1992) extensional outer rise earthquake was observed nearby. Here we examine the implications of recent seismicity for rupture of the Shumagin Islands/Alaska Peninsula segments and place this recent activity into the context of earlier published forecasts. We also describe the seismic release characteristics of a number of other fault segments along the Alaska-Aleutian seismic zone.

Historical Seismicity

Both the Alaska Peninsula and Shumagin Islands segments have relatively complete histories for large and great earthquakes that span the last 200+ years (see reviews in SYKES *et al.*, 1981 and DAVIES *et al.*, 1981). In addition to establishing recurrence intervals for these segments, the historic record also provides clear evidence for variable modes of earthquake rupture. Both segments ruptured simultaneously, or within a short time of one another during great earthquakes on 22 July and 7 August 1788 and again on 16 April 1847 or 1848. During this century, both segments ruptured independently of one another in a more complex sequence that includes events on 14 July 1899 (M_s 7.2), 31 May 1917 (M_s 7.4), and 14 May 1948 (M_s 7.5) for the Shumagin Islands, and 10 November 1938 (M_w 8.2) for the Alaska Peninsula. Both the 1899 and 1948 events are thought to have only filled in small portions of the Shumagin gap, with the 1948 event being confined to the deeper portions of the plate boundary (SYKES, 1971; DAVIES et al., 1981; BOYD et al., 1988).

The repeat times for those events which are thought to have ruptured the full width of the plate boundary range from 59 to 91 years and provide the primary, empirical constraint for long-term hazards estimates in this region. Based on these data, NISHENKO and JACOB (1990) estimated the probability for the recurrence of a large or great earthquake in the Shumagin and Alaska Peninsula segments to be 0.47–0.57 and 0.13–0.16, respectively, for the interval 1988–1998. Updating these earlier estimates for the interval 1993–2003 has increased the probabilities to 0.49–0.61 and 0.20–0.23, respectively. The joint probability of at least one gap-filling event with $M \geq 7.5$ in either the Shumagin or Alaska Peninsula segments during the 10-year interval 1993–2003 is 0.60–0.70. For comparison, the joint probability of at least one earthquake of $M \geq 7.0$ occurring on one of the four principal fault segments in the San Francisco Bay region during the 10-year interval 1990–2000 is 0.33 (WGCEP, 1990). While the Shumagin-Alaska Peninsula region has a significantly higher hazard during the next decade than the San-Francisco Bay region, the observed historic variability of the mode of rupture for this region does not allow more than a general specification as to the time and size of the next large event. For example, NISHENKO and JACOB (1990) estimate the 90 percent confidence interval for rupture of the Shumagin segment at ± 28 years. Both NISHENKO and JACOB (1990) and ESTABROCK and BOYD (1992) recognize that this region may rupture either in a single great earthquake or a series of large events In the sections to follow we attempt to more precisely define the magnitude and time frame of the next Shumagin Islands earthquake.

Adjacent to the Shumagin segment to the west, the Unimak Island segment, which last broke in 1946 (M_s 7.4, M_w 8.3), is also thought (NISHENKO and JACOB 1990) to have a high, though not as well constrained, probability of 0.58–0.77 of rupturing within this decade (1993–2003). In contrast to the long historic record of the Shumagin-Alaska Peninsula region, the short sample of 20th century seismic activity available for the region to the west along the Aleutian arc suggests that both the Fox and Delarof segments are characterized by the occurrence of large (M_s 7.0–7.4) earthquakes every 20 to 50 years. Both segments also apparently ruptured in conjunction with the occurrence of the great 1957 Central Aleutian earthquake.

The characteristic mode of failure for the Alaska-Aleutian region has been one of clustering or episodes of activity, where large portions of the Alaska-Aleutian arc

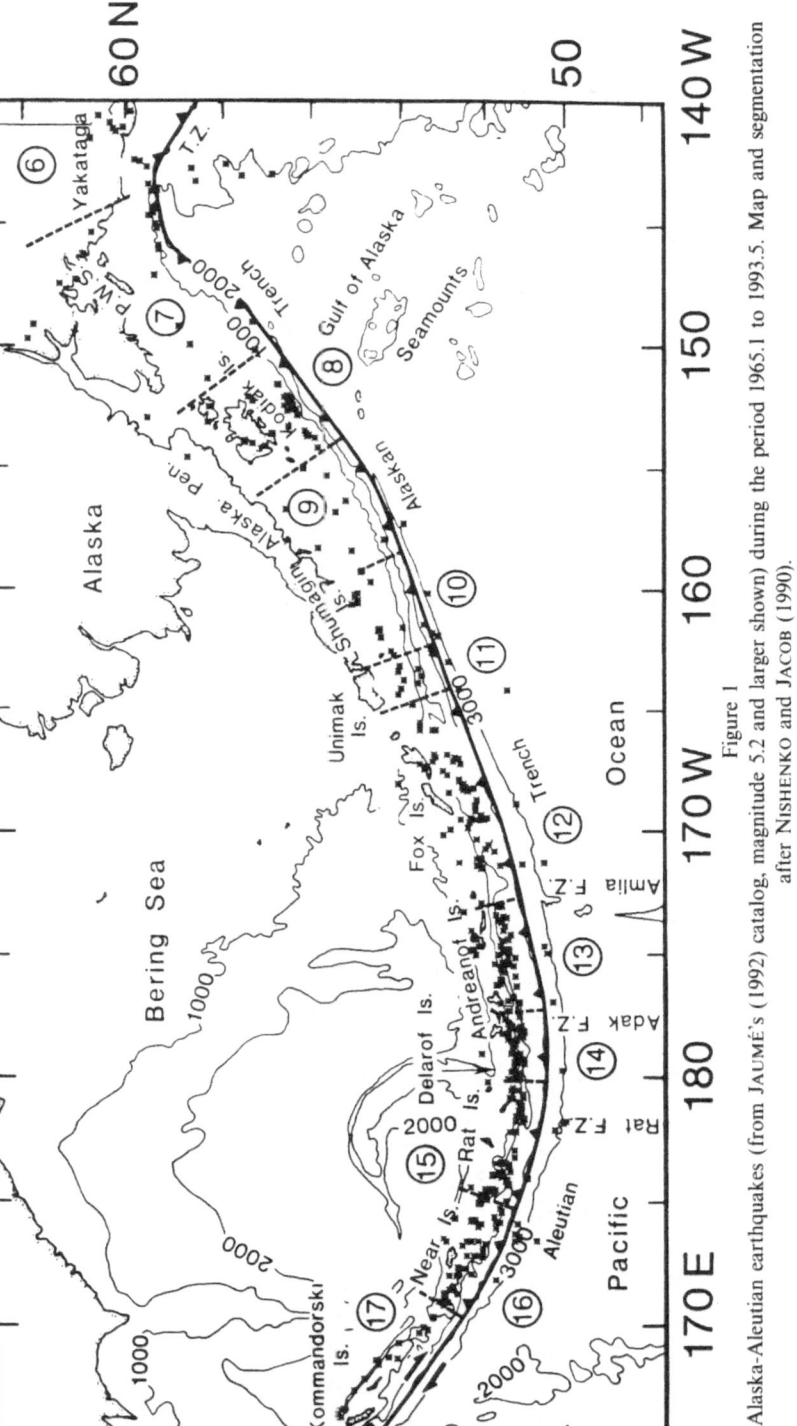

Figure 1

Alaska-Aleutian earthquakes (from JAUMÉ's (1992) catalog, magnitude 5.2 and larger shown) during the period 1965.1 to 1993.5. Map and segmentation after NISHENKO and JACOB (1990).

rupture in a series of events with inter-event times much shorter than the average recurrence interval for individual segments. During major tectonic episodes in the brief periods 1957–1958 and 1964–1965, more than half of the Aleutian-Alaska seismic zone ruptured in great earthquakes. In 1957 a great earthquake (M_w 8.7) originating in the Andreanof segment and its aftershocks ruptured the Andreanof, Delarof, and Fox Islands segments. The 1958 Lituya Bay earthquake (M_w 8.2) broke a large segment of the Fairweather fault. The 1964 Prince William Sound earthquake (M_w 9.2) ruptured the Prince William Sound and Kodiak segments. Finally, the 1965 Rat Islands event (M_w 8.7) and its aftershocks broke both the Rat and Near Island segments. In the nearly 30 years since 1965, the only great earthquake to occur in the region was an M_w 8.0 earthquake in 1986, again rupturing the Andreanof Islands segment.

Regional Seismic Release

Following BUFE and VARNES (1993), we use the term seismic release to denote measures of seismicity that can be derived or estimated from earthquake catalogs, specifically earthquake parameters (Ω's), such as seismic moment, event count, and square root of energy (or of moment). The Ω is usually estimated from earthquake magnitude (M) by an equation of the form:

$$\log \Omega = cM + d. \tag{1}$$

The coefficient c is 1.5 when Ω is moment or energy (KANAMORI, 1977; HANKS and KANAMORI, 1979), 0.75 for Benioff strain release (square root of energy) or square root of moment, and zero for the event count. Although seismic moment is the preferred parameter for most purposes, including application of the time-predictable model (BUFE et al., 1977; SHIMAZAKI and NAKATA, 1980), we have found Benioff strain release to be especially useful in time-to-failure analyses (VARNES, 1989, and BUFE and VARNES, 1993). Cumulative Benioff strain release is also useful in evaluating background seismic release rates where smaller events are of interest, but where some magnitude scaling is desirable. In contrast, cumulative moment release is typically dominated by the largest earthquake, and cumulative event count allows no magnitude scaling.

Cumulative Benioff strain release curves are shown in Figure 2 for individual segments of the Alaska-Aleutian seismic zone. The segmentation is after NISHENKO and JACOB (1990). We have placed the 1987–88 earthquakes (LAHR et al., 1988), which occurred in the Gulf of Alaska near the boundary between the Prince William Sound segment and the Cape Yakataga segment, into the Yakataga segment. The relation of these strike-slip events to the seismic cycles of either the Yakataga or Prince William Sound segments is not clear. For analysis we have extended the Yakataga segment eastward to 140 W.

Recent Seismicity Trends

Epicenters of earthquakes of M_s 5.2 and larger in the Alaska-Aleutian seismic zone are shown in Figure 1a for the period February 5, 1965 (following the occurrence of the M_w 8.7 Rat Islands earthquake on February 4) through May 25, 1993. Data shown are extracted from JAUMÉ's (1992) catalog of shallow (depth of

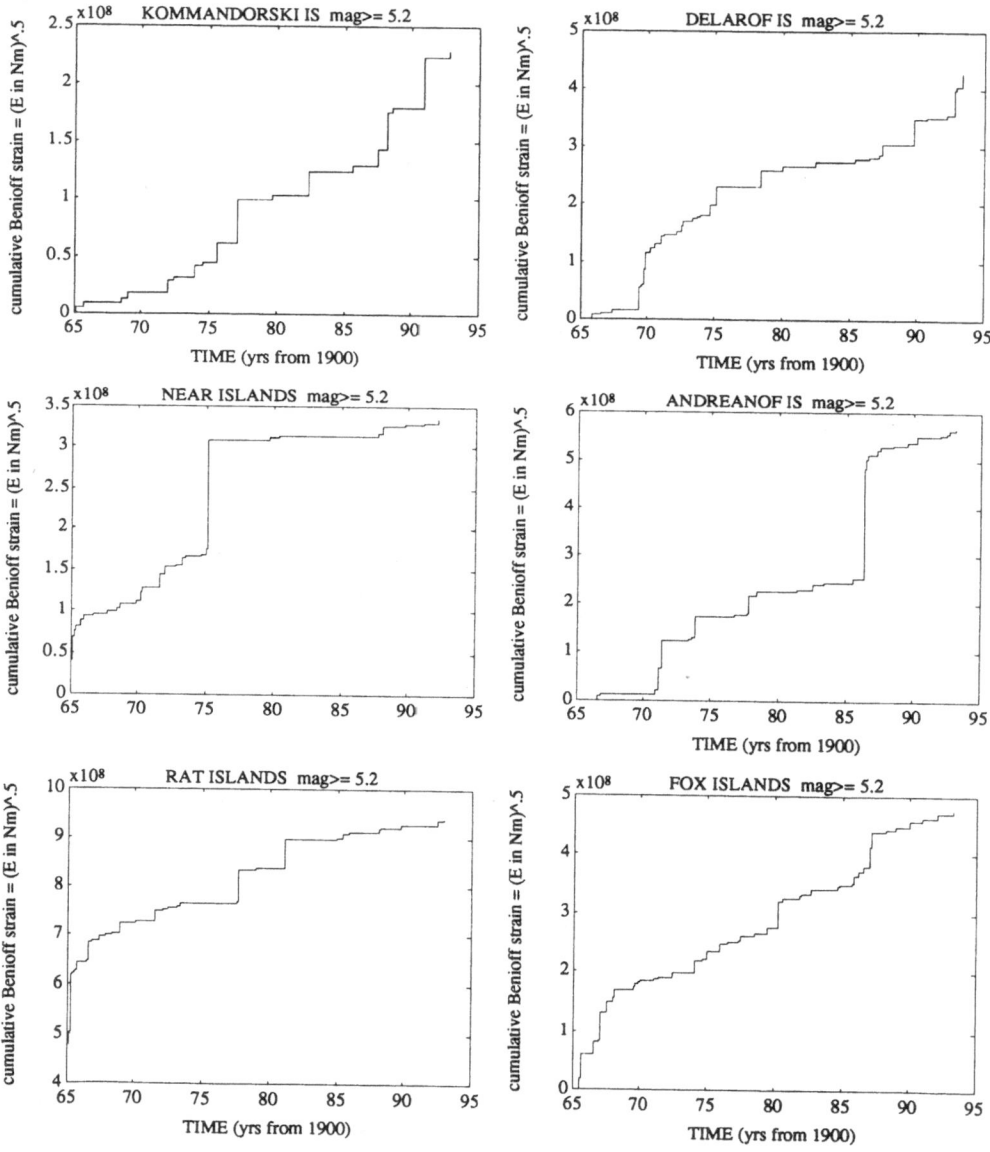

Figure 2(a)

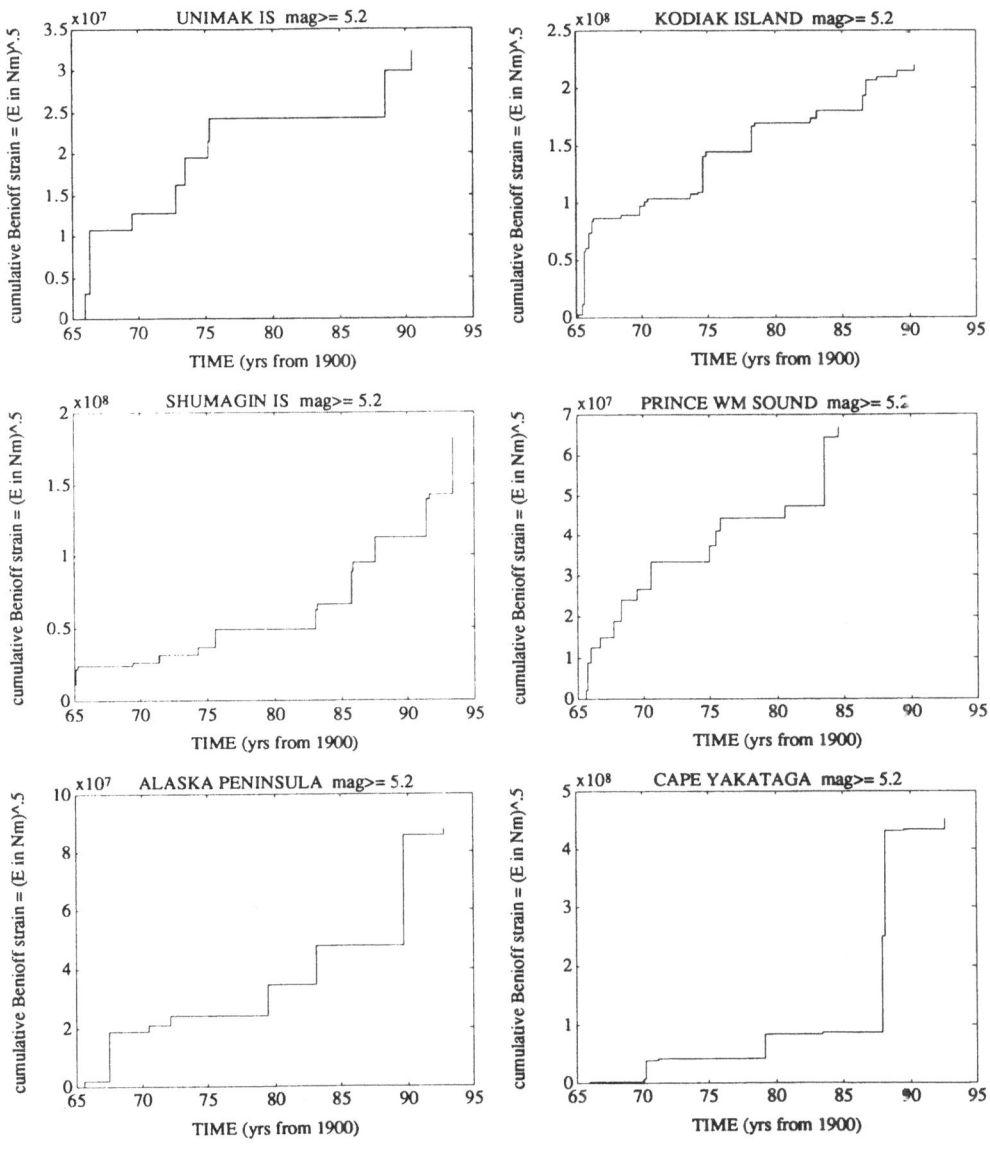

Figure 2(b)

Figure 2

Cumulative plots of Benioff strain release computed from magnitudes for each of the segments shown in Figure 1. Segments are: a) Left column, top to bottom; Kommandorski Islands, Near Islands, Rat Islands. Right column, top to bottom; Delarof Islands, Andreanof Islands, Fox Islands. b) Left column, top to bottom; Unimak Island, Shumagin Islands, Alaska Peninsula. Right column, top to bottom; Kodiak Island, Prince William Sound, Cape Yakataga.

60 km or less) earthquakes, updated using the U.S. Geological Survey National Earthquake Information Service's Preliminary Determination of Epicenters (PDE) data. The Jaumé catalog appears to be complete above about M_s 5.2. We use all earthquakes of M_s 5.2 and larger (see Figure 1), with no requirement that the earthquakes be along the plate interface or be of a particular mechanism. Our model of the earthquake cycle (see BUFE and VARNES, 1993) incorporates regional seismic release within a large volume, here incorporating both the subducting and overriding plates. We previously performed the analyses described below using data for earthquakes at all focal depths from the DNAG (ENGDAHL and RINEHART, 1991) catalog, updated using the PDE, with similar results.

The various segments of the Alaska-Aleutian seismic zone are listed from west to east in Table 1 and identified by name and number in Figure 1, using the segmentation and numbering of NISHENKO and JACOB (1990). We have quantitatively analyzed recent trends of seismic release for these segments along the Alaska-Aleutian seismic zone (including the Shumagin and Alaska Peninsula segments, both individually and jointly) to characterize the mode of strain release. Acceleration or deceleration is determined by analysis of the shape of cumulative Benioff strain release curves to determine whether the rate, or slope of the seismic release curve, is increasing with time (accelerating) or decreasing with time (decelerating). Where persistent acceleration is observed, we have applied time-to-failure analysis to estimate when gap-filling earthquakes may occur. Our analyses indicate that since about 1984 the Shumagin and Alaska Peninsula

Table 1

Seismic release trends (M ≥ 5.2) in Alaska and the Aleutian Islands

# Segment	Approx. Longitude	1967–1992 25-yr trend	1982–1992 10-yr trend	Segment type	Estimated failure time	Est. M
17 Kommandorski	171–165E	accel	accel	A	1995–2003	7.5–8.5
16 Near Islands	175–171E	decel	—	D		
15 Rat Islands	180–175E	decel	decel	D		
14 Delarof Islands	177–180W	decel	accel	A	1994–1996	7.3–8.2
13 Andreanof Isl.	173–177W	decel*	decel*	D		
12 Fox Islands	164–173W	decel	—	?		
11 Unimak Island	162–164W	decel	—	?		
10 Shumagin Isl.	159–162W	accel	accel	A	1994–1996	7.3–7.7
9 Alaska Pen.	155–159W	accel	accel	A	1994–1996	7.3–7.7
9 & 10 Combined	155–162W	accel	accel	A	1994–1996	7.5–8.2
8 Kodiak Island	150–155W	decel	—	D		
7 Prince Wm. Sd.	145–150W	decel	—	D		
6 Cape Yakataga	140–145W	accel	—	?		

\# Segment numbers and boundaries are after NISHENKO and JACOB (1990).
— No apparent trend.
* Deceleration observed both before and after the 1986 M 8 earthquake.

segments have both experienced accelerating seismic release (see Figures 3 and 4). Other segments of the Alaska-Aleutian seismic zone showing systematic accelera-tion are the Kommandorski and Delarof segments.

Although accelerating moment release has been observed preceding several large earthquakes in California (SYKES and JAUMÉ, 1990), and before two great earth-quakes in the Aleutians (JAUMÉ, 1992), not all of the segments along the Alaska-Aleutian seismic zone are characterized by accelerating seismic release preceding large or great earthquakes. The Andreanof segment, which most recently ruptured in great earthquakes in 1957 and 1986, has been characterized by decelerating seismicity (relative quiescence) prior to major failure (KANAMORI, 1981; JAUMÉ, 1992; and this paper, Figure 2). Although the 1957 earthquake originated in the Andreanof segment, its aftershocks extended into the adjacent Fox and Delarof segments. These segments experienced accelerating seismic release in the decade

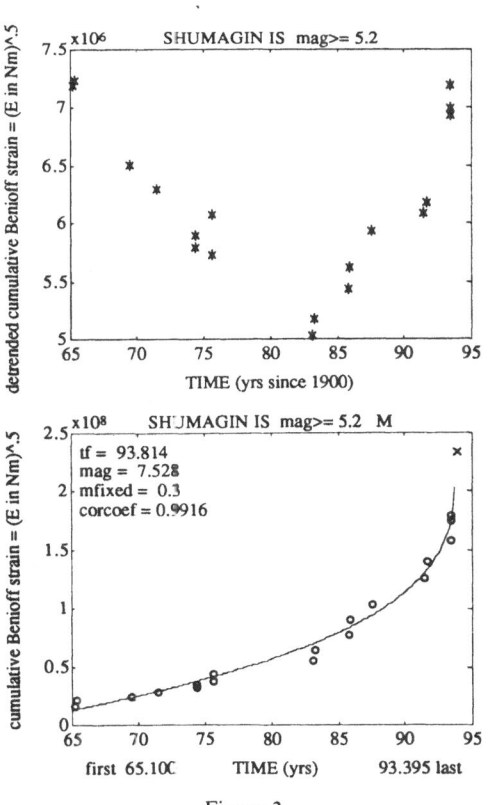

Figure 3
Detrended cumulative Benioff strain for Shumagin segment (approx. 159–162 W longitude) using data for $M_s \geq 5.2$ from JAUMÉ's (1992) catalog. Note change beginning in 1984. (bottom): Time-to-failure analysis for Shumagin segment; t_f is projected time of failure, mag is projected moment magnitude, m_{fixed} is exponent of time to failure in equation (3), and cor_{coef} is correlation coefficient for data fit.

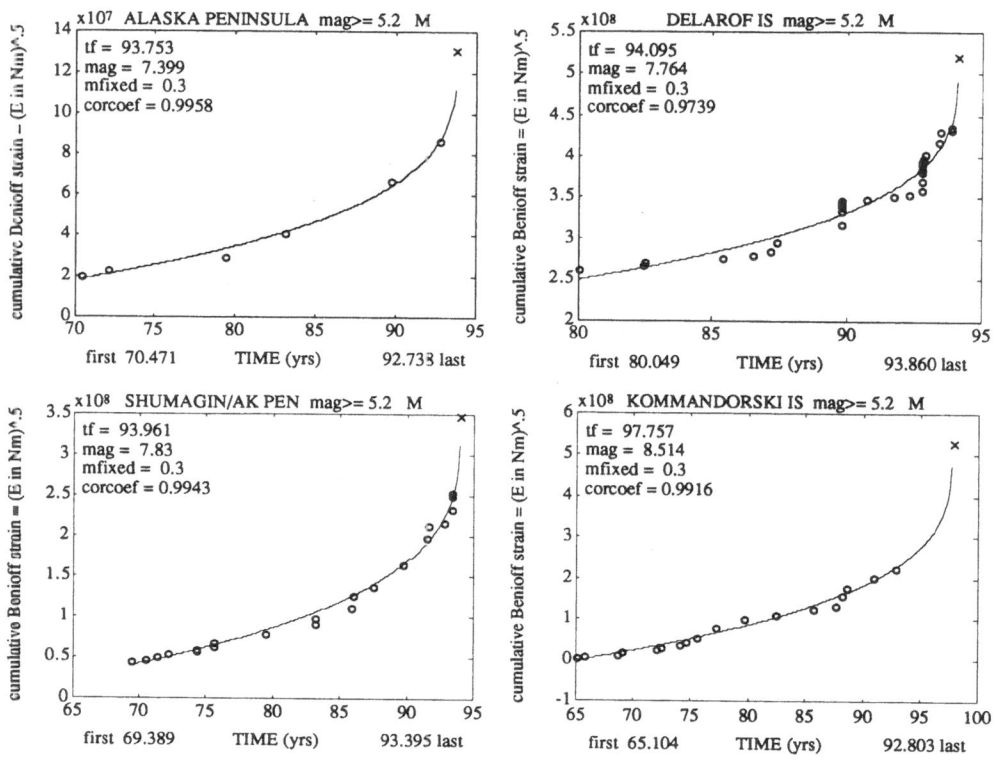

Figure 4

Time-to-failure analyses of cumulative Benioff strain for selected segments of the Alaska-Aleutian seismic zone, using data for $M_s \geq 5.2$ from JAUMÉ's (1992) catalog. t_f is projected time of failure, mag is projected moment magnitude, m_{fixed} is exponent of time to failure in equation (3), and cor_{coef} is correlation coefficient for data fit. Left column, top to bottom: 1) Time-to-failure analysis of the Alaska peninsula segment (155–159 W longitude); 2) time to failure for combined Shumagin Islands and Alaska Peninsula segments (155–162 W longitude). Right column, top to bottom: 1) Time-to-failure analysis of the Delarof Islands segment (177–180 W longitude); 2) time-to-failure analysis of the Kommandorski segment (165–171 E longitude).

before that event. Following the occurrence of the 1986 Andreanof Islands earthquake, the Delarof segment has shown acceleration of cumulative seismic release, continuing to the present. This pattern may result from transfer of stress and is interpreted as an indication of impending rupture of the Delarof zone itself. Immediately to the west, the Rat and Near Islands segments, which were quiescent (KANAMORI, 1981) prior to the great Rat Islands earthquake of 1965, show continuing deceleration from 1965 to the present.

Although the 1965–1993 data do not provide a sufficiently long baseline to firmly establish behavior of individual segments, these data, taken in combination with observations from earlier periods, suggest that some segments, such as the

Delarof Islands segment, are prone to exhibit acceleration of Benioff strain release, while others, such as the Rat Islands and Andreanof Islands segments, tend to show deceleration. The deceleration may indicate progressive locking of a strongly coupled zone or may simply be the result of decay of aftershock activity following a previous large earthquake. Acceleration may be the result of stress redistribution due to aseismic slip, possibly associated with progressive unlocking of a more weakly coupled fault segment. This behavior provides a means of classifying gaps as type A (accelerating or unlocking), as type D (decelerating or locking) or as neither. Some segments, such as the Delarof Islands, may show both A and D types of behavior at different times (see Figure 2). We have classified the Delarofs as predominantly A on the basis of acceleration observed before the 1957 event and acceleration occurring presently. The Kodiak Island segment is classified as D, based on recent seismicity, although JAUMÉ (1992) noted some acceleration in moment release there preceding the great 1964 Prince William Sound earthquake.

It is important to note that seismicity in type A gaps may be used to forecast large earthquakes using the time-to-failure analysis techniques discussed below. In some instances these large earthquakes may not originate within the type A segment showing the acceleration, but within an adjacent segment. Type D segments, on the other hand, will not show acceleration, but may be quiescent before large or great earthquakes. Epicenters of great earthquakes, such as the 1957 Andreanof Islands ($M_w = 8.7$), the 1964 Prince William Sound ($M_w = 9.2$), and the 1965 Rat Islands ($M_w = 8.7$) earthquakes appear to lie within type D segments. Although the absence of accelerating moment release within these segments is not indicative of low stresses, long-term forecasts (NISHENKO and JACOB, 1990) suggest that these great earthquakes are not due to repeat anytime soon.

Multi-segment ruptures which encompass both type A and D segments, such as the 1957 earthquake and its aftershocks, may be preceded by accelerating moment release within the type A segments. In the case of the 1957 earthquake, the premonitory acceleration of seismic release in the Delarof and Fox Islands segments and quiescence in the Andreanof segment were well developed. As noted by JAUMÉ (1992), the acceleration lies entirely within the Delarof and Fox Islands segments, with deceleration occurring in the Andreanof segment. It is also possible that type A segments may show acceleration preceding large earthquakes in type D segments that do not rupture the type A segment.

Time-to-failure Analysis

VARNES (1983) has shown that creep curves for various materials show decelerating (primary) and accelerating (tertiary) creep and that the most common form of accelerating creep is characterized by the INPORT relation, i.e., the *rate* is

proportional to the *IN*verse *P*ower *O*f *R*emaining *T*ime to failure. After VARNES
(1989) and BUFE and VARNES (1993):

$$d \sum \Omega/dt = k/(t_f - t)^n. \tag{2}$$

Integrated, this becomes

$$\sum \Omega = A + [k/(n - 1)](t_f - t)^m, \tag{3}$$

where Ω is seismic release calculated from magnitude, A, k, and n are constants,
$m = 1 - n$, $n \neq 1$, and t_f is time of failure (main shock). VARNES (1989) has shown
that Benioff strain release in precursory sequences often follows this relation, and
when it does, the time of the main shock may be predicted from the pattern of
accelerating release. SYKES and JAUMÉ (1990) investigated the distribution of
accelerating moment release on faults in the San Francisco Bay region preceding the
Loma Prieta earthquake. BUFE and VARNES (1993) have extended the concept of
accelerating seismic release in foreshocks to model the behavior of a type A segment
(the northern San Andreas fault) through a complete seismic cycle.

The progressive acceleration of seismic release in the Shumagin Islands segment
has been well established (BUFE *et al.*, 1990; DMOWSKA and LOVISON-GOLOB,
1991; JAUMÉ and ESTABROOK, 1992). BUFE *et al.* (1992) applied time-to-failure
techniques to analyze the accelerating seismic release which preceded the 1957
earthquake in the central Aleutians (their Figure 1) and to the current accelerating
release occurring in the combined Unimak, Shumagin, and Alaska Peninsula gaps
(their Figure 2). In this paper we provide time-to-failure analyses for the Shumagin
segment, the Alaska Peninsula segment, and the combined regions as well as for the
accelerating Delarof and Kommandorski Islands segments. The results of these
analyses are summarized in Table 1. Time-to-failure curves for the Delarof,
Kommandorski, Shumagin Islands, Alaska Peninsula, and combined Shumagin
Islands and Alaska Peninsula segments are shown in Figures 3 and 4.

Results and Discussion

Based on time-to-failure analysis of accelerating seismic release using magnitude
≥ 6 (maximum of M_s or m_b) from the extended DNAG catalog, BUFE *et al.* (1992)
predicted the occurrence of one or more large (M 7.4–8.3) earthquakes by 1997
(1992.5–1997.8) somewhere within the combined western Alaska Peninsula-Shu-
magin-Unimak segments of the Alaska-Aleutian subduction zone. (50–60 N, 156–
164 W), "assuming continued acceleration." The recent earthquakes in the
Shumagin gap (May 13, 1993, M_s 6.9; May 25, 1993, m_b 6.2) continue the
acceleration and are consistent with the projection (BUFE *et al.*, 1992, Figure 2) of
increasing cumulative seismic release approaching time of failure. However, the
Unimak segment does not show clear acceleration of seismic release (see Table 1)

and should probably not be included in the analysis. We have reanalyzed seismic release in the Shumagin and Alaska Peninsula segments, both individually and jointly, using time-to-failure analyses to estimate when gap-filling earthquakes may occur. Our analyses for these segments (Table 1) indicate that the Shumagin and Alaska Peninsula zones each show accelerating seismic release commensurate with the imminent occurrence of magnitude 7.3–7.7 main shocks. Another scenario is that these zones could rupture together in a single, larger event. The occurrence of the 1993 events continues the pattern of accelerating seismic release for the Shumagin Islands segment and for the Shumagin-Alaska Peninsula combined zone. Because the late-stage accelerating cumulative seismic release curve is steep, the estimate of time of failure is not very sensitive to the choice of the exponent of time to failure (see Figure 5). The magnitude estimate of 7.5 ± 0.2 for the Shumagin Islands segment is reasonably consistent with the moment estimate of ESTABROOK and BOYD (1992) for the 1917 Shumagin earthquake.

These results are also generally consistent with other evidence that conditions may be right for rupture of the Shumagin gap (BUFE et al., 1990; JAUMÉ and ESTABROOK, 1992; DMOWSKA et al., 1992). However, LISOWSKI et al. (1988) noted the absence of geodetic strain accumulation in the Shumagin gap during 1980–1987, suggesting weak coupling. DMOWSKA et al. (1992), estimate that while only 15 percent of the plate convergence takes place seismically, coupling at depths between 20 and 50 km is sufficient to permit the generation of large or great earthquakes.

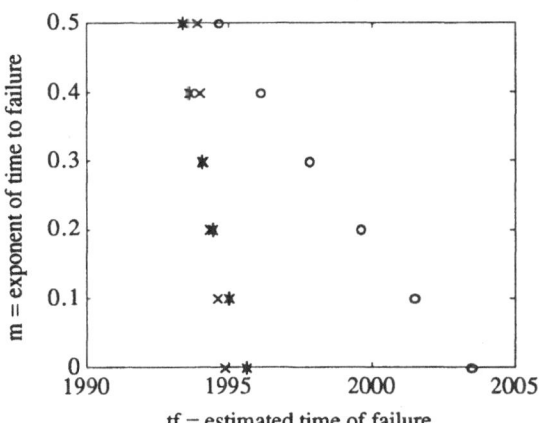

Figure 5

Stages of time to failure for segments of the Alaska-Aleutian seismic zone. Time of failure is shown as a function of exponent (m) of time to failure ($t_f - t$) in equation (3). Preferred value of m is 0.3 (range of 0.1 to 0.4) based on experience in ongoing postdiction studies of other events. Symbols are: * combined Shumagin Islands and Alaska Peninsula segments, × Delarof Islands segment, ○ Kommandorski Islands segment.

Other segments of the Alaska-Aleutian seismic zone showing systematic acceleration (see Figure 4) are the Delarof Islands segment (since 1980) and the Kommandorski Islands segment (since 1965 and possibly earlier). We are projecting the occurrence of a large earthquake ($M \geq 7.4$) in the Delarof segment (177–180 W) of the Aleutian arc by 1996. The recent Delarof Islands earthquake (May 15, 1993, M_s 6.6) occurred only two days following the Shumagin Islands event, and extends the acceleration curve for the Delarof segment, increasing the probability of a subsequently larger event. Interpretation of this acceleration is complicated by the possibility that the acceleration may be sympathetic or triggered, and the culminating earthquake, as in 1957, could originate in the adjacent type D Andreanof (or possibly Rat) Islands gaps and may or may not rupture the Delarof segment. In the Kommandorski segment (165–171 E), plate motion is nearly parallel to the plate boundary and no large earthquakes have recently occurred there. However, large or great tsunamigenic earthquakes may have occurred on this segment in the mid-1800s (SYKES *et al.*, 1981). The Kommandorski segment appears to be in an earlier stage of the failure process, with a larger uncertainty in estimated time of failure. Additional acceleration of seismic release in this segment will be required to narrow the uncertainty.

Time-to-failure analysis curves are shown in Figures 3 and 4 for the four segments (Kommandorski, Delarof, Shumagin, and Alaska Peninsula) showing clear, consistent acceleration of Benioff strain release. The results of time-to-failure analyses of accelerating segments are summarized in Table 1. The forecast time windows are somewhat larger than suggested in Figure 5 to account for the discontinuous nature of the seismicity used to determine time of failure. The computed times of failure cluster around 1994.0 for all but the Kommandorski segment. None of the events has occurred by the end of 1993, hence the expectation for their occurrence is asymmetrical, with the tail extending into the future. As additional smaller earthquakes occur in a given zone, the time-to-failure analysis should be updated. Failure of several of these zones is probable within the next few years.

Multiple large but discontinuous fault segments in the Alaska-Aleutian seismic zone have ruptured within relatively short (2-year) time periods in the past (BUFE *et al.*, 1992). The projected cluster of large earthquakes in the Alaska-Aleutian seismic zone within a period of a year or two is thus not without precedent and appears to be the normal mode of strain release in this region.

Conclusions

We have classified segments of the Alaska-Aleutian seismic zone as type A (accelerating) or type D (decelerating), based on cumulative seismic release histories for the time period 1965.1–1993.4. Data (through 1993.9) from four segments experiencing accelerating seismic release have been analyzed to estimate time of failure. The analyses indicate that the Shumagin segment and the Delarof segment are rapidly approaching failure (i.e., gap-breaking earthquakes of M 7.3 or greater are likely within 3 years). Accelerating seismic release has also been analyzed for the Kommandorski Islands segment, where culmination appears to be less imminent and time of failure not so well determined. It appears we are nearing the beginning of an episode of large or great earthquakes in the Alaska-Aleutian seismic zone, similar to the long-distance temporal clustering that has occurred in the past.

Tsunamis are one of the greatest hazards associated with large or great earthquakes along the Alaska-Aleutian seismic zone. These not only affect the epicentral region, but the entire circum-Pacific community. Numerical simulations of tsunamis generated along the Alaska-Aleutian seismic zone indicate that the west coast of Canada and the United States are more vulnerable to sources along the eastern portion of the seismic zone (i.e., the Shumagin Islands) than the central or western portion (i.e., the Delarof Islands) (HOUSTON et al., 1975). Hence, the recurrence of an earthquake the size of the 1788 event along the Alaska Peninsula would have significant economic consequences for the west coast of the United States and Canada. Additionally, models by ZOWALIK and MURTY (1989) indicate that the maximum energy for a Shumagin Islands tsunami would be directed towards the south and southeast (i.e., towards the Hawaiian Islands). Recognition of these facts is important for future hazards mitigation planning in these regions.

Acknowledgements

We especially thank William Spence, James Dewey, Steven Jaumé, Lynn Sykes, and Renata Dmowska for their helpful comments and criticisms.

REFERENCES

BEAVAN, J., GILBERT, L., and LARSON, K. (1993), *A GPS Reoccupation of the Shumagin Island, Alaska, EDM Network: Effects of the May 13, 1993 M_w 7.1 Earthquake*, EOS (Transactions, American Geophysical Union), 74, 92.

BOYD, T. M., and LERNER-LAM, A. L. (1988), *Spatial Distribution of Turn-of-the-century Seismicity along the Alaska-Aleutian Arc*, Bull. Seismol. Soc. Am. 78, 636–650.

BOYD, T. M., TABER, J. J., LERNER-LAM, A. L., and BEAVAN, J. (1988), *Seismic Rupture and Arc Segmentation within the Shumagin Islands Seismic Gap, Alaska*, Geophys. Res. Lett. 15, 201–204.

BUFE, C. G., HARSH, P. W., and BURFORD, R. O. (1977), *Steady-state Seismic Slip—A Precise Recurrence Model*, Geophys. Res. Lett. 4, 91–94.

BUFE, C. G., JAUMÉ, S. C., NISHENKO, S. P., SYKES, L. R., and VARNES, D. J. (1990), *Accelerating Moment Release in the Alaska Subduction Zone: Precursor to a Great Thrust Earthquake?* EOS (Transactions, American Geophysical Union), 71, 1451–1452.

BUFE, C. G., NISHENKO, S. P., and VARNES, D. J. (1992), *Clustering and Potential for large Earthquakes in the Alaska-Aleutian Region*, Proceedings, Wadati Conference on Great Subduction Earthquakes, University of Alaska, 129–132.

BUFE, C. G., and VARNES, D. J. (1993), *Predictive Modeling of the Seismic Cycle of the Greater San Francisco Bay Region*, J. Geophys. Res. *98*, 9871–9883.

DAVIES, J., SYKES, L., HOUSE, L., and JACOB, K. (1981), *Shumagin Seismic Gap, Alaska Peninsula: History of Great Earthquakes, Tectonic Setting and Evidence for High Seismic Potential*, J. Geophys. Res. *86*, 3821–3855.

DMOWSKA, R., and LOVISON-GOLOB, L. C. (1991), *Stress Transfer and Seismic Phenomena in the Shumagin Islands, Alaska*, EOS, Transactions American Geophysical Union, *72, Supplement*, 322.

DMOWSKA, R., and LOVISON, L. C. (1992), *Influence of Asperities along Subduction Interfaces on the Stressing and Seismicity of Adjacent Areas*, Tectonophysics *211*, 23–43.

DMOWSKA, R., ZHENG, G., RICE, J. R., and LOVISON-GOLOB, L. C. (1992), *Stress Transfer, Seismic Phenomena and Seismic Potential in the Shumagin Gap, Alaska*, Proceedings, Wadati Conference on Great Subduction Earthquakes, University of Alaska, 150–152.

ENGDAHL, E. R., and RINEHART, W. A., *Seismicity map of North America project*. In *Neotectonics of North America* (Slemmons, D. B., Engdahl, E. R., Zoback, M. D., and Blackwell, D. D., eds.) (Boulder, Colorado, Geological Society of America 1991) Decade Map Vol. 1, 21–27.

ESTABROOK, C. H., and BOYD, T. M. (1992), *The Shumagin Islands, Alaska, Earthquake of 31 May 1917*, Bull. Seismol. Soc. Am. *82*, 755–773.

HANKS, T. H., and KANAMORI, H. (1979), *A Moment Magnitude Scale*, J. Geophys. Res. *84*, 2348–2350.

HOUSTON, J. R., WHALIN, R. W., GARCIA, A. W., and BUTLER, H. L. (1975), *Effect of Source Orientation and Location in the Aleutian Trench on Tsunami Amplitude along the Pacific Coast of the Continental United States*, Res. Rep. H-75-4, 48 pp., U.S. Army Waterways Exp. Station Hydraul. Lab.

JACOB, K. (1984), *Estimates of Long-term Probabilities for Future Great Earthquakes in the Aleutians*, Geophys. Res. Lett. *11*, 295–298.

JAUMÉ, S. C. (1992), *Moment Release Rate Variations during the Seismic Cycle in the Alaska-Aleutians Subduction Zone*, extended abstract, Proceedings, Wadati Conference on Great Subduction Earthquakes, University of Alaska, 123–128.

JAUMÉ, S. C., and ESTABROOK, C. H. (1992), *Accelerating Seismic Moment Release and Outer Rise Compression: Possible Precursors to the Next Great Earthquake in the Alaska Peninsula Region*, Geophys. Res. Lett. *19*, 345–348.

JONES, L. M., and MOLNAR, P. (1979), *Some Characteristics of Foreshocks and their Possible Relationship to Earthquake Prediction and Premonitory Slip on Faults*, J. Geophys. Res. *84*, 3596–3608.

KANAMORI, H. (1977), *The Energy Release in Great Earthquakes*, J. Geophys. Res. *82*, 2981–2987.

KANAMORI, H., *The nature of seismicity patterns before large earthquakes*. In *Earthquake Prediction*, an International Review, M. Ewing Ser. 4 (Simpson, D., and Richards, P., eds.) (American Geophysical Union, Washington D.C. 1981) pp. 1–19.

KELLEHER, J. A. (1970), *Space-time Seismicity of the Alaska-Aleutian Seismic Zone*, J. Geophys. Res. *75*, 5745–5756.

KELLEHER, J. A., SYKES, L. R., and OLIVER, J. (1973), *Possible Criteria for Predicting Earthquake Locations and their Applications to Major Plate Boundaries of the Pacific and Caribbean*, J. Geophys. Res. *78*, 2547–2585.

LAHR, J. C., PAGE, R. A., STEPHENS, C. D., and CHRISTENSEN, D. H. (1988), *Unusual Earthquakes in the Gulf of Alaska and Fragmentation of the Pacific Plate*, Geophys. Res. Lett. *15*, 1483–1486.

LISOWSKI, M., SAVAGE, J. C., PRESCOTT, W. H., and GROSS, W. K. (1988), *Absence of Strain Accumulation in the Shumagin Seismic Gap, Alaska, 1980–1987*, J. Geophys. Res. *93*, 7909–7922.

NISHENKO, S. P. (1989), *Circum-Pacific Seismic Potential 1989–1999*, U.S. Geological Survey Open-File Report 89-85, map.

NISHENKO, S. P., and JACOB, K. H. (1990), *Seismic Potential of the Queen Charlotte-Alaska-Aleutian Seismic Zone*, J. Geophys. Res. *95*, 2511–2532.

SHIMAZAKI, K., and NAKATA, T. (1980), *Time-predictable Recurrence Model for Large Earthquakes*, Geophys. Res. Lett. *7*, 279–282.

SYKES, L. R. (1971), *Aftershock Zones of Great Earthquakes, Seismicity Gaps, and Earthquake Prediction for Alaska and the Aleutians*, J. Geophys. Res. *76*, 8021–8041.

SYKES, L. R., KISSLINGER, J. B., HOUSE, L., DAVIES, J. N., and JACOB, K. H. (1981), *Rupture zones and repeat times of great earthquakes along the Alaska-Aleutian arc, 1784–1980*. In *Earthquake Prediction*, an International Review, M. Ewing Ser. 4 (Simpson, D., and Richards, P., eds.) (Washington, D.C., American Geophysical Union 1981) pp. 73–80.

SYKES, L. R., and JAUMÉ, S. C. (1990), *Seismic Activity on Neighboring Faults as a Long-term Precursor to Large Earthquakes in the San Francisco Bay Area*, Nature *348*, 595–599.

TANIOKA, T., SATAKE, K., RUFF, L., and GONZALES, F. (1993), *Fault Parameters and Tsunami Excitation of the May 13, 1993 Shumagin Islands Earthquake*, EOS, Transactions, American Geophysical Union *74*, 92.

VARNES, D. J. (1983), *Time-deformation Relations in Creep to Failure of Earth Materials*, Proc. 7th Southeast Asian Geotechnical Conf. *2*, 107–130.

VARNES, D. J. (1989), *Predicting Earthquakes by Analyzing Accelerating Precursory Seismic Activity*, Pure Appl. Geophys. *4*, 661–686.

ZOWALIK, Z., and MURTY, T. S. (1989), *On Some Future Tsunamis in the Pacific Ocean*, Natural Hazards *1*, 349–369.

(Received August 6, 1993, revised January 4, 1994, accepted January 7, 1994)

PAGEOPH, Vol. 142, No. 1 (1994)

0033–4553/94/010101–71$1.50 + 0.20/0

Rupture Process of Large Earthquakes in the Northern Mexico Subduction Zone

Larry J. Ruff and Angus D. Miller[1,2]

Abstract — The Cocos plate subducts beneath North America at the Mexico trench. The northern-most segment of this trench, between the Orozco and Rivera fracture zones, has ruptured in a sequence of five large earthquakes from 1973 to 1985: the Jan. 30, 1973 Colima event (M_s 7.5) at the northern end of the segment near Rivera fracture zone; the Mar. 14, 1979 Petatlan event (M_s 7.6) at the southern end of the segment on the Orozco fracture zone; the Oct. 25, 1981 Playa Azul event (M_s 7.3) in the middle of the Michoacan "gap"; the Sept. 19, 1985 Michoacan mainshock (M_s 8.1); and the Sept. 21, 1985 Michoacan aftershock (M_s 7.6) that ruptured part of the Petatlan zone. Body wave inversion for the rupture process of these earthquakes finds the best: earthquake depth; focal mechanism; overall source time function; and seismic moment, for each earthquake. In addition, we have determined spatial concentrations of seismic moment release for the Colima earthquake, and the Michoacan mainshock and aftershock. These spatial concentrations of slip are interpreted as asperities; and the resultant asperity distribution for Mexico is compared to other subduction zones. The body wave inversion technique also determines the *Moment Tensor Rate Functions*; but there is no evidence for statistically significant changes in the moment tensor during rupture for any of the five earthquakes. An appendix describes the *Moment Tensor Rate Functions* methodology in detail.

The systematic bias between global and regional determinations of epicentral locations in Mexico must be resolved to enable plotting of asperities with aftershocks and geographic features. We have spatially "shifted" all of our results to regional determinations of epicenters. The best point source depths for the five earthquakes are all above 30 km, consistent with the idea that the down-dip edge of the seismogenic plate interface in Mexico is shallow compared to other subduction zones. Consideration of uncertainties in the focal mechanisms allows us to state that all five earthquakes occurred on fault planes with the same strike (N65°W to N70°W) and dip ($15 \pm 3°$), except for the smaller Playa Azul event at the down-dip edge which has a steeper dip angle of 20 to 25°. However, the Petatlan earthquake does "prefer" a fault plane that is rotated to a more east-west orientation — one explanation may be that this earthquake is located near the crest of the subducting Orozco fracture zone. The slip vectors of all five earthquakes are similar and generally consistent with the NUVEL-predicted Cocos-North America convergence direction of N33°E for this segment. The most important deviation is the more northerly slip direction for the Petatlan earthquake. Also, the slip vectors from the Harvard CMT solutions for large and small events in this segment prefer an overall convergence direction of about N20°E to N25°E.

All five earthquakes share a common feature in the rupture process: each earthquake has a small initial precursory arrival followed by a large pulse of moment release with a distinct onset. The delay time varies from 4 s for the Playa Azul event to 8 s for the Colima event. While there is some evidence of spatial concentration of moment release for each event, our overall asperity distribution for the northern Mexico segment consists of one clear asperity, in the epicentral region of the 1973 Colima earthquake, and then a scattering of diffuse and overlapping regions of high moment release for the

---

[1] Department of Geological Sciences, University of Michigan, Ann Arbor, MI 48109, U.S.A.
[2] Now at: The University of Durham, Durham, England.

remainder of the segment. This character is directly displayed in the overlapping of rupture zones between the 1979 Petatlan event and the 1985 Michoacan aftershock. This character of the asperity distribution is in contrast to the widely spaced distinct asperities in the northern Japan-Kuriles Islands subduction zone, but is somewhat similar to the asperity distributions found in the central Peru and Santa Cruz Islands subduction zones. Subduction of the Orozco fracture zone may strongly affect the seismogenic character as the overlapping rupture zones are located on the crest of the subducted fracture zone. There is also a distinct change in the physiography of the upper plate that coincides with the subducting fracture zone, and the Guerrero seismic gap to the south of the Petatlan earthquake is in the "wake" of the Orozco fracture zone. At the northern end, the Rivera fracture zone in the subducting plate and the Colima graben in the upper plate coincide with the northernmost extent of the Colima rupture zone.

Key words: Earthquake rupture process, asperities, moment tensor rate functions.

1. Introduction

1.1 Large Earthquakes and Tectonics along the Mexico Subduction Zone

The Cocos plate subducts beneath North America at the Mexico subduction zone. While this subduction zone has most of the features expected for subduction beneath a continent, it also has several unusual characteristics. There is a high level of shallow seismicity along the entire 1,000 km length of the zone, but a deep Wadati-Benioff zone is not present as the subducted plate apparently underplates the Mexico lithosphere. The associated volcanic arc is somewhat unusual as the trench-volcano distance varies from 200 km at the northern end of the subduction zone to more than 300 km toward the southern end of the Cocos-North America subduction zone. Subduction is characterized by the occurrence of large to great underthrusting earthquakes (M_s of 7 to 8), with a relatively short recurrence time of 30 years (RIKITAKE, 1976; MCNALLY and MINSTER, 1981). On a global basis, the characteristic largest earthquakes in Mexico of about magnitude 8 are smaller than "expected" (see RUFF and KANAMORI, 1983a). Compared to other subduction zones that subduct young lithosphere with a fast convergence rate, the plate contact interface in Mexico is apparently segmented into shorter rupture lengths. RUFF (1989a) speculated that a trench environment relatively devoid of sediments, such as the Mexico trench, might result in shorter along-trench seismic segments. BYRNE *et al.* (1988) emphasize the role of sediments in determining the up-dip edge of the seismogenic plate interface. Perhaps the lack of sediments in Mexico may allow the coupled zone to extend close to the trench axis. TICHELAAR and RUFF (1993) recently completed a global survey of the depth of the down-dip edge of the seismogenic plate interface in subduction zones. They show that the Mexico subduction zone is anomalous: while the depth of coupling in most subduction zones is about 40 ± 5 km, the down-dip edge of coupling in Mexico is just 25 ± 5 km. Although TICHELAAR and RUFF (1993) note that the shallow coupling depth may be related to the "small" size of Mexico earthquakes, there is no clear reason why depth extent of coupled interface would determine the along-trench seismic segmentation.

Although the Mexican earthquakes may be smaller than "expected," earthquakes such as the recent Michoacan earthquake of Sept. 19, 1985 (M_s 8 1) still represent terrible natural catastrophes. The Michoacan earthquake has undergone many studies, including detailed modern rupture process studies. In contrast, many of the earlier Mexico events have not undergone modern rupture studies. Figure 1 shows the northern segment of the Mexico subduction zone between the Cocos-North America-Rivera plates triple junction to the north, and the Orozco Fracture Zone and the Guerrero seismic gap to the south (see SINGH and MORTERA 1991). Five large earthquakes have ruptured this subduction zone segment since 1973. All five earthquakes have focal mechanisms determined from long-period surface waves. CHAEL and STEWART (1982) determined a time function of the 1979 event by forward modeling of body waves. SINGH and MORTERA (1991) show time functions for all five events as part of their study to examine the older pre-1963 events. Harvard CMTs (see DZIEWONSKI and WOODHOUSE, 1983) are available for four of the five events. As mentioned above, the 1985 mainshock received several rupture process studies (e.g., see EKSTRÖM, 1989; MENDOZA and HARTZELL, 1989, MENDEZ and ANDERSON, 1991; YOMOGIDA, 1988). However, the large aftershock of the Michoacan event, on Sept. 21, 1985 has not received as much attention (MENDOZA, 1993). One curious feature is that the 1985 aftershock is similar in size

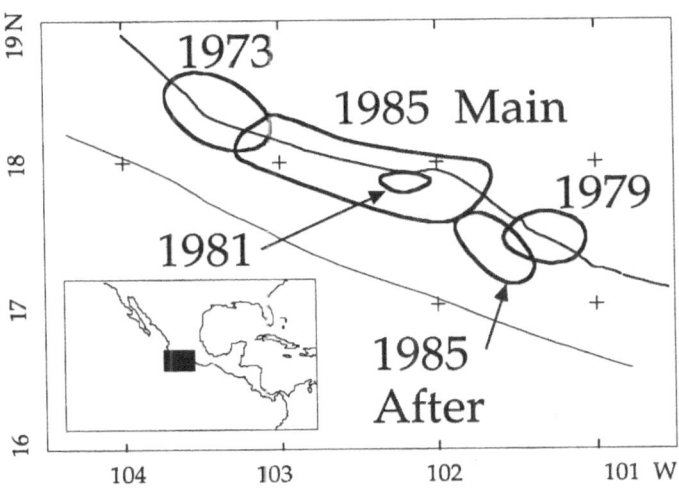

Figure 1

Base map of the northern Mexico subduction zone segment. Aftershock areas of the five large earthquakes (from UNAM, 1986), are shown with years of occurrence. Cocos plate subducts to the northeast beneath North America; the trench axis is plotted to the southwest of the rupture zones. The Mexico coastline is also shown. These basic features are plotted for reference in the following maps.

to the 1979 Petatlan event, but is apparently located up-dip of the 1979 event. Similar to other large events along the Mexico subduction zone, the 1979 event had an extensive foreshock sequence (GETTRUST et al., 1981). VALDES et al. (1982) show the detailed aftershock distribution of the 1979 event. REYES et al. (1979) conducted a special study of the 1973 Colima earthquake and associated seismicity. Given the basic interpretation that the Colima event is a subduction event, it is odd that the fault plane of their focal mechanism has a very steep dip of 30°, and the strike is 330°, more northerly than the trench strike. Even more strange is the fact that CHAEL and STEWART (1982) used the surface wave data from REYES et al. (1979), but they found a fault plane with an EW strike of 266°. Since the 1973 Colima earthquake is the northernmost large earthquake of the Cocos-North America plate interaction, an EW fault plane would have had major tectonic implications. Thus, it is important to reexamine the focal mechanism for the 1973 Colima earthquake. Of the five large events that we will consider, the 1981 Playa Azul earthquake has the smallest magnitude (M_s 7.3) and aftershock area (HASKOV et al., 1983). However, it is an interesting event because it occurred in the middle of the Michoacan gap four years before the 1985 event. Most rupture process studies of the 1985 mainshock conclude that the 1981 event ruptured the zone between two dominant asperities of the 1985 mainshock.

Earthquake slip vectors in subduction zones are used to determine relative plate motions. The NUVEL model (DEMETS et al., 1990) is the most recent global plate tectonic model, and the NUVEL data set includes many slip vectors from Harvard CMT solutions (see DZIEWONSKI and WOODHOUSE, 1983, for a description of the method, and the ISC bulletins for specific solutions). Overall, the Harvard CMTs are quite reliable; thus it is a curious observation that the Harvard CMT slip vectors for large earthquakes in northern Mexico seem to be systematically biased with respect to the NUVEL-predicted convergence direction. This observation motivates a reexamination of the moment tensors and focal mechanisms of the large events in Mexico.

The Orozco fracture zone subducts within the northern Mexico segment. The rupture zone of the 1985 Michoacan mainshock appears to straddle the down-dip extension of this fracture zone. Thus, another motivation to study these earthquakes is to see the effects of fracture zone subduction. The overall strike of the northern Mexico trench axis is about 295° (N65°W), but there is a subtle change from 290° to the south of the Orozco fracture zone to about 300° to the north. We will look for variations in the fault geometry that may be associated with the Orozco fracture zone.

1.2 Asperity Distributions

Rupture process studies of large and great earthquakes have either speculated or concluded that the "asperity distribution" along the plate interface controls the size of earthquakes (e.g., LAY et al., 1982: RUFF and KANAMORI, 1983b; SCHWARTZ and RUFF, 1987; RUFF, 1989a; BECK and RUFF, 1989; BECK and CHRISTENSEN, 1991; RUFF, 1992a). Many other rupture process studies of individual earthquakes have also referred to subregions of higher moment release as asperities. In the observational perspective of the above work, "asperities" are subregions of the rupture area that slip more than the average slip of the largest earthquake to break the plate boundary. In some rare cases, some arguments can be made that the "asperities" remained locked since the previous large earthquake (see BECK and RUFF, 1985). In a theoretical study, MADARIAGA (1977) showed that the simple mechanical interpretation of asperities as locked zones surrounded by freely slipping zones could result in coseismic moment release similar to some of the above observational results. KANAMORI (1981) provided a simple mechanical definition of asperities as regions of higher failure strength that he related to some aspects of seismicity during the seismic cycle. The above "asperity model" has evolved to become the asperity/barrier model (see BECK and RUFF, 1985; RUFF 1989a) where allowance is made for "geometric barriers" in the plane interface that can significantly affect the details of seismic wave radiation (see discussions of barrier model in DAS and AKI, 1977; AKI, 1979) The asperity/barrier model is most commonly described as the asperities having a higher failure stress level than the surrounding weaker regions, and it also assumed that there is an associated slip deficit at the asperities before the large earthquake (e.g., RUFF, 1989a). Since most subduction zones must have some fraction of aseismic slip in the plate interface, perhaps the most precise statement that can be made about the "weaker" regions is that aseismic slip is concentrated there over the seismic cycle. Unfortunately, various descriptions of the asperity model have not been completely self-consistent, nor have they been completely precise on the mechanical behavior of the plate interface. SCHOLZ (1990) offers a connection between the "asperity model" and theoretical concepts of the mechanical behavior of the plate interface. To return to the practical perspective of rupture process studies, we still define "asperities" as those subregions with larger coseismic slip in the largest earthquake. Many tests of the asperity model require determination of adjacent asperities along the subduction zone, i.e. the asperity distribution (e.g. RUFF, 1992a). To find the asperities, several large adjacent earthquakes must occur in the recent few decades to allow rupture process studies. The sequence of recent large earthquakes in the northern segment of the Mexico subduction zone (Fig. 1) provides the opportunity to add one more example to the global "catalog" of asperity distributions along subduction zone segments.

The scientific goals for this study can be summarized as follows: a systematic study of the rupture processes of the earthquakes; a detailed comparison of the slip vectors of these large Mexican earthquakes and the predicted slip directions from the NUVEL model (DEMETS et al., 1990); a teleseismic "view" of moment release and asperity distribution along the Mexico subduction zone; is there anything unusual about the 1981 event to indicate that it was an intra-asperity precursor; and, a seismotectonic comparison of the asperity distribution.

2. *Body-wave Inversion for Earthquake Source Process*

Progress in our understanding of earthquakes follows from the systematic determination of source parameters. While everyone acknowledges the complexity of earthquakes, only certain integral properties of the earthquake source can be reliably determined. The progression in source parameter determination has recently included the moment tensor (e.g., DZIEWONSKI and WOODHOUSE, 1983; SIPKIN, 1986a). The moment tensor gives the overall integral measure of earthquake size and faulting geometry. Moment tensors for earthquakes larger than magnitude 5 or so are now routinely listed in the NEIC and ISC global catalogs.

To return to the issue of earthquake complexity, rupture process studies seek to determine some spatial-temporal features of the moment release on the fault plane. Rupture process studies require special effort because resolution of spatial variations in moment release is quite difficult. For most earthquakes that occur in the world, the most detailed study of the source process can only determine the various "point source" parameters. Use of point source parameters does not mean that the earthquake is a point source, it means that we are describing the earthquake by integrals over the spatial variations in moment release. An example of a point source parameter is the moment tensor. Another example is the overall source time function, i.e., moment rate function, which can be deconvolved from body wave seismograms if the moment tensor is known.

Moment Tensor Rate Functions (MTRFs) are the most general description of a seismic source with no explicit spatial variations in moment release. The MTRFs include the moment tensor and source time function as specializations. The MTRF description of an earthquake allows the moment tensor to vary arbitrarily as a function of time. This variability could represent changes in faulting geometry during the rupture process. Alternatively, it could represent the simultaneous rupture of two or more different faulting geometries. Since there is a linear connection between seismograms and the MTRFs, the determination of the MTRFs is a linear inverse problem. Several investigators have taken advantage of this desirable characteristic (e.g., STUMP and JOHNSON, 1977; SIPKIN, 1986b; VASCO, 1989; RUFF and TICHELAAR, 1990). The primary problem with MTRFs is what to do with them after finding them. Most seismologists, while acknowledging the complexity of each earthquake, take the reasonable scientific approach that we should seek to determine the fewest number of earthquake parameters that will still provide an adequate explanation of the data. Thus, we should try to represent an earthquake by just one moment tensor and one time function, if possible. The full MTRF description should be kept only if the data "demand" a time-varying focal mechanism. The quantitative implementation of the above idea is tricky because extraction of a moment tensor and time function from the MTRFs is perceived to be a nonlinear inverse problem. Furthermore, due to errors and incompatibilities in the observed seismograms, inversion for the MTRFs always produces a time-varying moment tensor, thus some statistical test must be applied to the results to determine if the focal mechanism variations are significant. Thus, given the MTRFs, there are two questions to be answered: (1) how to extract the "best" moment tensor and time function from the MTRFs, and (2) how to assess whether a single moment tensor and time function are a statistically adequate representation of the earthquake? This paper answers the above two questions with a technique that may be systematically applied to earthquakes with magnitudes from 6 to 7.5 or so. The details are given in the appendix, here we only need to consider a single statistical parameter, ζ or the "MTRF parameter," that tells us if the MTRFs can be replaced by a single moment tensor and time function. If ζ is close to or less than 1, then the earthquake can be represented by a single moment tensor during the rupture process. Applications shown in this paper are for important and interesting earthquakes in the Mexico subduction zone.

The following sections give a brief overview of the body-wave inversion techniques that we use to find: the MTRFs (Moment Tensor Rate Functions); the best focal mechanism and time function; an image of the rupture process; calculation of radiated wave energy; and relative location of epicenters and rupture process features. Most of the details can be found in other papers, with the exception of the MTRF inversion and subsequent extraction of the best moment tensor and time function. These details are given in the Appendix.

2.1 Overview

Seismic source theory and elastodynamics provide a complete description of any internal source that generates seismic waves (see AKI and RICHARDS, 1980). The displacement along the ith direction observed at some location, denoted by Ω, is a convolution of the Green's functions with the source functions, over all components of the moment tensor, and integrated over the entire source volume:

$$u_i(t, \Omega) = \iiint_V G_{ij,k}(t, \Omega, V) * m_{jk}(t, V) \, dV \tag{1}$$

where G_{ij} are the Green's functions that depend on time (t), receiver location (Ω), and position within the source volume (V); $*$ denotes convolution with respect to time; and the seismic source is described by $m_{jk}(t, V)$, the moment tensor density functions that vary in time and space across the source volume. A key property of seismic source theory is that there is a linear connection between the source description and the observed wave displacements, hence the inverse is linear. Equation (1) is the most general description—several reductions must be made before we have an expression useful for applications. Also, seismologists have discovered that the Green's functions for certain parts of the seismograms can be reliably constructed. For example, teleseismic body waves have fairly simple

Green's functions that offer minimal distortion of the source time history. Let us make the first specialization of Eq. (1) by assuming that $u_i(t, \Omega)$ will be a teleseismic P- or S-wave seismogram, now denoted by $s_i(t, \Omega)$,

$$s_i(t, \Omega) = \iiint\limits_V g_{ij,k}(t, \Omega, V) * \dot{m}_{jk}(t, V) \, dV \tag{2}$$

where g_{ij} are modified versions of G_{ij} that now include the seismographic instrument response and attenuation effects, plus reflected phases for shallow earthquakes; and $\dot{m}_{jk}(t, V)$ are now the moment tensor rate density functions. A completely general source description is retained in Eq. (2). However, unless the source region is quite large, the observed seismograms will be unable to resolve any spatial variation in the moment release. As discussed by many previous investigators (e.g, AKI and RICHARDS, 1980; RUFF, 1987), resolution of spatial variations in moment release depends on the combination of good distribution of stations about the source and precise temporal resolution of coherent features across the stations. An approximate measure of horizontal spatial resolution is T/p, where p is the ray parameter in s/km for P or S waves, and T is the shortest period of temporal coherence across seismograms. Optimistic values of p and T would be $p = 0.10$ s/km and $T = 1$ s, while pessimistic values might be $p = 0.05$ s/km and $t = 3$ s. Consequently, for earthquakes with fault zones that are smaller than 10 to 60 km, teleseismic waves can only resolve fault-averaged source properties. The above fault dimensions translate into earthquake magnitudes as large as 7 to 7.5 for underthrusting subduction events. Thus for certain circumstances, even large subduction earthquakes may be "seen" as a point source by teleseismic waves. It may sometimes be desirable to treat an earthquake as a point source even if spatial resolution is available in the recorded seismograms. This is easily accomplished by filtering the seismograms to remove the higher frequencies. When the seismograms consist of wave periods substantially longer than the directivity time shifts, then the earthquake can be modeled as a point source.

Directivity time shifts appear in the seismograms due to the travel-time shifts in g_{ij} across the fault zone (see AKI and RICHARDS, 1980; or RUFF, 1983). When these directivity time shifts cannot be resolved in the recorded seismograms, then the Green's functions can be taken outside the integral over the source region. Thus, we rewrite $g_{ij}(t, \Omega, V)$ as $g_{ij}(t, \Omega, V_0)$, where V_0 implies that we evaluate g_{ij} at a specified fixed location within the source region. Two reasonable choices are the hypocenter or centroid locations. To retain a completely linear formulation, the hypocentral location is used. Thus, Eq. (2) becomes

$$s_i(t, \Omega) = g_{ij,k}(t, \Omega, V_0) * \iiint\limits_V \dot{m}_{jk}(t, V) \, dV. \tag{3}$$

We now see that the seismograms can only see the spatial integration of $\dot{m}_{jk}$. Define the Moment Tensor Rate Functions (MTRFs) as this integral

$$\dot{M}_{jk}(t) = \iiint_V \dot{m}_{jk}(t, V)\, dV. \tag{4}$$

Equation (3) then becomes

$$s_i(t, \Omega) = g_{ij,k}(t, \Omega, V_0) * \dot{M}_{jk}(t). \tag{5}$$

The basic units of $\dot{M}_{jk}$ are moment rate, each function gives the time history of moment rate of a particular moment tensor component. The above development shows that the MTRFs represent the spatial integration of moment release throughout the source volume, which could even contain a network of faults. Since the only restriction to go from Eq. (2) to Eq. (5) is the lack of resolution of spatial variations, the MTRFs are the most general "point source" description of an earthquake. They allow the source to display arbitrary changes in the moment tensor as a function of time. Of course, there is still a direct linear connection between the seismograms and the MTRFs, thus the inverse problem allows full use of linear inverse methodology, including calculation of the model covariance matrix.

2.2 A priori Reduction to Moment Tensor and Time Function

Most seismological investigators have further reduced Eq. (5). Recall that the MTRFs contain both faulting geometry and time history information, hence the reductions typically emphasize one source aspect or the other. If we use wave periods that are longer than the source duration, then we cannot distinguish the time history of the source. For this case, we can characterize the time history of all MTRFs by a single source time function, $f(t)$, and the MTRFs are reduced to

$$\dot{M}_{jk}(t) \Rightarrow M_{jk} f(t) \tag{6}$$

where $f(t)$ has units of $1/time$, such that $\int f(t)\, dt = 1$. The simplest a priori choice for $f(t)$ is the delta function, though one could use a more complicated time function, if desired. Substitute (6) into (5) to find:

$$s_i(t, \Omega) = [g_{ij,k}(t, \Omega, V_0) * f(t)] M_{jk} \tag{7}$$

where the functions in the bracket are specified a priori. We see from Eq. (7) that when the coherent wave periods are longer than the total source duration, the moment tensor is the remaining source characteristic.

Suppose that we have a good estimate of M_{jk} from long-period surface waves, then we can estimate $f(t)$ from body waves. In this case, Eq. (7) becomes

$$s_i(t, \Omega) = [g_{ij,k}(t, \Omega, V_0) M_{jk}] * f(t) \tag{8}$$

where the function in the bracket is specified a priori. Based on Eq. (8), we can deconvolve the source time function from a single seismogram, or simultaneously from several seismograms. In some cases, seismologists wish to determine both the moment tensor and the source time function from Eq. (7). Unfortunately, the reparameterization of the source description from $\dot{M}_{jk}(t)$ to $(M_{jk}f(t))$ now makes the simultaneous estimation of M_{jk} and $f(t)$ a nonlinear inverse problem. One can assume an initial simple $f(t)$ and then linearly invert for M_{jk}. Then one can seek changes to $f(t)$ and M_{jk} that reduce the error between observed and synthetic seismograms (e.g., BARKER and LANGSTON, 1981; NABELEK, 1984). Recall that the source description is fundamentally a linear inverse problem; the above nonlinear problems result from the reparameterization of the source description to reduce the five functions of MTRFs to the product of five numbers and one function. The number of unknowns is less, but the inverse problem is more burdensome, and formal error estimates are not possible. The methodology that we follow employs linear inversion to obtain the MTRFs, and then tests whether or not the MTRFs can be reduced to a single moment tensor and time function (see Appendix).

2.3 Inversion for Spatial Variations in Moment Release

If we think that we can resolve information about the spatial distribution of moment release from our collection of seismograms, then we must return to Eq. (2). If one is willing to invert an underdetermined linear system, then the integral of Eq. (2) can be discretized, and various methods will give some results for the space and time variable moment tensor rate density functions. On the other hand, any a priori information that we can incorporate into the problem would allow us to reduce the number of unknowns and solve a better posed inverse problem. One of the most useful reductions is to assume that we know the focal mechanism of the earthquake. While there are many examples of changes in focal mechanism during earthquakes, it is typically assumed that large subduction earthquakes rupture with a nearly constant focal mechanism. With this assumption, we need not invert for the five components of the moment tensor at each space-time grid point, we only need to invert for the displacement rate. Thus, Eq. (2) becomes

$$s_i(t, \Omega) = \iiint_V [M_{jk}g_{ij,k}(t, \Omega, V)] * \dot{m}(t, V) \, dV. \tag{9}$$

The next most useful simplification is to realize that the waveshape of Green's functions for teleseismic waves does not vary significantly in the horizontal direction across the fault plane, except for the relative travel time $(T(V) - T_0)$. Since underthrusting earthquakes are characterized by fault planes with small dips, the change in hypocentral depth across the rupture area is small for many earthquakes. Thus, under these conditions (see AKI and RICHARDS, 1980; RUFF, 1983, 1987; for details) we can rewrite Eq. (9) as

$$s_i(t, \Omega) = [M_{jk}g_{ij,k}(t - T_0, \Omega, V_0)] * \iiint_V \dot{m}(t - (T(V) - T_0), V) \, dV. \tag{10}$$

If we further assume that the moment release is concentrated along one direction, say the x direction with its directivity parameter Γ, then Eq. (10) can be reduced to

$$s_i(t, \Omega) = [M_{jk}g_{ij,k}(t - T_0, \Omega, x_0)] * \int_{x_{min}}^{x_{max}} \dot{m}(t - \Gamma x, x)\, dx. \qquad (11)$$

The above approximation is quite useful for strike-slip earthquakes, and can be used for large underthrusting events that have an elongated rupture area. To invert Eq (11) for the moment release, we use the iterative "tomographic imaging" technique as discussed in RUFF (1987) and applied in SCHELL and RUFF (1989) and BECK and RUFF (1989). For underthrusting events, we can determine the best overall rupture directions by pointing the x axis in all possible directions. Also, to further reduce the effective number of unknown parameters, we use the a priori notion that most of the moment release occurs close to a rupture front. Thus, we can try all possible rupture azimuths and rupture velocities and choose the "best" model from one that best fits the data. The iterative solution to the tomographic imaging problem uses a variant of the conjugate gradients technique. This method guarantees convergence, and in practice only a small number of iterations, typically less than 10, is required to achieve convergence. The tomographic imaging technique used here parallels that used by SCHELL and RUFF (1989), except for the "preprocessing" of the input time functions. SCHELL and RUFF (1989) studied the 1972 Sitka (M_w 7.6) strike-slip earthquake, and there was an abundance of on-scale P-wave seismograms before the shadow zone. In the case of large underthrusting events, we typically will use some seismograms from the shadow zone, thus their absolute amplitudes are diminished. We found that the best method for preprocessing the resultant time functions was: (1) first normalize the time functions to zero moment by subtraction of half-sine function of the duration of the time function; and (2) rescale the peak amplitudes of all the time functions to the peak amplitudes from the nondiffracted stations. We invert these rescaled time functions to find the best rupture direction and velocity. To produce a final estimate of the moment release along this rupture direction, we add a third step to the above scheme: (3) the seismic moment is added to the time functions with a half-sine function of the same duration as the time functions. Also, in contrast to SCHELL and RUFF (1989), we find a single best time function for a group of stations in the same azimuthal sector by multistation inversion for the time function with an assumed focal mechanism (method of RUFF, 1989b). Thus, there are fewer time functions that are used in the tomographic imaging, but each time function is less "noisy," and there is a better balance between different azimuths than in the original azimuthal distribution of stations.

2.4 Radiated Wave Energy

Earthquake energy is a fundamental characteristic of earthquakes, but it is quite difficult to reliably measure. Most estimates of earthquake energy use empirical formulas that relate magnitude to energy (i.e., GUTENBERG and RICHTER, 1956; KANAMORI, 1977). The formula from KANAMORI (1977) can be written as the ratio of energy to moment is a constant, that is, $E/M_0 = 5 \times 10^{-5}$. It is possible to determine the radiated wave energy, a lower bound on total earthquake energy, by two methods: (1) directly summing the wave energy in many seismograms that completely sample the focal sphere; or (2) determine the rupture process as the space-time history of moment release, and then use the computer to "radiate" the P and S waves in all directions and integrate the wave energy. For large earthquakes, teleseismic rupture process studies should image all of the significant wave energy. KIKUCHI and FUKAO (1988) applied method #2 to several great and large earthquakes, and they found that the "observed" wave energy is about one-tenth the amount expected from the Kanamori energy-moment formula.

The energy calculation is fairly straightforward and is based on established procedures. The basic relation for radiated wave energy from HASKELL (1964) is

$$E_P = \frac{1}{16\pi^2 \rho \alpha^5} \iiint R_P^2 [\ddot{M}(t, \theta, \phi)]^2 \, dt \, \sin\theta \, d\theta \, d\phi$$

$$E_S = \frac{1}{16\pi^2 \rho \beta^5} \iiint [R_{SH}^2 + R_{SV}^2][\ddot{M}(t, \theta, \phi)]^2 \, dt \, \sin\theta \, d\phi \, d\theta \tag{12}$$

where E_P and E_S are the P- and S-wave energies; ρ, α, and β are the density, P wave and S wave velocities; R_P, R_{SH}, and R_{SV}, are the radiation pattern factors for P, SH, and SV waves; θ and ϕ are the spherical coordinates; and $\ddot{M}$ is the second derivative, with respect to time, of the moment accumulation as seen from all directions around the focal sphere. Henceforth, we will just explicitly write the S-wave energy equation. Our tomographic imaging method idealizes the moment release to occur just along one direction, i.e., a line source. Thus, the moment rate function, $\dot{M}$, depends only on time and the directivity parameter, $\Gamma_S = \cos(\theta)/\beta$, as follows

$$\dot{M}(t, \Gamma_S) = \int_{x_{\min}}^{x_{\max}} \dot{m}(t - \Gamma_S x, x) \, dx \tag{13}$$

where $\dot{m}$ is the moment rate density function, and the faulting extends from $x_{\min}$ to $x_{\max}$ along the x direction, $\theta = 0°$. Note that $\dot{M}(t, \Gamma_S)$ does not depend on ϕ. Substitute the above specialization of $\dot{M}$ into the basic energy equation (12), integrate over ϕ, and change the integration variable to $\chi = \cos(\theta)$ to find the radiated S-wave energy

$$E_S = \frac{1}{16\rho\beta^5} \left\{ \int_{-1}^{1} R(\chi) \left[\int_0^\infty \ddot{M}(t, \Gamma_S(\chi)) \, dt \right] d\chi \right\} \tag{14}$$

where $R(\chi) = (4\chi^4 - 3\chi^2 + 1)$ for fault slip parallel to the x direction, and $R(\chi) = (1 - \chi^4)$ for slip perpendicular to the x direction. Given our tomographic inversion for $\dot{m}(t, x)$, the above double integral is numerically evaluated by generating $\ddot{M}(t, \Gamma_S)$ for the full range of χ, i.e., the entire focal sphere. Unfortunately, we have discovered that the energy estimate is sensitive to various parameters of the source imaging that are poorly determined; in particular, the assumed rupture velocity can exert a strong influence on the energy estimate. Thus, we will show a range of energy estimates that correspond to rupture velocities from the lower bound up to the S-wave velocity.

2.5 Epicentral Relocations and "Hand-picked" Directivity Analysis

We have tested various epicentral locations by using linearized relative epicentral relocation. This simple technique is well-known and has been used by many seismologists; we briefly describe it here for the sake of completeness. Suppose we want to locate event "A" with respect to event "o" based on the observed time delays, Δt, at various stations. From the observational perspective, the time delay at the ith station is: $\Delta t_i = t_{Ai} - t_{oi}$, where t_{Ai} and t_{oi} are the arrival times of events "A" and "o". The event-to-station azimuth (counterclockwise from East) and ray parameter of the wave from event "o" to the ith station are ϕ_i and p_i. We place an (x, y) coordinate system at the fixed epicentral location of event "o", with $+x$ pointed East and $+y$ pointed North. For the teleseismic case, any hypocentral depth difference between events "o" and "A" is absorbed into the origin time shift, T_A. The linearized expansion of the travel-time function then produces the theoretical connection between the origin time shift and epicental perturbations $(T_A, x_A,$ and $y_A)$ of event "A" with respect to event "o" and the observed travel-time delays as follows

$$\Delta t_i = T_A - (p_i \cos \phi_i) x_A - (p_i \sin \phi_i) y_A. \qquad (15)$$

With a total number of observed travel-time delays of N, we seek the least-squares solution of the N equations for the three unknowns: T_A, x_A, and y_A. Error estimates are important for relocation assessment, thus we use the resultant *rms* misfit between observed and best-fit time delays, together with the model covariance matrix, to calculate the (x, y) error ellipse about x_A and y_A. Note that this relocation procedure gives only the relative location of event "A" with respect to event "o"—we must assume the absolute location of event "o". Also, this procedure uses only the common stations—this restriction makes the procedure quite simple and yet it implicitly corrects for any station effects. We have used this relocation procedure to test the mainshock locations with respect to the ISC catalog locations, and also to relocate mainshocks with respect to each other.

The above procedure is also used to study the rupture process of large earthquakes. From an observational perspective, we "pick" the arrival times at each station of an easily identified feature in the source time functions. This feature, event "A", is relocated with respect to event "o" which is typically the point of rupture initiation, i.e., the epicenter. Thus, this "hand-picked" directivity analysis is simply a relative epicentral location problem, and possesses the advantage over other techniques of rupture imaging in that we obtain statistical estimates of the reliability of the feature location. It is desirable to use this "hand-picked" directivity whenever possible, and it is especially useful for multiple-event rupture processes where we can clearly pick the beginning of each subevent (e.g., BECK and RUFF, 1987). This simple and reliable method can be used for *any* coherent feature in the time functions. The usefulness of the solution depends on whether: (1) the origin of the observed feature is spatially concentrated within the rupture zone, and (2) the seismologist can "pick" the same feature in all seismograms. Fortunately, any departure from ideal circumstances should be seen in the calculation of the error ellipse.

3. Results of Seismological Analysis

Teleseismic rupture process studies can locate moment release with respect to a reference point, typically the epicenter. To compare results from adjacent earthquakes, the relative epicentral locations must be correct. To compare moment release to tectonic features, then the absolute locations should also be reliable. Seismicity in Mexico presents some special problems. Given the lack of numerous large aftershocks, the rupture zones of large earthquakes are mostly based on the small aftershocks located by local and regional networks. Thus epicentral locations must be consistent with the local aftershock locations. SINGH and LERMO (1985) document a strong systematic bias in the ISC and PDE/NEIC epicenters with respect to the local epicenters. We shall use epicentral locations determined by various local networks, and thus the teleseismic rupture process studies will be referenced to these locations. While the epicentral locations based on local networks are the best estimates, the depths and origin times from the local network may not be compatible with the global data set. Teleseismic waveform modeling can find the "best" depth for the overall moment release, for an average velocity between the surface and hypocenter. Given this "best" depth, we can recalculate the origin time such that the teleseismic arrival times are well-predicted. Since the spatial location of moment release is sensitive to the start times of the body waves, these start times will be listed for all body-wave phases used in the inversions for all events. In addition, we shall also list the seismogram scale factors that result from the omnilinear inversion of the seismograms. An additional special problem exists for the 1985 Michoacan mainshock since local accelerograms have been used to determine the rupture process—these records do contain information on the absolute location of the moment release. Hence, we must use a compatible epicenter to mix teleseismic and local records in rupture process studies.

3.1 The 1973 Colima Earthquake

3.1.1 Epicenter and aftershock area

REYES et al. (1979) presented a detailed study of the 1973 Colima earthquake and its aftershocks. They deployed a temporary seismic network within days after the mainshock, and thus recorded the large Feb. 10, 1973 (M 6.2) aftershock. The epicenter for this aftershock is probably the most reliable absolute location in the Colima region of an event large enough to be teleseismically recorded. This large aftershock is near the northwestern end of the aftershock zone of REYES et al. (1979). The main cluster of aftershocks is located at the southeastern end of the aftershock zone, about 55 km to the southeast of the Feb. 10 event. REYES et al. (1979) also show a location for the 1973 mainshock that is displaced about 30 km to the southwest from the ISC and PDE epicenters. This systematic shift is documented for other large Mexico earthquakes in SINGH and LERMO (1985). Thus, the REYES et al. (1979) mainshock location is more reliable than the ISC or PDE location, but their epicenter is still down-dip of the main aftershock cluster.

We locate the mainshock epicenter with respect to the Feb. 10 aftershock by using the P-wave arrivals at 18 common stations that range in epicentral distance from 13.9° to 48.9°. The azimuthal range of the stations covers only the eastern sector from 322° around to 149°. The relocation places the mainshock epicenter 80 km along an azimuth of 117° from the Feb. 10 event, with an error estimate of about 11 km. This azimuth of 117° coincides with the local trend of the coastline and also points towards the southeastern aftershock cluster, but the 80 km distance places the mainshock epicenter beyond the southeastern aftershock cluster. If the mainshock epicenter of REYES et al. (1979) is shifted by 15 km further to the southwest, then it falls on the 117° azimuth from the Feb. 10 event, and also is at the center of the aftershock cluster. We use this mainshock epicenter as our compromise location; the geographic coordinates are: 18.20°N, 103.18°W.

3.1.2 MTRF inversion for focal mechanism and depth

As noted by CHAEL and STEWART (1982), the P waves for the Colima earthquake show a "complication" of an emergent arrival. While CHAEL and STEWART (1982) were unable to find a time function with their forward modeling technique, there is no difficulty with a deconvolution or inverse method. On the other hand, finding the proper start time of the P waves can be difficult for an emergent arrival for seismograms at large epicentral distance. After careful comparison of seismograms, we have used the arrival times listed in Table 1. We have used the naturally rotated SH wave at AKU in the MTRF inversion. Usually, it is not possible to use SH at a distance of 72° for a large earthquake because the ScS arrival interferes with S. However, we noticed that both the S and ScS are essentially nodal on the AKU SH record—this special case allows the use of this record.

Table 1

Station data and some results for the 1973 Colima earthquake

Sta	Net	Comp	Phase	Epicentral params Δ,°	Az,°	Start time 21 hrs +	obs− pre	a factor	direct. Δt, s peak	duration	M_0
KBS	W	Z	P	76.9	10	13: 1.3	−2.2	0.93 ⎤			
KTG	W	Z	P	70.0	21	12:20.4	−2.9	1.27 ⎟			
NUR	W	Z	P	91.1	23	14:11.9	−3.4	1.54 ⎟	13	31	0.9
AKU	W	Z	P	71.6	26	12:30.7	−1.7	1.27 ⎦			
ESK	W	Z	P	80.8	35	13:22.0	−2.4	1.00 ⎤			
STU	W	Z	P	90.4	38	14: 9.6	−2.4	1.62 ⎟			
VAL	W	E	P	77.7	39	13: 4.9	−2.8	0.94 ⎟	15	32	0.8
TOL	W	Z	P	85.3	50	13:46.2	−1.4	1.24 ⎦			
NAT	W	Z	P	71.0	102	12:26.4	−3.0	0.98	15	35	0.6
ARE	W	Z	P	46.5	136	9:37.8	−0.6	0.91 ⎤			
SOM	W	Z	P	76.6	160	13: 2.0	+0.5	0.13 ⎦	17	31	0.6
WEL	W	Z	P	96.0	229	14:42.1	+4.2	1.65* ⎤			
RAR	W	Z	P	67.9	237	12:13.3	+3.4	1.88 ⎟	13	27	0.5
AFI	W	Z	P	74.8	250	12:52.7	+1.4	1.24 ⎦			
KIP	W	Z	P	51.5	283	10:17.8	+0.8	0.24	13	26	
MAT	W	Z	P	100.5	314	14:57.1	−1.0	2.17*	13	31	
AKI	W	N	SH back az.	71.6	26 (273)	21:41.2	−8.0	1.01			

"Net" code: W, WWSSN. "Comp": Z, vertical; N, north; E, east. "Phase": P wave. "Epicentral params": distance and azimuth. "Start time": onset of observed waves, 21 hours plus the listed minutes:seconds. "obs−pre": arrival time residual. "a factor": the seismogram scale factors from omnilinear analysis, based on MTRF inversion with AKU SH and at the best depth of 25 km, * diffracted P waves. "direct. Δt": times of features in source time functions, in s, "peak" is the first peak in moment release of the time functions, "duration" is duration of the pulse. "M_0": seismic moment of the source time functions, in 10^{20} Nm.

For the initial body wave inversions, the seismograms are filtered such that the MTRF inversion can be used to find the overall best depth and focal mechanism. We use a "zero-phase yet-causal" time domain symmetric triangle filter with 10 s duration. Using different combinations of stations, allowing a variable or fixed focal mechanism, and with different damping values, the overall best point source depth is 20 to 25 km. To show a particular example, using 17 P waves and the AKU SH wave, with 70 s duration, we invert for MTRFs of 60 s duration with a 4 s sampling, a damping of 0.10, and at a depth of 25 km (Figure 2). The MTRFs are plotted in the left column, and the observed (solid) and synthetic (dashed) seismograms are plotted in the right column. Two of the P-wave phases are diffracted, so the final value of seismic moment uses only the nondiffracted station amplitudes. In Figure 2, the time functions from the reduction to the major double

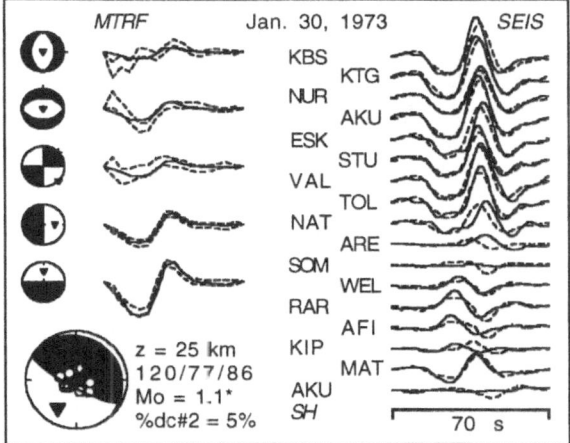

Figure 2

Inversion of seismograms for MTRFs for the 1973 Colima earthquake. The five MTRFs are shown on the left, while the observed (solid) and synthetic (dashed) seismograms are shown to the right. The inversion is for a source depth of 25 km. Each MTRF is identified by its equivalent double-couple focal mechanism. The dashed traces show the ± uncertainty of the individual MTRFs obtained by omnilinear inversion. A single source time function and double-couple focal mechanism are extracted from the MTRFs (see Appendix for details). The solid traces are the source time function rescaled by the moment tensor components corresponding to the double-couple mechanism shown below. The focal mechanism parameters are listed as: strike of 120°, dip of 77°, and rake angle of 86°. The seismic moment is 1.1, in units of 10^{20} Nm, The second double couple is 5% of the size of the major double couple. The synthetic seismograms are calculated for the final model of one time function and one double couple. Station distribution is plotted on the focal sphere, and the time scale is the same for both the MTRFs and seismograms.

couple are plotted with the error bounds about the original MTRFs. Visual inspection of Figure 2 shows that the individual MTRFs are consistent with a single moment tensor and double couple, but the MTRF parameter is the best overall quantitative measure of the adequacy of reduction. The left side of Figure 3 shows the fit between the synthetic and observed seismograms as a function of assumed point-source depth. We find the best point-source depth to be 25 km, but with low resolution, expected for a large event (see TICHELAAR et al., 1992). The minimum in the MTRF parameter gives an indication of the "best" solution. Figure 3 also shows the statistical parameters for MTRF inversion without the *SH* seismogram; the best fit for a double couple and time function is now at a depth of 15 km. For the inversion with the *SH* wave, the MTRF parameter obtains a minimum value of less than 1 at 25 km. This means that the MTRFs can be replaced by a single time function and moment tensor. We select the inversion run with *SH* and a depth of 25 km as the overall best solution, plotted in Figure 2. As seen in Figures 2 and 3, the best double-couple focal mechanism has a fault strike of 317°, fault dip of 14°, and a slip vector angle of 30° (N30°E) that is nearly pure thrust. This fault strike

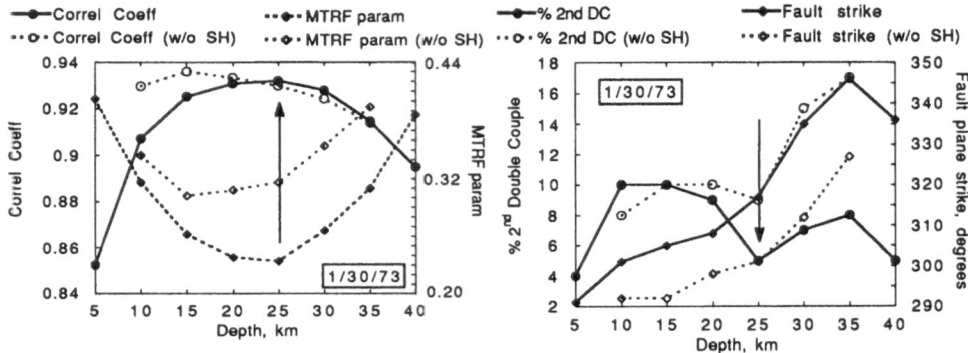

Figure 3
Various MTRF inversion and earthquake parameters as a function of assumed depth for the 1973
Colima earthquake. The match between observed seismograms and the synthetics for one time function
and double couple is measured by the correlation coefficient (plotted in the left graph). The MTRF
parameter (ζ in the Appendix) is also plotted in the left graph. For the Colima event, we show the curves
for inversion runs with and without the *SH* wave. We choose 25 km as the overall best point-source
depth, shown as the arrow. The graph at right shows the fault strike of the major double couple and the
percent size of the second double couple for all depths. We use this figure to estimate the allowed range
in fault plane strike.

is close to the estimate of REYES *et al.* (1979), but our dip of 14° is shallower than
their preferred values. Recall that the fault plane strike of CHAEL and STEWART
(1982) is 266°. On the other hand, the auxiliary plane from CHAEL and STEWART
has a strike of 122° and a dip angle of 76°; our auxiliary plane has a strike of 120°
and a dip of 77°. Thus, the auxiliary plane agrees with CHAEL and STEWART, but
their fault plane is rotated to an east-west orientation. In the right side of Figure 3,
we plot the fault strike of the major double couple extracted at each depth. We see
that if we choose a depth less than 25 km, or if we use the *P* wave-only inversion
results, the fault strike is rotated to a more east-west orientation; but is never less
than 290°. Thus, the body waves prefer a fault strike between 290° and 320°, more
or less parallel with the local trench orientation; the fault strike of CHAEL and
STEWART is not consistent with this result. The auxiliary plane strike is quite stable
at the various depths. For the MTRF inversion with *SH*, the auxiliary plane strike
is between 120° and 123° over the entire 5 to 40 km depth range. For the MTRF
inversion without the *SH* phase, the auxiliary plane strike only varies from 130° to
133° for a depth range of 10 to 25 km. This basic character of a stable auxiliary
plane strike hence slip vector strike, is seen for all the Mexico earthquakes.

3.1.3 Directivity and rupture

The basic history of moment release is a single pulse of about 30 s duration,
with a small precursor 4 to 8 s before the main pulse. To test for directivity, we
invert for time functions of the station groups indicated in Table 1. We use a depth

of 20 km, seismogram duration of 80 s, time function duration of 60 s with 2 s sampling, triangle filter duration of 3 s. We do a "hand-picked" directivity analysis of two features: (i) the first peak of moment release, and (ii) the termination of moment release, i.e., the zero-crossing of the time functions. An example of these features in the time functions for the North Atlantic stations is shown in Figure 4. We relocate these features with respect to the epicenter, as shown in Figure 5. The peak in moment release is located 32 km to the northwest along an azimuth of 315° with a time delay of 14 s; the apparent rupture velocity is 2.2 km/s. The error ellipse for this location is ±11 km, along a strike of 345°, and ±8 km along a direction of 75°. The time function zero-crossing is located at a distance of 64 km along an azimuth of 257° from the epicenter, with a delay time of 30 s; this gives an apparent rupture velocity of 2.1 km/s. One interpretation of the azimuthal difference between the peak and termination of moment release is that a rupture front propagated from the southeastern corner of the aftershock both up-dip and along strike to the northwest.

As an additional test, we use the one-dimensional tomographic imaging technique to find the overall average azimuth of rupture propagation and moment

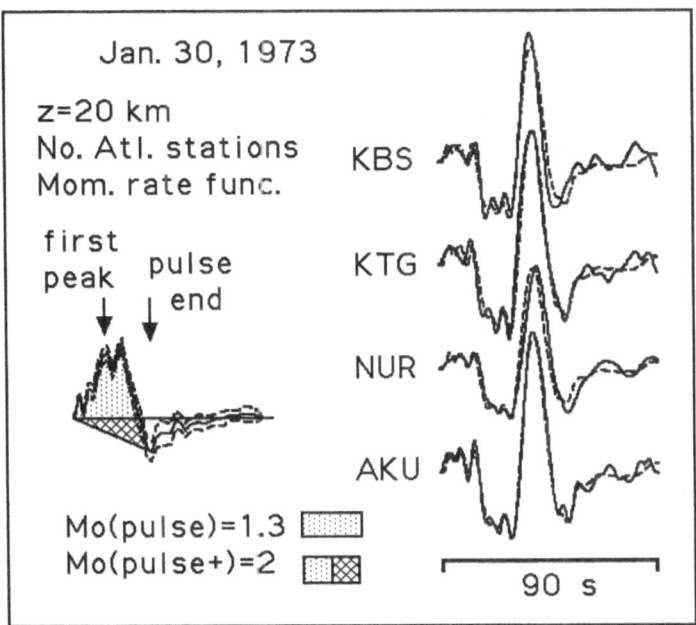

Figure 4

Inversion for the source time function as seen by the "North Atlantic" station group. With a depth of 20 km and the focal mechanism from Figure 2, the four seismograms are inverted for the time function shown to the left. The dashed traces are the resultant synthetic seismograms. The arrows show the two features that are used in the "hand-picked" directivity analysis. Two different values of seismic moment are shown, the larger value of 2×10^{20} Nm adds a baseline correction to the main pulse.

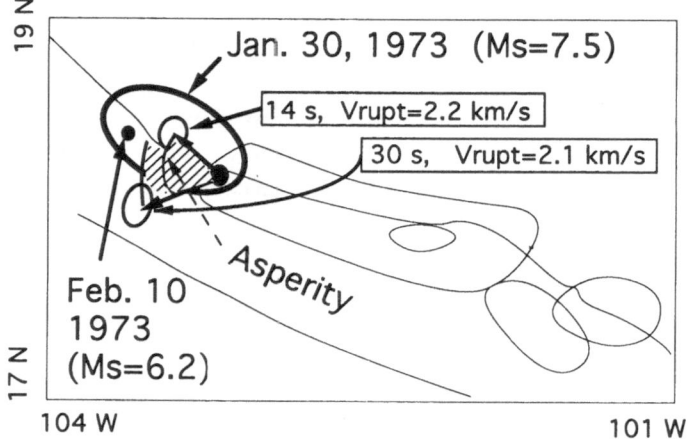

Figure 5

Mexico base map with the rupture process features of the 1973 Colima earthquake. Solid dot in the southeast corner of the 1973 aftershock zone shows our best estimate of the mainshock rupture initiation epicenter. The aftershock epicenter is from REYES *et al.* (1979). The arrows and error ellipses show the results from the directivity analysis, which is corroborated by tomographic imaging of the moment release. The hachured region is presumed to be the main pulse asperity.

release. We find that irrespective of rupture velocity, the best average rupture direction is 280°, which is intermediate to the above two directions. The seismic moment contained in this first pulse is about 1.3×10^{20} Nm (Fig. 4). If we assume that the inversion "damps out" a longer period component that would make the first 30 s entirely positive, then we can boost the seismic moment in Figure 4 to about 2×10^{20} Nm. The long-period surface wave estimates of REYES *et al.* (1979) and CHAEL and STEWART (1982) are about 3×10^{20} Nm, where REYES *et al.* (1979) also estimate the uncertainty as $3 \pm 2 \times 10^{20}$ Nm. Thus, we can conclude that about half to most of the seismic moment was released in the southeast corner of the aftershock area. Following the interpretations of BECK and RUFF (1985) and SCHWARTZ and RUFF (1987), this southeast corner is identified as the dominant asperity of the 1973 Colima earthquake. If we take the asperity area to be a quarter-circle of radius 50 km, then the average displacement will be 3.4 m. There is only about 1.6 m of tectonic displacement accumulation between the 1973 Colima earthquake and the previous large earthquake in 1941 (see NISHENKO, 1991). Thus, the asperity of the 1973 earthquake certainly qualifies as an "official" asperity in the context of the asperity model, where the coseismic slip should equal the total accumulated tectonic displacement. This asperity occupies only about half the aftershock area. While the aftershock cluster at the southeastern corner is in the asperity region, the large Feb. 10 aftershock is located just beyond the asperity edge. See MENDOZA and HARTZELL (1988) for other examples of the relationship between an asperity and aftershocks.

3.2 The 1979 Petatlan Earthquake

3.2.1 Epicenter and aftershock area

Due to fortunate circumstances, a temporary local network recorded the fore-shocks, mainshock, and aftershocks of the 1979 Petatlan earthquake (GETTRUST et al., 1981; VALDES et al., 1982). While the foreshock sequence is interesting and important, we are mostly concerned with the aftershock area and mainshock hypocenter. We use the mainshock epicenter reported by the above two papers: 17.46°N, 101.46°W. They also report a hypocentral depth of 15 km and an origin time of 11 hr 7 min 11.2 s. The above mainshock epicenter is displaced by about 40 km southwest of the PDE epicenter, similar to the bias seen from the 1973 Colima earthquake. VALDES et al. (1982) show the distribution of 22 aftershocks that occur up to 2.5 days after the mainshock, all with magnitude less than 4. With the exception of one event, these aftershocks define a narrow zone that is 27 km wide in the down-dip direction, and extends 60 km along the coastline. The mainshock epicenter is more or less at the center of this aftershock region. The one exception is an aftershock that is 15 km further inland than the down-dip edge of the narrow zone; inclusion of this one aftershock greatly increases the aftershock area. As documented in VALDES et al. (1982), the aftershock area expands with time, primarily in the inland, or down-dip, direction. In contrast to most other large subduction earthquakes, it seems that the 1979 rupture initiated in the center of the rupture area, rather than at the down-dip edge. The detailed relocations of HSU et al. (1985) indicate a fault plane dip angle of about 14°; they also further discuss a possible foreshock migration pattern that they claim is consistent with the model of DMOWSKA and LI (1982). The slip front in the DMOWSKA and LI (1982) model progresses upward from the down-dip edge; thus the conclusion of HSU et al. (1985) implies that the entire sequence started at the down-dip edge, though the mainshock epicenter is displaced up-dip.

3.2.2 Body-wave inversion for MTRFs, depth, and rupture process

CHAEL and STEWART (1982) modeled the P and surface waves of the Petatlan event, and a Harvard CMT solution is also available. We use a total of 14 WWSSN long-period P waves as listed in Table 2. We first perform the MTRF inversion for seismogram duration of 60 s, filtered with a 10 s duration triangle, MTRF duration of 60 s with 4 s sampling, and damping of 0.1. The MTRF inversion fits the P waves quite well for a point-source depth anywhere between 10 and 25 km (Fig. 6); again, this lack of depth resolution is expected, given the smooth source time function duration of 20 s. Recall that our depths are determined for a local P-wave velocity of 6.7 km/s, hence our depth would be less than the above values if we used the slightly slower velocity above a depth of 25 km in the velocity model of VALDES et al. (1982). The MTRF parameter is considerably less than 1 (Fig. 6), thus the MTRFs can be replaced by a single moment tensor and time function, as will be the case for all the Mexico earthquakes.

Table 2

Station data and some results for the 1979 Petatlan earthquake

Sta	Net	Comp	Phase	Epicentral params Δ,	Epicentral params Az,	Start time 11 hrs+	obs– pre	a factor	direct. Δt, s peak	direct. Δt, s duration	M_0
DAG	W	E	P	71.4	14	18:30.8	−2.2	0.94			
NUR	W	E	P	91.1	24	20:15.3	−1.3	1.18			
AKU	W	Z	P	71.5	26	18:31.5	−1.9	0.80			
ESK	W	E	P	80.4	35	19:22.6	−1.2	0.70			
IST	W	Z	P	105.9	37	21:21.3	−2.1	4.70*	15	21	0.8
STU	W	Z	P	89.9	38	20: 9.4	−1.9	0.98			
VAL	W	E	P	77.2	39	19: 4.2	−2.0	0.58			
TRI	W	E	P	94.1	40	20:28.9	−1.7	1.28			
SJG	W	E	P	33.6	83	13:50.8	−2.2	0.86	16	25	1.4
LPB	W	N	P	47.1	134	15:43.1	−1.6	0.79			
ARE	W	Z	P	44.8	137	15:25.8	−0.6	0.52	15	22	0.7
LPA	W	Z	P	66.5	142			0.58			
RAR	W	Z	P	68.9	238	18:16.7	−0.6	0.70	13	18	1.4
MAT	W	Z	P	102.2	315	21: 5.9	−1.0	3.29*	18	22	0.3

"Net" code: W, WWSSN. "Comp": Z, vertical; N, north; E, east. "Phase": P wave. "Epicentral params": distance and azimuth. "Start time": onset of observed waves, 11 hours plus the listed minutes:seconds. "obs–pre": arrival time residual. "a factor": the seismogram scale factors from omnilinear analysis, based on MTRF inversion at best depth, * diffracted P waves. "direct. Δt": times of features in source time functions, in s. "peak" is the peak moment release just before truncation of pulse, "duration" is duration as measured by baseline-crossing time. "M_0": seismic moment of the source time functions, in 10^{20} Nm.

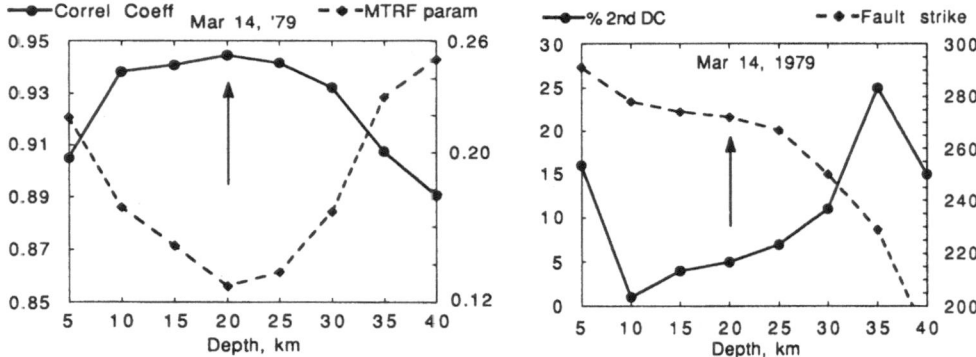

Figure 6

Various MTRF inversion and earthquake parameters as a function of depth for the 1979 Petatlan earthquake. Same as Figure 3. Arrow shows the best point-source depth of 20 km. The MTRF parameter is less than 1, thus a single source time function and single moment tensor is a statistically valid representation of the MTRFs. Fault strike is stable for depths between 10 and 25 km.

The focal mechanisms that result from the MTRF inversion are quite similar for depths between 10 and 25 km: the fault plane strike varies from 278° to 257° and fault dip varies from 15° to 16°; the auxiliary plane strike varies from 99° to 107°. At the best depth of 20 km—best in the sense of the final fit to the data—the corresponding double couple (Fig. 7) from the MTRF inversion is: strike 104°, dip 74°, slip rake angle of 93°, which gives a fault plane strike and dip of 272° and 16°. While the Harvard CMT yields a similar auxiliary plane strike of 106°, the CMT fault plane strike is 306°. CHAEL and STEWART (1982) have a fault plane strike of 293° and an auxiliary plane strike of 116°.

To strive for better depth resolution, we fix the focal mechanism, and then invert for the source time function at different depths: the seismograms are filtered with a 3 s triangle, time function is sampled at 2 s, and the damping is 0.01. The best point-source depth is anywhere between 15 to 20 km.

We then invert for individual time functions for the station groups shown in Table 2. An interesting feature of the time functions is that the earthquake starts with a small precursor, followed 6 s later by a single large pulse that peaks at about 15 s and ends at about 20 to 25 s (see MTRFs in Fig. 7). This basic time history is similar to that of the 1973 Colima earthquake. However, for the 1979 Petatlan earthquake, we could not distinguish any coherent directivity in the "hand-picked" features of the time functions. There is no resolvable directivity for the initiation of the main pulse at about 6 s; it is difficult to reliably pick the main pulse initiation

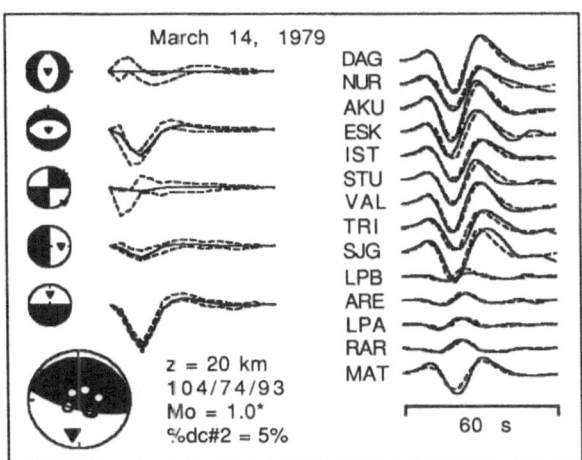

Figure 7

Inversion of seismograms for MTRFs for the 1979 Petatlan earthquake. Same as Figure 2. The inversion is for the best depth of 20 km. Note that for the first moment tensor component at top, the reconstructed MTRF from the time function and double couple (solid trace) falls outside the error bounds of the original MTRF for this component; yet Figure 6 shows that in fact this reconstructed MTRF is statistically acceptable. The MTRF parameter accounts for the full covariance matrix of the MTRFs, while a visual inspection of the above agreement/disagreement does not.

in the observed seismograms due to interference from the depth phases from the precursor. The overall duration varies from 18 s at RAR, which is the only station to the south and west, to a range of 21 to 25 s at all other azimuths. Unfortunately, this difference does not appear to be significant. The lack of apparent directivity is consistent with a rupture front that spreads out in all directions to fill the elliptical aftershock area. The peak in moment release can then be interpreted as the rupture front reaching the edges of the high moment release region. Furthermore, if we associate the moment release peak with a distance of ± 30 km, then we have an apparent rupture velocity of about 2 km/s, but this is all based on assumptions. CHAEL and STEWART (1982) used a simple trapezoid-shaped time function of 17 s duration in their forward modeling of the P waves; but they note that there is "... *greater complexity*" in the onset of this earthquake.

Our overall estimate for the moment in the main pulse from MTRF inversion is about 1×10^{20} Nm (Fig. 7). If we use the maximum value for the first pulse from the individual time functions, then the moment is 1.4×10^{20} Nm. From the seismic moment resolved by the body waves, the displacement in the assumed elliptical area around the epicenter is approximately 2.9 m. Since the Harvard CMT moment estimate is 1.7×10^{20} Nm and CHAEL and STEWART (1982) found 2.7×10^{20} Nm in their surface wave analysis, it is likely that additional moment release occurred in the region around the epicentral asperity over a longer time scale. Since the previous large Petalan earthquake occurred in 1943 (NISHENKO, 1991), the accumulated tectonic displacement is only about 2 m; hence we can consider the epicentral elliptical area to be an asperity.

3.3 The 1981 Playa Azul Earthquake

3.3.1 Epicenter and aftershock area

HASKOV et al. (1983) located the mainshock and aftershocks up to 6 days after the mainshock with local and regional stations. Just as for the other events, their mainshock epicenter is about 35 km to the southwest of the PDE location. The aftershocks fall into two clusters at the far ends of a 40 km long region that is oriented nearly east-west. The north-south width is about 20 km, with the main-shock epicenter at the extreme up-dip, i.e. southwestern, edge of the aftershock area. Again, it is unusual to see the epicenter at the up-dip edge of the aftershocks for a large subduction event, though the Playa Azul is a relatively small event.

3.3.2 Body wave inversion for MTRFs, depth, and rupture process

A Harvard CMT mechanism is available for this event, and the focal mechanism is nearly pure underthrusting. The fault plane strikes at 287° with a dip of 20°, the auxiliary plane strike is 115° with a dip of 70°, and the moment is 0.7×10^{20} Nm at a centroid depth of 32 km. In a later study ASTIZ et al. (1987) find a solution very similar to the Harvard CMT, except that their fault dip is just 11°. TICHELAAR

Table 3

Station data and some results from the Oct. 25, 1981 Playa Azul earthquake

Sta	Net	Comp	Phase	Epicentral params Δ,°	Epicentral params Az,	Start time 3 hrs +	obs– pre	a factor	main– pre	direct. top	Δt, s durat.	M_0
NUR	W	Z	P	91.1	23	35:17.1	−0.7	1.16	4.3 ⎤			
AKU	W	Z	P	71.6	26	33:33.6	−1.2	0.88	4.3 ⎦	15	19	0.6
STU	W	Z	P	90.2	38	35:14.3	+0.9	0.99	4.8 ⎤			
VAL	W	Z	P	77.4	39	34: 7.5	−1.2	0.54	5.5 ⎬	13	15	0.7
PTO	W	E	P	81.2	50	34:28.3	−0.7	0.58	5.5 ⎦			
SJG	W	N	SH	34.3	84	34:22.6	−3.1	0.27		16	20	3.6
CAR	W	Z	P	35.0	97	29: 5.9	+0.1	1.16	5.6	15	18	0.8
ARE	W	Z	P	45.5	137	30:36.9	−0.9	1.79	6.0 ⎤			
NNA	W	N	P	38.8	138	29:41.5	+3.7	2.94	⎬	13	20	0.8
LPA	W	Z	P	67.2	142	33: 9.7	+1.9	1.38	5.6 ⎦			
RAR	W	Z	P	68.4	238	33:20.6	+5.3	0.91	6.0 ⎤			
AFI	W	Z	P	75.4	250	34: 1.7	+4.3	1.56	6.8 ⎦	14	16	0.4

"Net" code: W, WWSSN. "Comp": Z, vertical; N, north; E, east. "Phase": P or SH wave. "Epicentral params": distance and azimuth. "Start time": onset of observed waves, 3 hours plus the listed minutes:seconds. "obs–pre": arrival time residual. "a factor": the seismogram scale factors from omnilinear analysis, based on inversion at best depth and focal mechanism. "main–pre": time difference, in s, between start and main pulse onset as measured on the seismograms. "direct. Δt": times of features in source functions, in s, "top" is peak moment release, "durat." is duration as measured by baseline-crossing time. "M_0": seismic moment of the source time functions, in 10^{20} Nm.

and RUFF (1993) used the Playa Azul earthquake in their global study of earthquakes at the down-dip edge of the seismogenic zone; they concluded that the best estimate of point-source depth is 18 km, but that the bootstrap uncertainty allowed the depth to be between 16 to 32 km. SINGH and MORTERA (1991) find a precursor in the source time function derived from single-station deconvolution. MENDOZA (1993) performed a detailed rupture process study of the Playa Azul event. he also notes a precursor to the main rupture process, and concludes that most of the moment release is concentrated just down-dip of the epicenter.

We invert 40 s of 12 body waves phases, 11 P waves and 1 SH waves at SJG, for the MTRFs with 30 s duration and a damping of 0.10. Given the reduction of the MTRFs to the best double couple and time function, we find that the resultant match between synthetic and observed seismograms allows the depth to be anywhere between 20 and 30 to 35 km. On the other hand, the MTRF parameter shows a fairly sharp minimum at 20 km depth, consistent with the depth estimate of TICHELAAR and RUFF (1993). Since a depth of 20 km was acceptable in all tests, we use this depth for rupture process inversions. From the right side of Figure 8, we see that the fault plane strike changes rapidly from 276° at 20 km depth to 296°

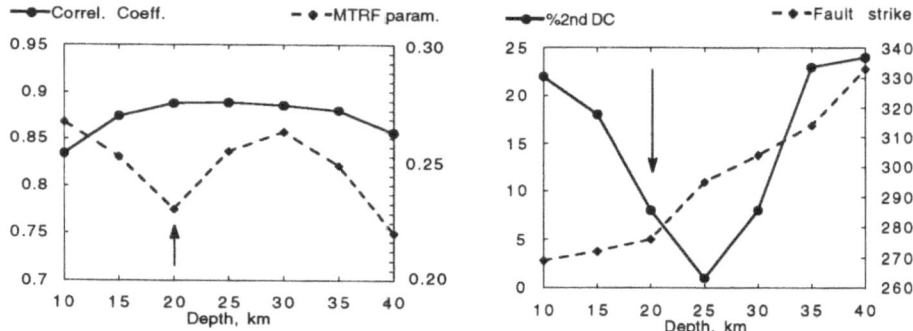

Figure 8
Various MTRF inversion and earthquake parameters as a function of depth for the 1981 Playa Azul
earthquake. Same as Figure 3.

at 25 km. We must conclude that the fault plane strike can be anywhere in this range. The auxiliary plane strike has a smaller range of 120° to 125°, and the dip is between 67° to 72°.

The basic character of the source time function in Figure 9 is a weak initiation, followed 4 to 6 s later by the single main pulse of moment release that peaks at about 12 to 15 s and then has a fairly sharp truncation of moment release centered at 15 to 18 s. Thus, once again we see a time history similar to the 1973 Colima and 1979 Petatlan earthquakes, except that the main pulse is slightly shorter in duration with less seismic moment. As for the 1979 Petatlan event, the time shifts of the

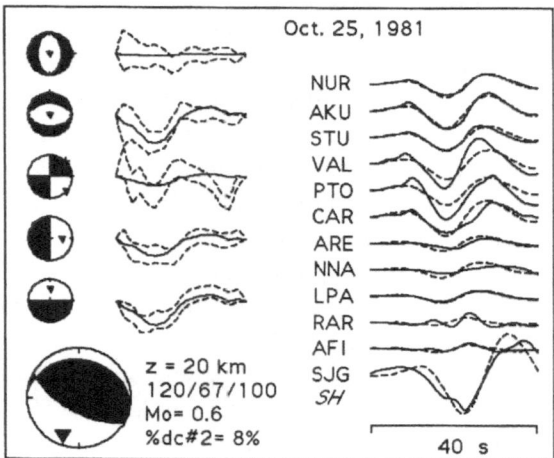

Figure 9
Inversion of seismograms for MTRFs for the 1981 Playa Azul earthquake, at a depth of 20 km. Same as Figure 2.

"hand-picked" features in the time history show no coherent directivity effects. In fact, the times of the truncation of moment release range over just one sampling interval. Thus, this lack of directivity implies that: (1) the termination of moment release was quite close to epicenter, within about 30 km; and/or (2) the rupture termination occurred close to the same time over several parts of the rupture area. If we assume a rupture velocity of about 2 km/s, then the pulse duration of about 16 s allows the rupture front to cover all of the aftershock region of HASKOV et al. (1983). Since we cannot resolve any spatial details, we simply assume that the moment was released uniformly over the aftershock zone. This assumed moment distribution is consistent with the shape and duration of the time history, and is also quite similar to the model presented in MENDOZA (1993).

There is no distinct additional pulse of moment release after this main pulse, which has a seismic moment of 0.6×10^{20} Nm (Fig. 9), essentially the same as the overall moment release. The average displacement in the 20×40 km^2 rupture area is then about 2.9 m. The pulse of moment release returns to the baseline at times that range from 15 to 20 s. Thus, the total duration of moment release appears to be no longer than 20 s. If we suppose that the rupture went down-dip for 30 km (see above), then the depth range of moment release would be 10 km for a 20° fault dip (Harvard CMT), or 13 km for 25° fault dip (the MTRF results in Fig. 9). This latter number seems consistent with the bootstrap error estimates for depth. Thus, we do not see anything highly unusual in the rupture process of the Playa Azul event; except that it initiated at the up-dip edge of the aftershock area, the fault dip is somewhat steeper, and the time function duration is somewhat long for the earthquake size.

3.4 The 1985 Michoacan Aftershock

We discuss the Sept. 21, 1985 aftershock before the Sept. 19, 1985 mainshock. Although this is a departure from the chronological order, we need to use the epicenter of the Sept. 21 event to relocate the Sept. 19 mainshock.

3.4.1 Epicenter and aftershocks

The Sept. 21, 1985 event was recorded by a local temporary network that was installed soon after the Sept. 19 mainshock (UNAM, 1986). The Sept. 21 epicentral location should be reliable. Furthermore, since UNAM (1986) used the same crustal structure as VALDES et al. (1982) used for the Petatlan aftershocks, the relative locations of the aftershocks between these two sequences should also be reliable. The UNAM (1986) location of the Sept. 21 event is: 17.62°N, 101.82°W; this location is about 30 km southwest of the ISC location.

Seismicity that occurred after the Sept. 21 earthquake fills a region that is southeast of the presumed rupture zone of the Sept. 19 mainshock. UNAM (1986) considers the Sept. 21 earthquake as a separate "mainshock" in the sense that it extended the rupture zone beyond the limit of the Sept. 19 earthquake. As a caveat to this interpretation, there is one aftershock that occurred Sept. 20 within the UNAM aftershock zone of the Sept. 21 event.

The aftershock regions of UNAM (1986) show some overlap of the Sept. 21 and 1979 Petatlan regions. In detail, the Sept. 21 aftershocks occur in two main clusters, and the down-dip cluster does overlap with the inner aftershock zone of VALDES *et al.* (1982). STOLTE *et al.* (1986) show aftershocks that occurred from Sept. 22 to Oct. 5; their epicentral maps also show some overlap of the Sept. 21 region with the 1979 Petatlan aftershock zone.

We have picked the *P*-wave arrivals on a total of 33 analog and digital records. The *P*-wave arrival has an emergent character with an initial ramp before the main pulse. Although difficult for some records, we consistently picked the time of the initial ramp. If we use the travel-time residuals between our observed arrivals and the predicted times from the UNAM location, then the globally-derived epicenter is 40 to 50 km northwest of the UNAM epicenter—the typical bias seen for other Mexico events. We use the UNAM epicenter as the best location for the Sept. 21 event.

3.4.2 Body-wave inversion for MTRFs, depth, and rupture process

The Harvard CMT for the Sept. 21 event has a moment of 2.5×10^{20} Nm at a centroid depth of 21 km, and the CMT yields an underthrusting focal mechanism on a fault plane with a strike and a dip of 296° and 17°. The auxiliary plane strikes at 121° with a dip of 73°. *P*-wave first motions at some stations require a steeper dip of the auxiliary plane, perhaps a dip of 78°. We inverted different combinations of stations with several different trial focal mechanisms to find the best point-source depth. The main conclusion is that depth is poorly resolved, and the best point-source depth is between 10 and 25 km. The MTRF inversion uses a damping of 0.1, 60 s of seismogram, filtered with a 10 s duration triangle, and the MTRFs are sampled at 4 s. WWSSN and digital seismograms are inverted together, the normalization is discussed in the Appendix. The MTRF inversion also produces a good fit to the seismograms over a broad depth range (Fig. 10); note that the MTRF parameter is less than 0.2 for depths between 15 to 35 km. As seen in Figure 10, the fault strike is fairly constant for depths between 10 and 30 km, it varies between 303° and 294°. The auxiliary plane is even more stable, with a strike of 111° to 114°. The overall best MTRF inversion, as reduced to a double couple, is for a depth of 25 km, and is shown in Figure 11. The resultant focal mechanism has a fault plane strike and dip of 299° and 12°, and the auxiliary plane strikes at 114° with a dip of 78°. Note that the simple time history fits both the WWSSN long period and GDSN intermediate period seismograms in Figure 11. The seismic moment is about 1.1×10^{20} Nm, based on the nondiffracted *P* waves.

Source time functions are determined for the station groups depicted in Table 4. MENDOZA (1993) reports that a short-period precursor is observed 4 to 5 s before

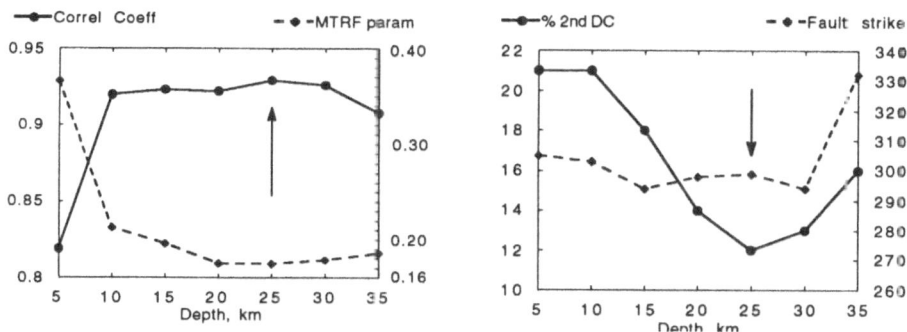

Figure 10
Various MTRF inversion and earthquake parameters as a function of depth for the Sept. 21, 1985 Michoacan aftershock. Same as Figure 3. Note that depth is poorly resolved.

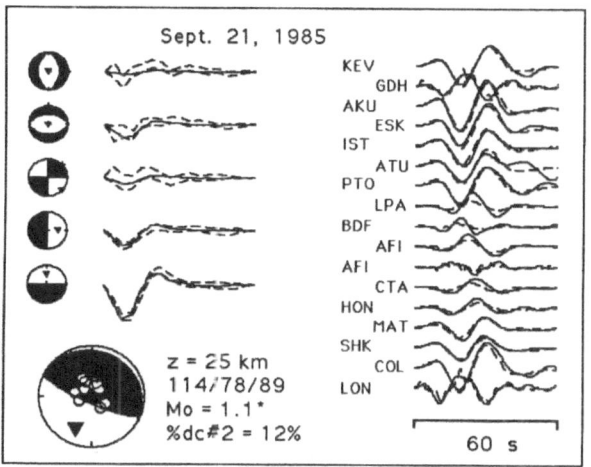

Figure 11
Inversion of seismograms for MTRFs for the Sept. 21, 1985 Michoacan aftershock, at a depth of 25 km. Same as Figure 3.

the main pulse. Indeed, this precursor is responsible for the emergent P-wave arrival on long-period records, and it is expressed as an initial ramp in the time functions with a distinct increase in moment release at 4 s after initiation. This basic character is illustrated in Figure 12 with the time functions from the European and the combined Japan-North America station groups. The seismograms are smoothed with a 3 s triangle, time functions are sampled at 2 s, with a damping of 0.10. The impact of adding a digital "broad-band" station is to sharpen the features of the time function, though the basic features are the same as when inverting just the WWSSN seismograms. The moment release peaks at about 10 s, and ends at

Table 4

Station data and some results for the Sept. 21 Michoacan earthquake

Sta	Net	Comp	Phase	Epicentral params Δ,°	Az,°	Start time 1 hr+	obs– pre	a factor	direct. Δt, s duration	M_0
KEV	W	N	P	85.8	16	49:50.7	−0.3	0.42		
GDH	G	IZ	P	59.5	18	47:13.2	−3.1	0.44	15.6	1.1
AKU	W	Z	P	71.5	26	48:32.3	−1.1	0.60		
ESK	W	Z	P	80.5	35			0.51		
IST	W	Z	P	105.9	37	51:22.7	−0.9	3.38*		
ATU	W	E	P	104.7	42	51:17.4	−0.6	1.50*	17.4	0.4
PTO	W	E	P	81.0	50	49:25.3	−1.4	0.31		
LPA	W	Z	P	66.8	142	48: 6.6	+2.1	0.78		
BDF	G	IZ	P	62.6	119	47:36.2	−0.9	0.54	16.2	1.8
AFI	W	Z	P	75.8	250	48:49.5	−8.8	1.18		
AFI	G	IZ	P	75.8	250			1.79	17.7	0.4
CTA	W	Z	P	115.9	256	52:13.8	+5.2	6.89*		
HON	W	Z	P	52.9	284	46:30.7	+2.0	0.78	16.4	
MAT	W	Z	P	101.8	315	51: 5.4	+0.2	2.11*		
SHK	W	Z	P	106.7	315	51:34.0	+6.9	3.07*	17.6	0.3
COL	W	E	P	56.3	338	46:52.0	−1.7	0.43		
LON	G	IZ	P	33.4	335			0.85	16.9	1.1

"Net" code: W, WWSSN. G, GDSN. "Comp": Z, vertical; N, north; E, east; IZ, intermediate-period vertical. "Phase": P wave. "Epicentral params": distance and azimuth. "Start time": onset of observed waves, 1 hour plus the listed minutes:seconds. "obs–pre": arrival time residual. "a factor": the seismogram scale factors from omnilinear analysis, based on MTRF inversion at best depth, * diffracted P waves. "direct. Δt": times of features in source time functions, in s, "duration" is duration as measured by baseline-crossing time. "M_0": seismic moment of the source time functions, in 10^{20} Nm.

around 20 s. This basic time history is quite similar to that observed for all four earthquakes discussed above. Clearly, rupture initiation along the Mexico subduction zone is characterized by a low level of moment release for 4 to 6 s, followed by a single simple pulse of moment release. The maximum moment resolved by the body wave inversion is about 1.5×10^{20} Nm.

Our directivity study includes "hand-picked" times for the termination of moment release as measured by the baseline-crossing time. This time was distinctly less at the northern European stations. The relocation places this feature only 8 km to the northeast, but the error ellipse includes the reference epicenter. Thus, there is no resolvable directivity in the "hand-picked" times. Next, we try tomographic imaging to test if there is some overall preferred rupture direction. We tested for all azimuths between 80° and 240°, and found that a rupture direction of 100° provides the best fit to the time functions (shown in Fig. 13). The rupture velocity is poorly

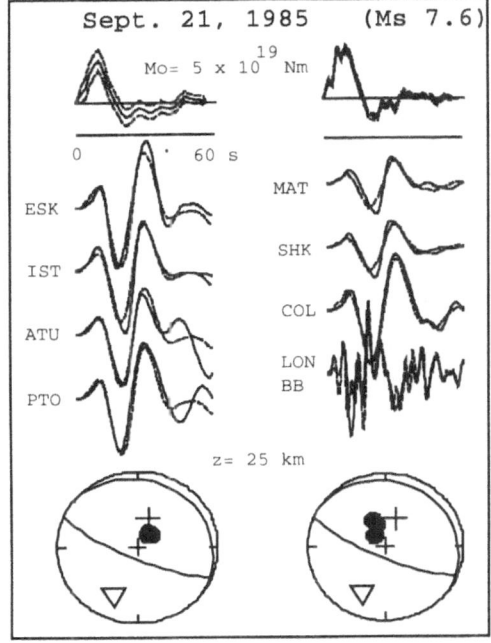

Figure 12
Inversion for source time functions for two different station groups, at a depth of 25 km. Time functions are at top, observed (solid) and synthetic (dashed) seismograms are shown below each time function. Note the distinct precursor before the main pulse. Position of stations on the focal sphere is shown at bottom.

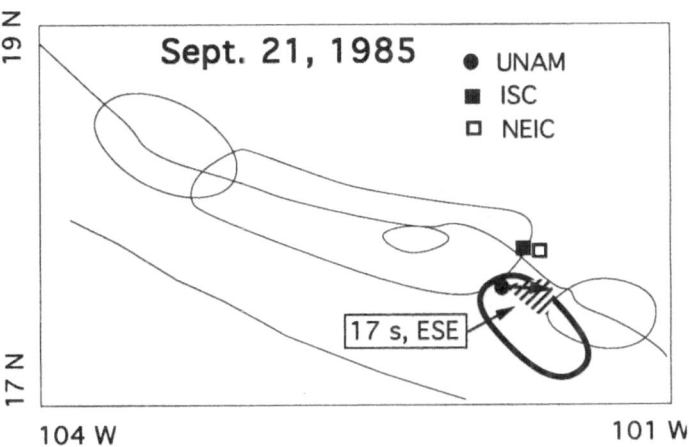

Figure 13
Base map with summary of location information for the Michoacan aftershock. Different epicenters are plotted, as is our best estimate of the region of concentrated moment release (hachured area).

resolved. Nonetheless, the overall rupture direction of 100° offers some corroboration of the detailed rupture model of MENDOZA (1993), where he places most of the moment release to the southeast of the epicenter. If we use a seismic moment of 2.5×10^{20} Nm and a rupture area of 30×60 km^2 from UNAM (1986), then the average displacement is about 4.5 m; distinctly larger than any of the previous earthquakes. One implication is that the major moment release of the Sept. 21 event does overlap the inner aftershock region of the 1979 Petatlan event.

3.5 Sept. 19, 1985 Michoacan Mainshock

3.5.1 Epicenter and aftershocks

Determination of the Sept. 19 Michoacan mainshock epicenter is a difficult problem for two reasons: (1) the typical bias for teleseismically-determined epicenters along Mexico; and (2) due to the emergent nature of rupture initiation, it is possible that the arrival data are not all measuring the same feature. The ISC epicenter is clearly too far inland, but it is curious that the PDE epicenter is displaced to the southwest from the ISC location. The most reliable location is from UNAM (1986). They fix the depth at 16 km, and used four arrivals from triggered strong-ground motion stations, including P and S waves from the station CAL that is only about 10 km from their epicenter. Most recent rupture process studies have used the UNAM (1986) epicenter as the reference point (e.g., EKSTRÖM, 1989).

We examine all possible P-wave records from the WWSSN and global digital networks to "pick" the arrival times. We soon discovered that the P wave arrives at different times for the short-period and long-period channels at the same station. There is a relatively small precursor about 5 s before the sharp onset of major moment release. From our experience with the above-described Mexico earthquakes, we are not surprised to see some type of emergent rupture initiation. Local stations and most teleseismic short-period records see the presursor as the first arrival, but other stations might pick the main pulse onset as the first arrival. Indeed, when we compare our arrival times to those reported in the ISC bulletin, there are many "picks" that are the precursor, a few that are the main pulse onset, and several that are somewhere in between. Thus, we speculate that the poor epicenter estimate from ISC is due to a data set with mixed arrivals. Many WWSSN long-period vertical records are off-scale for the Sept. 19 event, but these records offer reliable identification of both the precursor and main pulse onset arrivals. The overall time difference between the precursor and main onset is about 5 s, the times vary from 4 to 6 s (Table 5). There was no significant directivity in our measured times, hence the relative location of the precursor and main onset could be the same or differ by up to 20 km, given the scatter in our times. In contrast to our choice of start times for all the above earthquakes, the inversion of the Sept. 19 seismograms will use the main pulse onset as the start time.

Table 5

Time delay between precursor and main pulse onset for the Sept. 19, 1985 Michoacan event

Sta	Net	Epicentral params		Delay time, s
		$\Delta,°$	Az,°	
JAS	G	25.0	324	4.2
COL	W	55.3	338	5.1
	G	55.3	338	3.3
RSNT	T	44.7	352	4.3
RSON	T	33.0	10	4.5
KEV	W	85.0	16	5.5
SSB	Geo	88.8	42	6.0
SJG	W	34.3	85	4.9
TRN	W	40.3	95	5.7
BDF	G	63.4	119	5.5
LPB	W	48.5	134	5.7

"Net" code: W, WWSSN; G, GDSN; T, regional test network; Geo, GEOSCOPE. "Epicentral params": distance and azimuth. "Delay time", in s.

We do not know if the UNAM (1986) epicenter is the location of the precursor or main pulse onset, but we speculate that it is probably the precursor location. Recall that the UNAM (1986) epicenter for the Sept. 21 event is well determined. Thus, we use our time-picks of the Sept. 19 and Sept. 21 events at common stations to locate the main pulse onset of the Sept. 19 event relative to the Sept. 21 epicenter. The relocation is fairly stable with respect to the use of times from either analog or digital stations. The results from the digital readings (Table 6) place the Sept. 19 main pulse onset 124 km to the northwest of the Sept. 21 event along an azimuth of 288° (alternatively, 118 km to the west and 39 km to the north). The error ellipse is essentially circular with a radius of 10 km. Times read from analog records place the epicenter 103 km to the northwest along an azimuth of 290°. The main pulse onset epicenter is 24 km southwest of the UNAM (1986) epicenter, or just 12 km on the oceanward side of station CAL.

Aftershocks of the Sept. 19 event are small and infrequent (UNAM, 1986). The rupture area of the Sept. 19 event is mostly defined by events after Sept. 21. At the northwestern end, the aftershocks abut the presumed epicenter for the 1973 Colima earthquake. There is a cluster of aftershocks just a few km northwest of the main pulse onset location. Most of the aftershocks in the southeastern end are in the Sept. 21 aftershock zone of UNAM (1986). There are a few scattered aftershocks that are down-dip of the 1981 Playa Azul earthquake. The southeastern down-dip corner of the UNAM (1986) aftershock area is defined by several aftershocks, including one of magnitude greater than 4 that occurred within one day after the Sept. 21 event.

Table 6

Main pulse P arrival times at broad-band digital stations, used to relocate the Sept. 19, 1985 main pulse onset with respect to the Sept. 21, 1985 epicenter

Sta	NET	Epicentral params Δ,	Epicentral params Az,	Sept. 19 arrival time 13 hrs +	Sept. 21 arrival time 1 hr +
RSON	T	33.0	10	24:29.1	43:57.2
KEV	G	85.0	16	30:26.4	49:54.9
GDH	G	58.8	18	27:50.7	47:18.6
RSNY	T	34.8	36	24:45.2	44: 9.6
RSCP	T	22.5	38	22:55.3	42:19.1
SCP	G	30.5	38	24: 6.1	43:30.8
SSB	Geo	88.8	42	30:43.6	50: 8.0
NE17	Nar	84.4	50	30:27.0	49:50.3
BDF	G	63.4	119	28:19.9	47:37.5
AFI	G	75.7	250	29:36.1	49: 9.8
LON	G	32.4	335	24:23.8	43:56.3
COL	G	55.3	338	27:26.0	46:57.2
RSNT	T	44.7	352	26: 7.4	45:37.2
RSSD	T	25.5	357	23:25.3	42:55.2

"Net" code: W, WWSSN; G, GDSN; T, regional Test network; Geo, GEOSCOPE; Nar, NARS. "Epicentral params": distance and azimuth. "Sept. 19 arrival time", 13 hours plus the listed minutes:seconds. "Sept. 21 arrival time", 1 hour plus the listed minutes:seconds.

EKSTRÖM (1989) modeled two magnitude 5 aftershocks that are located near the down-dip edge in the Sept. 19 epicentral region. The depth of these two events is 15 to 16 km, for a crustal structure modified from VALDES et al. (1982). These depths place the down-dip edge of the Sept. 19 mainshock within the depth range of TICHELAAR and RUFF (1993). For some final comments on the 1985 aftershocks, the largest "aftershock" with M_s of 7.0 occurred April 30, 1986, and is located at the northwestern end of the aftershock zone; and an event occurred at the southeastern end on Sept. 28, 1985 with a centroid depth of 41 km and a strike-slip mechanism.

UNAM (1986) estimates the aftershock area to be 170 × 50 km². With a seismic moment of 11×10^{20} Nm (EKSTRÖM, 1989), and a crustal rigidity, the average displacement across the aftershock area is 4.3 m. The previous rupture in the central part of the Michoacan zone was in 1911, hence the accumulated tectonic displacement is less than 4 m. Therefore, any concentration of moment release within the aftershock area would qualify as an "official" asperity. The estimates of coseismic displacement can be decreased if we assume a mantle rigidity; but this change is difficult to justify given the shallow depth of the Mexico earthquakes. The convergence rates of 49 to 55 mm/year are from NUVEL; global constraints make it difficult to change these numbers by a factor of two or so. Overall, it would seem that the coseismic displacement of the recent sequence of earthquakes is greater than required to keep up with plate motions. A simple mechanical model of a subduction zone segment with interacting asperities can produce "synthetic earthquake" sequences where the coseismic displacements are greater than tectonic motions for several events, but there must be an occasional "unusual" event that resets the sequence (RUFF, 1992a).

3.5.2 Body-wave inversion and rupture process

Due to the large rupture length and strong directivity effects in the body waves, long-period surface waves provide the most reliable average moment tensor and focal mechanism for this event. We found that the focal mechanism of EISSLER et al. (1986), with the fault dip changed to 17°, can provide a good match of the body waves. While the fault strike of EISSLER et al. (1986) is 288°, the Harvard CMT has a fault strike of 301°. As expected, the best equivalent point-source depth is poorly resolved and can be chosen anywhere between 10 to 27 km. The stations are separated into azimuthal groups as indicated in Table 7, and the source time

Table 7

Station used for P-wave inversion and some directivity results for the 1985 Michoacan mainshock

Sta	Net	Comp	Epicentral params. Δ,°	Epicentral params. Az,°	Start time 13 hrs+	Directivity Δt, s 1st pulse end	Directivity Δt, s 2nd pulse peak	Directivity Δt, s 2nd pulse end
KEV	G	IZ	85.0	16	30:28.0			
GDH	G	IZ	58.8	18				
AKU	W	E	70.9	26	29:12.9			
IST	W	Z	105.5	36	32: 6.4	19.6	31.0	41.0
SCP	G	IZ	30.5	38	24:36.3			
ATU	W	E	104.3	41	32: 0.0			
TOL	G	IZ	84.4	50	30:19.8			
WIN	W	Z	123.5	105	33:26.4			
BDF	G	IZ	63.4	119	28:20.7	20.4	30.7	39.3
LPA	W	Z	67.8	142	28:48.3			
AFI	W	Z	75.7	250	29:37.2			
AFI	G	IZ	75.7	250	29:36.5			
CTA	W	Z	115.7	256	32:47.3	25.0	37.1	48.6
HNR	W	E	100.2	263	31:38.2			
PMG	W	Z	112.3	267	32:34.0			
GUA	W	Z	106.5	291	32: 5.1	25.5	38.7	50.5
ANP	W	Z	119.1	314	33: 6.5			
MAT	W	E	100.9	314	31:43.0			
SHK	W	Z	105.7	315	32: 8.2	23.2	35.5	47.1
HKC	W	Z	126.0	317	33:35.7			
LON	G	IZ	32.4	335	24:23.4			
COL	G	IZ	55.3	338	27:25.4			

"Net" code: W, WWSSN; G, GDSN. "Comp": Z, vertical; N, north; E, east; IZ intermediate-period vertical. "Epicentral params": distance and azimuth. "Start time": onset of observed waves, 13 hours plus the listed minutes:seconds. "directivity Δt": times of features in source time functions, in s.

functions are determined. As described by many previous investigators, the basic time history is a double event with strong directivity placing the second event to the southeast (see Fig. 14). For example, the "hand-picked" peak of the second pulse of moment release follows the main pulse onset by 31 s to the southeast azimuth, and by 37 to 39 s for western azimuths. We could consistently identify three features for the "hand-picked" directivity analysis: end of the first pulse of moment release; peak of the second pulse; and end of the second pulse. Examples of our "hand-picked" times for the ends of the first and second pulses are shown in Figure 14. The locations of these features are shown in Figure 15, all relative to our epicenter for main pulse onset ("LP rupture initiation"). The peak of moment release of the second pulse is always located half-way between the truncations of the two pulses. All three features are located at the extreme down-dip edge in the eastern part of the aftershock area. Tomographic imaging of the source time functions finds the overall best rupture direction to be 100°, with a preferred rupture velocity of 2.0 km/s. The "hand-picked" directivity of the end of the second pulse gives a rupture direction that is the same as the tomographic imaging, but with an apparent rupture velocity of 2.9 km/s. EKSTRÖM (1989; also see EKSTRÖM and DZIEWONSKI, 1986) also obtains an overall rupture velocity direction and velocity for his body-

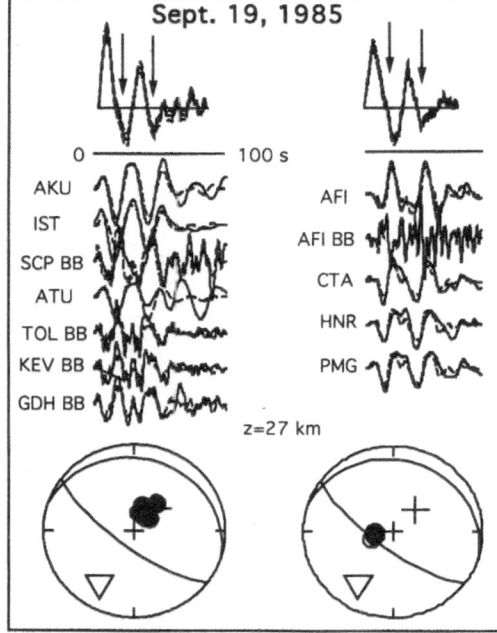

Figure 14
Inversion for source time functions for two different station groups, at a depth of 27 km, for the Michoacan mainshock. Same as Figure 12. The double event nature is clearly seen, and the arrows show our directivity time picks.

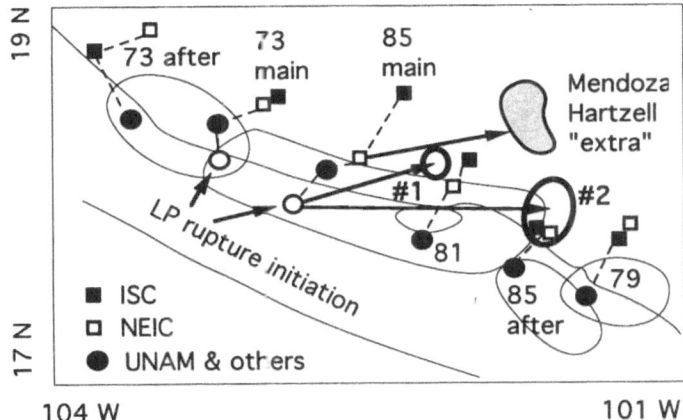

Figure 15

Map of Mexico that summarizes directivity results for the Michoacan mainshock and epicentral bias for all earthquakes. We use the epicenters labeled "LP rupture initiation" for the 1973 Colima and 1935 mainshock events, and we use the regional determinations of epicenter for the other events (solid dots, labeled as "UNAM & others"). The ISC and NEIC epicenters generally agree with each other, but are systematically biased with respect to the regional epicenters. There is an additional problem for the 1935 mainshock (see text). The arrows and error ellipses for features #1 and #2 show our directivity results for the 1985 mainshock; we place the moment release of the second event along the down-dip edge of the aftershock zone between #1 and #2. MENDOZA and HARTZELL (1989) found an "extra" asperity; if we moved our epicenter down-dip to the NEIC epicenter, then we would move the second event down-dip close to the "extra" asperity. Conversely, perhaps the "extra" asperity can be moved up-dip to our location of the second event.

wave inversion; he finds a rupture direction of 97° and rupture velocity of 2.4 km/s. Thus, although the details of the techniques used by EKSTRÖM (1989) and ourselves are quite different, and we used a total of 22 analog and digital stations while EKSTRÖM (1989) used 12 "broad-band" digital records, the two teleseismic body-wave studies give essentially the same result for moment release projected onto a one-dimensional "rupture ray." As concluded by nearly all the previous teleseismic studies, the temporal gap in moment release between the double events corresponds spatially to the 1981 Playa Azul event, and moment release is concentrated in the southeastern corner of the UNAM aftershock region. Our results do not make a clear statement as to whether there was no significant moment release in the 1981 Playa Azul zone, but we do locate the moment release of the second event in the southeastern corner. We see no evidence to continue the rupture process beyond the southeastern limit of the UNAM (1986) aftershock zone. Figure 15 also shows a summary of the epicentral location bias for all earthquakes. We show the epicenters used in our analysis, and related information is listed in Table 8.

The remaining key aspect of the rupture process of the Sept. 19 event is the additional "extra" asperity (Fig. 15) of MENDOZA and HARTZELL (1989; also see HOUSTON and KANAMORI, 1986 for a somewhat different interpretation). The

Table 8

Summary of origin times, epicenters, and depths for the five earthquakes. These parameters are used to calculate the theoretical arrival times in the previous tables

Earthquake	Date	M_s	Origin Time hr:min:sec	Epicenter Lat.	Lon.	Depth km
Colima	Jan. 30, 1973	7.5	21:01:13.8	18.20	−103.18	20
Petatlan	Mar 14, 1979	7.6	11:07:11.2	17.46	−101.46	15
Playa Azul	Oct. 25, 1981	7.3	03:22:13.0	17.75	−102.25	20
Michoacan	Sept. 19, 1985	8.1	13:17:53.7	17.97	−102.85	27
Michoacan	Sept. 21, 1985	7.5	01:37:12.7	17.62	−101.82	25

All hypocentral parameters for the Petatlan event are from GETTRUST *et al.* (1981). All hypocentral parameters for the Playa Azul events are from HASKOV *et al.* (1983). Michoacan aftershock epicenter is from UNAM (1986). Other parameters are described in text.

MENDOZA and HARTZELL study used the local strong-ground motion records with a few SRO-type long-period records to invert for the two-dimensional spatial distribution of slip across a specified fault plane. They argue that the deep extra asperity is resolved by the data. The local records presumably provide the resolution to split the second event into two separate locations along the dip direction. Our tomographic imaging method woud find the average location between their shallow and deep high-slip regions. We wish to point out that MENDOZA and HARTZELL used the PDE epicenter as the reference point to start the rupture front in their model (see Fig. 15). Although one might suppose that the local records could constrain the absolute location of moment release, we offer the possibility that their spatial locations may depend on the choice of a reference point. Now note that if we were to shift our reference epicenter down-dip to the UNAM or PDE location in Figure 15, our spatial location of the second pulse would be about 20 km down-dip of the aftershock zone edge, and approximately halfway between the separate high-slip zones of MENDOZA and HARTZELL (1989). Hence, we would have to conclude that our teleseismic results were compatible with the model of MENDOZA and HARTZELL (1989). If we are allowed to shift their pattern toward the trench by about 30 km, then the center of their down-dip extra asperity is at the down-dip edge of the UNAM (1986) aftershock region. We speculate that this trenchward shift of their slip pattern would reconcile their high-slip regions with all the other indications that seismic coupling goes no further down-dip than the UNAM (1986) aftershock zone.

4. Discussion of Results

Our discussion starts with some detailed seismological aspects of these earthquakes, and we end with some seismological and tectonic generalizations.

4.1 Radiated Wave Energy

Recall from the earlier discussion the question of whether radiated wave energy estimates are systematically less than expected from the energy-moment formula of KANAMORI (1977). Since we have determined some aspects of the space-time history of moment release for three earthquakes, we can also determine the radiated wave energy of the 1973 Colima event, and the 1985 Michoacan mainshock and aftershock. The source time function duration for these earthquakes is at least 20 s, thus the teleseismic body waves with periods of a few seconds and greater should contain most of the radiated wave energy (see KIKUCHI and FUKAO, 1988, for more discussion of this point). As KIKUCHI and FUKAO (1988) give an energy estimate for the 1985 Michoacan mainshock, we provide an independent estimate based on: a different teleseismic data set; a different inversion procedure; and different assumed fault finiteness. In addition, we will add energy estimates for two additional earthquakes. The radiated wave energy is presented as the ratio of energy to the seismic moment.

For the 1985 Michoacan mainshock, we can see from Figure 16 that our upper bound on radiated wave energy is the same as the KIKUCHI and FUKAO (1988) estimate, while our lower bound is a factor of four less. Our independent calculation supports the result of KIKUCHI and FUKAO (1988). The wave energy estimates for the 1973 Colima and the 1985 Michoacan aftershock are 0.1 to 0.2×10^{-5}, similar to the 1985 mainshock. Thus, these earthquakes add two more examples to the list in KIKUCHI and FUKAO (1988) of large subduction earthquakes with a E/M_0 ratio less than one-tenth the "expected" value of 5×10^{-5}. As discussed in KANAMORI (1977) and KIKUCHI and FUKAO (1988), the "expected" energy/moment ratio is

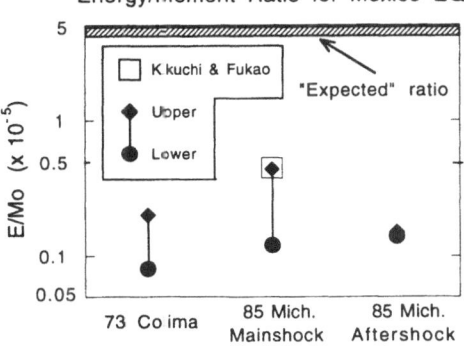

Figure 16

The energy/seismic moment ratio for the Colima and Michoacan earthquakes. Given the uncertainty in some of the rupture parameters, we calculate upper and lower bounds on the radiated wave energy. The value of this ratio calculated by KIKUCHI and FUKAO (1988) for the Michoacan mainshock is also plotted. The "expected" value of the ratio is 5×10^{-5}. Our calculations show that the radiated wave energy is only about 2% to 10% of the expected value.

based on two assumptions: (1) the stress drop is a typical value of about 30 bars; and (2) the seismic efficiency is 100%. Hence, the E/M_0 ratio can be lower if the stress drop is lower, the seismic efficiency is lower, or a combination of these two. Seismic efficiency is a measure of the frictional properties during the rupture process, and an efficiency of 100% results from the dynamic frictional stress equal to the final stress on the fault. If the average dynamic frictional stress is greater than the final stress, then the efficiency is less than 100%. KIKUCHI and FUKAO (1988) and KIKUCHI (1990) argue that the observed values of the energy/moment ratio are explained by a seismic efficiency of only 1% to 10%. Since our estimates of the static stress drop of the 1973 Colima and 1985 Michoacan aftershock are in the range of 30 to 40 bars, then we are also compelled to argue that the seismic efficiency of the large Mexico earthquakes is only about 2% to 10%. However, other evidence exists that the above conclusions may not be correct. KANAMORI *et al.* (1993) have used the "direct" method, described as method #1 in Section 2.4, to estimate energy of California earthquakes, and they find that the observed energy agrees with the "expected" value. Even more relevant to our study are the results of SINGH and ORDAZ (1993), who apply the "direct" method to Mexico earthquakes, including the 1985 Michoacan earthquakes. They find far more energy in the radiated waves than we do—their results are roughly compatible with the "expected" value. How to reconcile these results? As one possibility, some recent numerical experiments that we have performed indicate that inversion for the rupture process of large earthquakes may "lose" much of the wave energy (see, e.g., RUFF, 1992b). This important question must be resolved in future studies.

4.2 Fault Geometry, Slip Vectors, and Plate Motions

Figure 17 shows estimates of the strike and dip of the fault plane and the slip vector direction for the five earthquakes. For our results, we have indicated the acceptable ranges in these parameters by the boxes. Given the scatter in the results, it seems that a fault strike of 290° to 295° would be acceptable for all five events. If we focus on fault strike determinations by the same investigator, the Harvard CMT fault strikes are between 287° and 306°, while our preferred MTRF fault strikes range from 272° to 317°. The only earthquake for which the range in MTRF results do not include the overall value of 290°–295° is the 1979 Petatlan earthquake, with a preferred strike of 272°. However, since two other investigators found a fault strike of 290° or greater, we should conclude that all five events allow the same fault strike.

Fault dip angles from the Harvard CMT solution and our study are shown in the middle part of Figure 17. The MTRF results give a fault dip of 12° to 16°, except for the 1981 Playa Azul event which has a distinctly steeper dip of 25°. The Harvard CMT solution for the Playa Azul event yields a dip angle of 20°. Thus, it appears that the Playa Azul event does have a steeper fault plane dip as compared

Fault Plane & Slip Vector Geometry

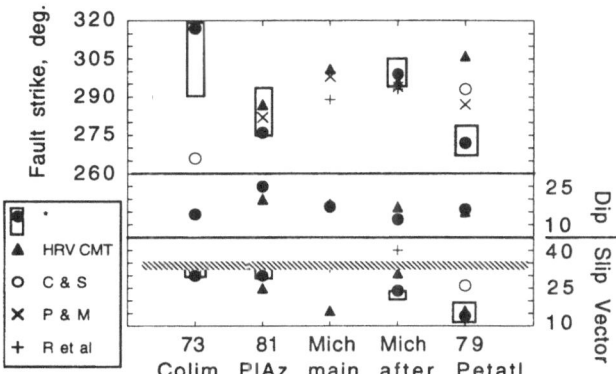

Figure 17

Summary of different estimates of focal mechanism geometry for the five earthquakes. The legend shows the symbols for different investigators: *, our results; HRV CMT, Harvard CMT results; C & S, CHAEL and STEWART (1982); P & M, PRIESTLEY and MASTERS (1986); R et al, RIEDESEL et al. (1986). The open bars show the allowed range from our MTRF study. The upper panel shows the fault plane strike, the middle panel shows fault plane dip, and the lower panel shows slip vector direction. The hachured line through the lower panel is the predicted direction from NUVEL (DeMETS et al., 1990).

to the other larger Mexico earthquakes. Since the Playa Azul rupture area is located at the down-dip edge of the coupled zone, this steeper dip could reflect a general steepening in fault dip across the seismogenic plate interface. Alternatively, the specific plate interface area ruptured by the Playa Azul event is geometrically anomalous as compared to the adjacent segments along the down-dip edge. Perhaps future detailed studies of microseismicity can discriminate between these two possibilities.

The lower part of Figure 17 shows the slip vector strike for the five events from different investigators. As previously discussed, the slip vector strikes from the Harvard CMTs are less (i.e., more northerly) than the predicted NUVEL strike of 33° to 34° across the subduction zone segment (the horizontal bar in Fig. 17). However, the results from RIEDESEL et al. (1986) for the 1985 mainshock and aftershock are consistent with the NUVEL direction. Our results for the 1973 Colima and 1981 Playa Azul events agree quite closely with the NUVEL direction. On the other hand, all three values plotted in Figure 17 for the 1979 Petatlan earthquake are less than the NUVEL direction. Either the seismological results are biased, or the slip vector for the Petatlan earthquake is rotated to a more northerly direction with respect to NUVEL. The Harvard CMTs in northern Mexico for smaller events (see compilation in DeMETS et al., 1990) show that the overall preferred slip vector strike is in the range of 20° to 25°. The Harvard CMTs from earthquakes further south *are* consistent with the NUVEL direction. To accommodate this difference of 8° in slip vector direction, between 25° to 33°, requires a significant change in the location of the Cocos-North America rotation pole, and this change would not be compatible with the global constraints on the location of the rotation pole. At this time, the most parsimonious solution is to accept the NUVEL rotation pole, and ascribe the slip vector bias to seismological "noise."

4.3 Time Functions and Rupture Initiation

One common characteristic of all earthquakes is the emergent rupture initiation. All five earthquakes show a precursor, followed 4 to 8 s later by the main pulse of moment release. After identification in the time functions, the precursor timing can be seen in the seismograms. Figure 18 shows simplified versions of the basic time functions for all five earthquakes. These time functions are all plotted at the same amplitude and time scales. In detail, the precursor is seen either as a small ramp, or as some "glitches" superimposed on a ramp—these details have been simplified in Figure 18. The precursor duration varies from 4 to 8 s, while total time function duration varies from 20 s to 65 s. The precursor duration is 4 s for the Michoacan aftershock and Playa Azul event, which both have a total duration of about 20 s. The Petatlan event has a 6 s precursor and a total duration of about 25 s, and the precursor duration is 8 s for the Colima earthquake with a total duration of 30 s. The Michoacan mainshock has a precursor duration of 5 s. Although the overall duration of this event is at least 65 s, the duration of the two individual pulses is

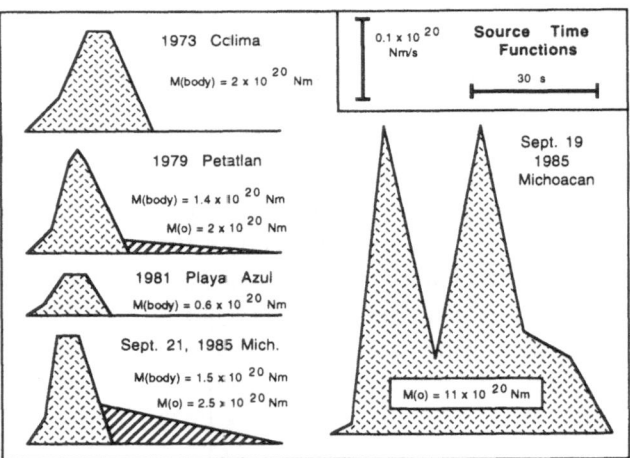

Figure 18
Schematic summary of the source time functions for each earthquake. The stylized time functions retain the duration and amplitude information; amplitude and time scales are at top. The resolved body-wave moment is listed. If additional moment release is required to match the long-period surface wave moment, it is added as a later ramp for the Petatlan event and Michoacan aftershock, and as an overall baseline correction for the Michoacan mainshock (also see EKSTRÖM, 1989). An important feature is the small precursor for all earthquakes.

about 20–25 s. Thus, it is possible that the precursor duration is linked to the duration of the main pulse that follows, though there are too few examples to convincingly prove this idea. If we accept some linkage between precursor and main pulse duration, then the precursor duration would be related to the size of the "asperities" since longer pulse duration is usually related to a larger region of higher moment release. At this point, we move into the realm of speculation by suggesting that the precursors are an intrinsic part of the rupture process of Mexico earthquakes, and that the time delay between precursor and main pulse onset is determined by the size of the asperity that is about to break. There are at least two theoretical models for rupture initiation that might explain this speculative interpretation of the precursors. One model is presented in OHNAKA (1992), where the rupture process is characterized by a slow precursory growth of the rupture front until the crack obtains a critical size, then a typical earthquake rupture occurs. The characteristics of this precursory phase depend on the unknown frictional properties of fault zones. Another model is that of KOSTROV and DAS (1988), who snow the results of a theoretical study of how an isolated asperity would fail. They discuss a curious geometric effect in which the rupture front initially propagates around the edge of the asperity, and then eventually a rupture front can sweep across the entire asperity. They use the term: "double encircling pincer," to describe this geometric effect. This style of rupture produces a source time function with a precursory ramp, followed by a main pulse as the rupture front breaks the interior of the asperity. Note that the duration of the precursor should be linked to the overall size of the asperity. Thus, the KOSTROV and DAS (1988) model for asperity failure appears to offer a good explanation of our interpretation of the Mexico earthquakes rupture process. This explanation has the appealing feature that the linkage between the precursor and the main pulse is a purely geometric effect. Precursors have been noted by many previous investigators for many subduction zone earthquakes. The results from Mexico certainly stimulate the desire to study these precursors in a more systematic fashion.

4.4 Asperity Distribution and Subduction of the Orozco Fracture Zone

One goal of this study is to determine the asperity distribution along the northern Mexico subduction zone. Our overall conclusion is that the rupture process studies do *not* give us a clear picture of distinct asperities separated by extensive "weak" regions. We were able to resolve a distinct subregion of higher moment release for the 1973 Colima earthquake, shown by the triangular region in Figure 19. Also, there is a concentration of moment release in the epicentral area of the 1979 Petatlan earthquake, shown as the oval region in Figure 19. There is also a concentration of moment release in the epicentral area of the Sept. 21 Michoacan aftershock, but it is poorly defined and is represented in Figure 19 by the open triangular area. Our results for the Sept. 19 Michoacan mainshock place bounds on

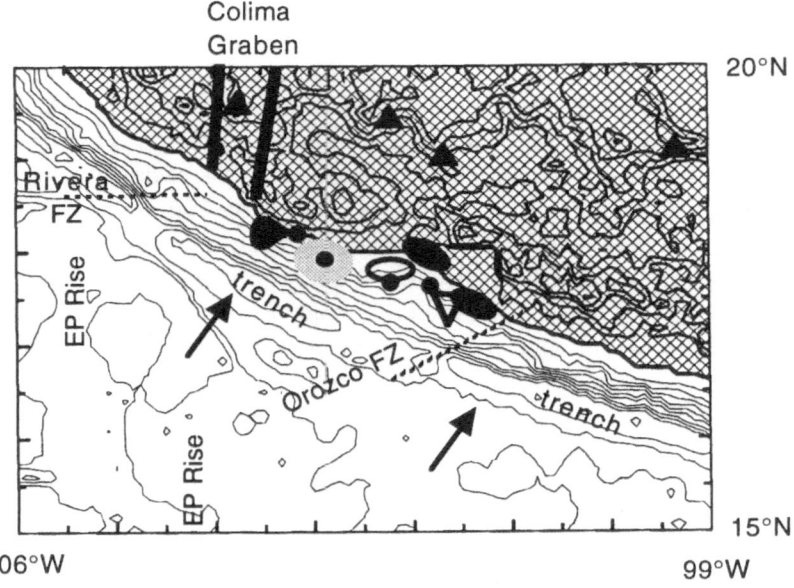

Figure 19
Detailed map of the northern Mexico subduction zone segment with the results from the rupture process studies. Contour interval is 500 m. Land is indicated by the cross-hatching. Oceanic features include the East Pacific Rise, the Orozco and Rivera fracture zones, and the Mexico trench. Arrows show the convergence direction of the Cocos plate with respect to North America; the convergence rate varies from 49 to 55 mm/year from Colima to Petatlan. Triangles are active volcanos, and the Colima graben is bordered by bold lines. Dots are large earthquake epicenters. Asperity candidates are shown as regions in the seismogenic zone; more confidence is attached to the solid regions. The asperity distribution along this segment can be characterized as "indistinct," and it is possible that the asperities overlap each other (see text). Most of the moment release is closely associated with the subducting Orozco Fracture Zone. Note the physiographic effects in the upper plate above the Orozco Fracture Zone.

the along-strike location of the second pulse of moment release; this "asperity" is shown as the oval region down-dip of the Sept. 21 aftershock. In detail, the moment release of this pulse could be distributed both up-dip and down-dip from the oval "asperity." Recall the work of MENDOZA and HARTZELL (1989) who do split this asperity into two distinct regions of higher slip. Our study does not place useful bounds on the moment release of the first pulse of the Michoacan mainshock; this fact is indicated by the dotted region about the mainshock epicenter in Figure 19. Finally, the moment release of the 1981 Playa Azul event is probably contained within the open oval attached to the epicenter in Figure 19, but recall that the common interpretation of this region is a "weak" area between the two dominant asperities of the 1985 Michoacan mainshock. The 1981 Playa Azul rupture area could be an intervening "weak" region, but it is also possible that significant moment release occurred in this area during the 1985 Michoacan mainshock. The collection of results in Figure 19 would allow us to postulate a nearly continuous distribution of "asperities" along the subduction zone segment. This map of "asperities" looks quite different from the asperity map for the Kuriles Islands (SCHWARTZ and RUFF, 1987) with clear distinct asperities separated by distances of 50 or more km. Given the relatively smaller size of the Mexico earthquakes, one explanation could be that the asperity lengths and separations are so small that our teleseismic rupture process techniques may not resolve the distinct asperities. Use of acceleragrams, such as in the study of MENDOZA and HARTZELL (1989), might provide better spatial resolution. However, since acceleragrams are not available to study the adjacent 1973 Colima and 1979 Petatlan earthquakes, we still cannot conclude that there are distinct asperities along the subduction segment. To summarize, our rupture process studies are consistent with a continuous chain of "asperities" from the 1979 Petatlan earthquake to the 1973 Colima asperity. Given some evidence from the other studies, we speculate that in fact there is heterogeneity in moment release along the subduction segment.

Subduction of the Orozco Fracture Zone causes a shallow trench to the southwest of the Petatlan, Playa Azul, and Sept. 21, 1985 earthquakes. Although the Orozco Fracture Zone is a complex bathymetric feature (see SINGH and MORTERA, 1991), its overall trend is clearly rotated to a more easterly azimuth than the convergence direction of the Cocos plate with North America (see Fig. 19). The southern edge of the Orozco Fracture Zone projected down the seismogenic zone would follow a trend similar to the dashed line in Figure 19. The "width" of the Orozco Fracture Zone is conveniently determined at the trench axis by shallowing of the trench to less than 4500 m. Thus, the 1979 Petatlan earthquake and the 1985 Michoacan aftershock are located more or less in the middle of the Orozco Fracture Zone. Also, the southern part of the Michoacan mainshock is located over the Orozco Fracture Zone. On a global basis, the subduction of bathymetric features like the Orozco Fracture Zone is usually associated with reduced seismic coupling (see McCANN and HABERMANN, 1989). However, for the case of the Orozco Fracture Zone in Mexico, we see earthquakes that are typical in size for Mexico occurring on the crest of the subducted feature. It appears that the Playa Azul earthquake initiated at the northern edge of the Orozco Fracture Zone. Furthermore, the slip associated with the second pulse of moment release of the Michoacan mainshock is located somewhere along the extension of the Orozco Fracture Zone. Thus, a significant portion of the moment release in the northern Mexico subduction is associated with the Orozco Fracture Zone.

It is interesting to note the topographic effects in the North America plate that may be associated with the subduction of the Orozco Fracture Zone. The trend of the mountain front, i.e., the 500 m contour, north of the Orozco Fracture Zone is nearly east-west (Fig. 19). Along the projection of the Orozco Fracture Zone there is a sharp embayment with a steep mountain front. The trend of the mountain front south of the Orozco Fracture Zone more nearly parallels the trend of the trench axis. McCANN and HABERMANN (1989) present a model in which the subduction of bathymetric features such as the Orozco Fracture Zone has a profound effect on the physiography and geology of the upper plate. Two predictions of their model are that: (i) there will be coastal uplift and terraces over the subducting features; and (ii) the volcanic arc will be displaced inland behind the feature (also see NUR and BEN-AVRAHAM, 1983). In Mexico, there is clearly a topographic anomaly over the ridge, with possible uplift at the southern, i.e., trailing, edge of the fracture zone. Also, a simple extension of the Orozco Fracture Zone inland shows that it corresponds with a change in the trench-volcano distance, and possibly with a gap in the most recent volcanic activity (see Fig. 19).

Given the difference in azimuths between the Orozco Fracture Zone and the Cocos-North America plate motion, the intersection of the Orozco Fracture Zone with the Mexico trench migrates north as plate motions proceed. The Guererro seismic gap (SINGH and MORTERA, 1991) is in the region just behind the passage of the Orozco Fracture Zone, though we note that it takes several million years for the Orozco Fracture Zone to traverse the seismic gap region. Would this tectonic interaction change the character of the plate interface in the Guererro region for a few million years? It is important to know the answer to this question.

The northern limit of the 1973 Colima earthquake appears to coincide with the extensions of the Rivera Fracture Zone and the Colima graben (Fig. 19). This northern limit of the seismogenic plate interface could be defined by the subducted boundary between the Cocos and Rivera plates, or the fragmentation of the upper plate, or both. The small seismic gap between the 1973 Colima and 1932 Jalisco earthquake (see NISHENKO, 1991) would appear to coincide with the Colima graben, and hence may not be capable of generating a large underthrusting earthquake.

In the spirit of "comparative subductology" (UYEDA and KANAMORI, 1979; JARRARD, 1986), we propose that there are seismotectonic similarities between northern Mexico and two other subduction zone segments: the Santa Cruz Islands segment (see TAJIMA et al., 1990); and the central Peru segment (BECK and RUFF, 1989). As fully described in TAJIMA et al. (1990), the Santa Cruz Islands segment is characterized by overlapping aftershock areas of two large earthquakes in 1966 (M_s 7.9) and 1980 (M_s 7.7). This is similar to the overlapping rupture areas of the 1979 Petatlan event and 1985 Michoacan aftershock. In the Santa Cruz Islands segment, it is also difficult to identify distinct asperities (see Fig. 20). The central Peru segment also offers an interesting comparison. As presented in BECK and RUFF (1989), the central Peru subduction segment extends for about 500 km between the Mendana Fracture Zone to the north and the Nazca Ridge to the south (segment shown in Fig. 20). This segment was ruptured by three large earthquakes in 1940, 1966, and 1974; however it seems that the entire segment was ruptured by a single great earthquake in 1746. The 1974 Peru earthquake (M_s 7.8) shares many similarities with the 1985 Michoacan mainshock: (1) seismic moment is the same; (2) it was a double event; (3) the first event in the northern part displayed some

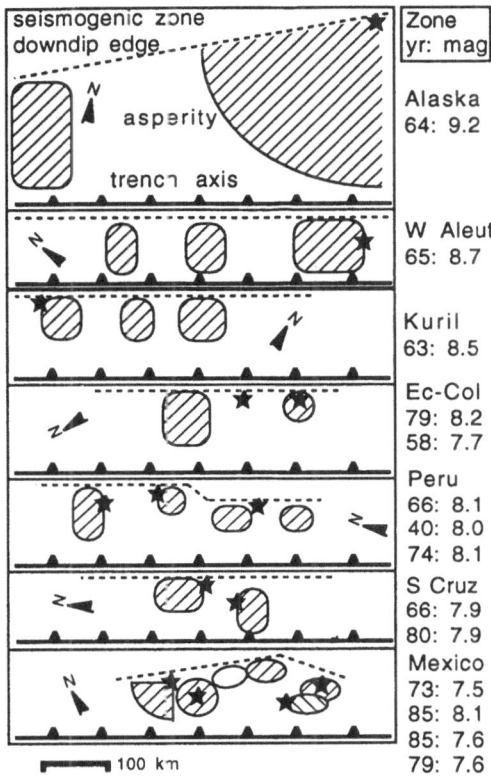

Figure 20

Schematic map views of asperity distributions determined for several subduction zones from rupture process studies of great earthquakes. For each subduction zone segment: asperities are the hachured regions; trench axis and down-dip edge of the seismogenic zone are plotted as the bold and dashed lines, respectively; north arrow is given for approximate orientation; stars plot the great earthquake epicenters; for more than one star, the list corresponds to left-to-right ordering of epicenters. Earthquake list to the right of each zone gives the year, e.g., "64" is the March 28, 1964 Alaska earthquake, and magnitude is either M_w or or M_s. Abbreviations and references are: Alaska, CHRISTENSEN and BECK (1994) this issue; W Aleut (Western Aleutians), BECK and CHRISTENSEN (1991); Kuril (Kurile Islands), Ec-Col (Ecuador-Colombia), and Peru, BECK and RUFF, (1989); S Cruz (Santa Cruz Islands), TAJIMA et al. (1990); and Mexico, results from this paper. Same scale is used for all plots, 100 km is shown at bottom. Asperity size varies from large in the subduction zones plotted at top of figure, to small (yet close together) in the zones plotted at bottom.

component of bilateral rupture; (4) the second event ruptured to the southern edge of the segment, more or less near the crest of the Nazca Ridge; (5) there is a precursory rupture before the first main pulse onset; and, (6) there is even a strange cluster of aftershocks located significantly down-dip of the main rupture area of the 1974 Peru event (DEWEY and SPENCE, 1979). While there are many similarities between the rupture process of the 1974 Peru and 1985 Michoacan earthquake, the overall asperity distribution in central Peru appears to be more distinct than

Mexico. One speculative aspect to consider is whether the entire northern Mexico segment could rupture in a single great earthquake, just as central Peru has in previous earthquake cycles. The combined moments of all five earthquakes in Mexico would be about twice as large as the 1985 Michoacan mainshock. An event of this size is contrary to the known earthquake history along the Mexico subduction zone (SINGH et al., 1984). On the other hand, a simple mechanical model of asperity interaction can produce synthetic earthquake sequences with a long succession of smaller "single" events, punctuated by an occasional larger "multiple" event (RUFF, 1992a). We emphasize that it is pure speculation to suggest that this type of variable rupture mode, as observed in central Peru, is relevant to the northern Mexico subduction segment. The facts of earthquake history in Mexico offer a stronger suggestion that "indistinct" asperity distributions cause a spatially-overlapping sequence of large earthquakes, as opposed to great earthquakes that rupture the entire segment.

5. Conclusions

We have studied the focal mechanism and rupture process of five large earthquakes in the northern Mexico subduction zone. The rupture process of all five earthquakes shares a common feature: rupture begins with a low-level of moment release for the first 4 to 8 s, followed by the main pulse onset with a duration of about 20 s. In the case of the 1985 Michoacan mainshock, the rupture continued with a second main pulse of 20 s duration, with a total duration of 70 s or more. Rupture directivity and an epicentral region asperity are resolved for the 1973 Colima earthquake. There are various difficulties in resolving the spatial extent of rupture for all other events. As found by previous investigators, there is a strong directivity for the second pulse of the 1985 Michoacan mainshock. Our results for the Michoacan mainshock are most similar to those of EKSTRÖM (1989); he also used a one-dimensional rupture model. Our results are also consistent with the MENDOZA and HARTZELL (1989) model where the second pulse is split into a shallow asperity and an "extra" deep asperity—as long as we are able to shift their entire slip pattern trenchward. One unusual feature is the overlap in rupture areas of the 1979 Petatlan event and the 1985 Michoacan aftershock. Moment release for the 1981 Playa Azul earthquake can be contained within the aftershock area.

Detailed considerations of the focal mechanisms allow us to choose an acceptable overall fault strike of 290° to 295° for all events. The fault dip angle can be chosen as 16° for all events, except for the 1981 Playa Azul earthquake which has a dip angle of 20° to 25°. Given the freedom to combine results from different investigators, the slip vector directions of four earthquakes are consistent with the NUVEL (DE METS et al., 1990) direction of 33° for Cocos-North America convergence. The slip vectors of the 1979 Petatlan earthquake are more northerly than the NUVEL direction. The Harvard CMT solutions for large and small earthquakes in northern Mexico prefer an overall slip vector direction of 20° to 25°.

The northern boundary of this subduction segment is defined by significant features in both the subducting and upper plates. The southern boundary of this segment coincides with the southern edge of the subducting Orozco Fracture Zone. Most of the seismic moment release in this segment is associated with the Orozco Fracture Zone. We speculate that the subducting Orozco Fracture Zone influences the physiography and volcanism in the upper plate. The Guererro seismic gap is located just to the south of the Orozco Fracture Zone. We cannot speculate as to what effect the geologically recent passage of the Orozco Fracture Zone has had on the plate interface in the Guererro region.

There are several seismotectonic similarities between the northern Mexico segment and the subduction zone segments in central Peru and the Santa Cruz Islands. The small separation of asperities is a common feature for all three segments, with the northern Mexico and Santa Cruz Islands characterized by nearly continuous "asperities" and overlapping rupture areas. This asperity distribution and earthquake behavior is in contrast to that of Alaska, where the very large and distinct epicentral asperity generates "truly great" earthquakes when it breaks (see Fig. 20). The 1985 Michoacan mainshock and the 1974 Peru earthquake appear to have quite similar rupture processes and tectonic environments. The central Peru segment is characterized by a variable rupture mode, where the previous earthquake cycle consisted of a single great event that ruptured the entire segment. Significant differences between central Peru and northern Mexico include the age of subducting lithosphere and the fact that the known earthquake history of Mexico does not include the occasional great earthquakes that occur in Peru.

Acknowledgments

Thanks to C. Mendoza, S. K. Singh, S. Hartzell, and J. Anderson for materials and preprints. Thanks to Bart Tichelaar and Roland LaForge for stimulating discussions, and to the reviewers for their helpful comments. Yuichiro Tanioka assisted with the preparation of Figure 19. Also, special thanks to those "Harvard CMT" people! These research efforts are supported by the National Science Foundation (NSF90–19003 to LJR).

Appendix: MTRF Inversion and Extraction of Source Time Function and Moment Tensor

This section presents a thorough and detailed discussion of inversion for the MTRFs. The first subsection covers the forward problem and "standard" linear inversion. The second subsection covers omnilinear inversion for MTRFs. The final section addresses the inverse problem for reduction to the best source time function and moment tensor together with a statistical measure of the adequacy of this reduction.

A1. MTRF Notation and Linear Inversion for MTRFs

We can write the MTRFs explicitly as:

$$\dot{M}_{jk}(t) = \begin{pmatrix} \dot{M}_{11}(t) & \dot{M}_{12}(t) & \dot{M}_{13}(t) \\ & \dot{M}_{22}(t) & \dot{M}_{23}(t) \\ & & \dot{M}_{33}(t) \end{pmatrix} \tag{A1}$$

where the moment tensor is required to be symmetric. M_{11} represents a source dipole along the x_1 direction of a Cartesian coordinate system; positive M_{11} corresponds to a "dilatational" dipole. The orientation of the Cartesian system used here is: $+x_1$ points "East"; $+x_2$ points "North"; and $+x_3$ points "Up"; at the source location.

An isotropic seismic source, i.e., an explosion or implosion, is represented by the trace of the moment tensor. Since we shall be focused on earthquake studies, the isotropic source component will be eliminated by setting the trace to zero. This constraint is linear, hence the inverse problem for the five independent moment tensor components is linear. There are several different choices that can be made to define the five independent moment tensor components from the six components of Eq. (A1). Recall that a double-couple source corresponds to a moment tensor for which one eigenvalue is zero, and hence the other two are equal and opposite . Following the formulation in RUFF and TICHELAAR (1990), the moment tensor is constructed from five double-couple sources as follows:

$$\dot{M}_{jk}(t) = \begin{pmatrix} 1 & 0 & 0 \\ 0 & 0 & 0 \\ 0 & 0 & -1 \end{pmatrix} \dot{M}_1(t) + \begin{pmatrix} 0 & 0 & 0 \\ 0 & 1 & 0 \\ 0 & 0 & -1 \end{pmatrix} \dot{M}_2(t) + \begin{pmatrix} 0 & 1 & 0 \\ 1 & 0 & 0 \\ 0 & 0 & 0 \end{pmatrix} \dot{M}_3(t)$$

$$+ \begin{pmatrix} 0 & 0 & 1 \\ 0 & 0 & 0 \\ 1 & 0 & 0 \end{pmatrix} \dot{M}_4(t) + \begin{pmatrix} 0 & 0 & 0 \\ 0 & 0 & 1 \\ 0 & 1 & 0 \end{pmatrix} \dot{M}_5(t). \tag{A2}$$

For a given seismogram, the Green's functions for each of the above five components are synthetic seismograms calculated for the five pure double couples with a delta function time history and unit moment. The MTRFs are identified by their focal mechanisms (e.g., Fig. 2 in main text). Substitution of Eq. (A2) into Eq. (5) in the main text then yields

$$s_i(t, \Omega) = g_{fm1} * \dot{M}_1(t) + g_{fm2} * \dot{M}_2(t) + \cdots + g_{fm5} * \dot{M}_5(t) \tag{A3}$$

where g_{fm1} through g_{fm5} are the Green's functions for the first through fifth focal mechanisms in Eq. (A2) (written in full, they are: $g_{i,fm1}(t, \Omega, V_0)$). To streamline the notation of Eq. (A3), let $s_i(t, \Omega)$ simply be the nth seismogram, $s_n(t)$; and g_{fm1}, becomes g_{nm}, the Green's function for the nth seismogram from the mth focal mechanism and MTRF. Then, Eq. (A3) is written as the contraction of a row vector with the Green's functions and convolution and a column vector with the MTRFs:

$$\overbrace{g_{n1}^* \quad g_{n2}^* \quad g_{n3}^* \quad g_{n4}^* \quad g_{n5}^*} \begin{pmatrix} \dot{M}_1(t) \\ \dot{M}_2(t) \\ \dot{M}_3(t) \\ \dot{M}_4(t) \\ \dot{M}_5(t) \end{pmatrix} = s_n(t) \qquad (A4)$$

where * represents the convolution operation. We cannot expect to determine the five unknown functions from one observed seismogram. We now write down in symbolic form the simultaneous equations for N seismograms:

$$\begin{bmatrix} g_{11}^* & g_{12}^* & \cdots & g_{15}^* \\ g_{21}^* & g_{22}^* & \cdots & g_{25}^* \\ \vdots & \vdots & & \vdots \\ g_{N1}^* & g_{N2}^* & \cdots & g_{N5}^* \end{bmatrix} \begin{bmatrix} \dot{M}_1(t) \\ \dot{M}_2(t) \\ \vdots \\ \dot{M}_5(t) \end{bmatrix} = \begin{bmatrix} s_1(t) \\ s_2(t) \\ \vdots \\ s_N(t) \end{bmatrix} . \qquad (A5)$$

The above system corresponds to the equation in RUFF and TICHELAAR (1990).

We must now discretize the above system. The unknown MTRFs are stacked into the model column vector, and the observed seismograms are stacked into the data vector. It is not necessary for the duration or averaging intervals of the seismograms and time functions to be the same. The $g*$ operation becomes a submatrix (see RUFF and KANAMORI, 1983b for details).

Our preference is to invert the seismograms as recorded, rather than preprocess the seismograms to equalize instrument responses. Digital seismograms are rescaled from digitial counts to ground displacement or velocity. Amplitudes are normalized by dividing both the Green's functions and seismograms by the average spectral level of the instrument response in the pass-band of 10 to 30 s. This procedure allows us to simultaneously invert WWSSN seismograms with digital seismograms from various instruments and networks. The relative amplitudes as plotted are meaningful in the sense that nodal seismograms will have smaller amplitudes, etc. The discretized version of Eq. (A5) is then written as

$$Am = ? \, d \qquad (A6)$$

where "$=?$" indicates that Eq. (A6) is a statistical estimation "equation." As in RUFF and KANAMORI (1983b) and RUFF and TICHELAAR (1990), we use the damped least squares inverse to obtain the model estimate

$$m = (A^T A + \delta J)^{-1} A^T d = A^* d \qquad (A7)$$

where the A^T matrix is the transpose of the A matrix, δ is the model damping parameter, which is specified by the user as a fraction of the average value of the diagonal elements of A^TA; J is a "sawtooth" diagonal matrix that increases from zero to the average diagonal value of A^TA for each MTRF (see RUFF and KANAMORI, 1983b, for statistical explanation of this choice). A^* is commonly referred to as the generalized inverse.

At this point, we can invert seismograms for the MTRFs. However, a problem appears due to the scatter in the amplitudes of long-period body waves. This scatter introduces unwanted noise into the model results. We now discuss a method that eliminates this problem.

A2. Omnilinear Inversion of Seismograms for MTRFs

The underlying reasons, numerical experiments, and some philosophical notions of body-wave amplitude scatter and omnilinear inversion are discussed in RUFF (1989b) and TICHELAAR and RUFF (1991). Here we shall simply discuss the technical details of seismogram scaling factors and omnilinear analysis. The data vector in Eq. (A6) contains the N seismograms. An explicit representation of this is

$$Am = ? \begin{bmatrix} \dfrac{s_1}{\quad} \\ \dfrac{s_2}{\quad} \\ \vdots \\ \dfrac{\quad}{s_N} \end{bmatrix} \tag{A8}$$

where the horizontal lines within the data vector separate the subvectors for each seismogram. We now modify this equation by explicitly including the unknown scale factors $a_1, \ldots, a_N$, one for each seismogram:

$$Am = ? \begin{bmatrix} \dfrac{s_1}{\quad} & a_1 \\ \dfrac{s_2}{\quad} & a_2 \\ \vdots & \vdots \\ s_N & a_N \end{bmatrix} \tag{A9}$$

where a_i multiplies every component of seismogram s_i. We want to determine the set of scale factors that will make the data vector most compatible with the model description. Assemble the a factors into the vector a by placing the seismograms in a matrix:

$$Am = ? \begin{bmatrix} \dfrac{s_1 \cdot a_1}{s_2 \cdot a_2} \\ \vdots \\ s_N \cdot a_N \end{bmatrix} = \begin{bmatrix} [s_1] & & & \mathbf{0} \\ & [s_2] & & \\ & & \ddots & \\ \mathbf{0} & & & [s_N] \end{bmatrix} \begin{bmatrix} a_1 \\ a_2 \\ \vdots \\ a_N \end{bmatrix} = Sa. \tag{A10}$$

The S matrix has N columns; the ith column vector contins the ith seismogram shifted to the ith position, and all other elements are zero. The above formulation shows that the simultaneous estimation of m and a can be treated as a linear problem, but it is a different problem than the simple linear form of Eq. (A8). First notice that Eq. (A10) is now always exactly solved by all a's and all m's equal to zero. This trivial solution must be eliminated. As discussed in RUFF (1989b), the physical problem provides the constraint equation for a: the product of scale factors is one, i.e., $\prod_i a_i = 1$. Other constraint equations on a for different physical problems could be $\sum_i a_i = N$, for example.

We shall briefly outline the solution of the above described problem. First rewrite Eq. (A10) as:

$$e = [A \mid -S]\left(\frac{m}{a}\right) \tag{A11}$$

where e is the error vector; the difference between the rescaled observed seismograms, Sa, and the synthetic seismograms, Am. We now minimize the squared length of the error vector with respect to both m and a to find the normal equations

$$\left[\begin{array}{c|c} A^TA & -A^TS \\ \hline -S^TA & S^TS \end{array}\right]\left(\frac{m}{a}\right) =?\left(\frac{0}{0}\right) \tag{A12}$$

where 0 is a vector with all zeros, and " $=?$ " symbolizes the fact that this equation might not be exactly solved due to constraints that may be placed on m or a. The goal is to find some acceptable nontrivial solution of the above system. Note that the upper-half of Eq. (A12) is

$$A^TAm = A^TSa \tag{A13}$$

which represents the normal equations to find the least-squares estimate for m, while the lower-half of Eq. (A12) are the normal equations to find the least-squares estimate for a. One style of solution of the above problem is known as canonical correlation analysis, as invented by HOTELLING (1936); however, this solution has undesirable properties for the model constraints and the rescaling of the data. The new solution adopted by RUFF (1989b) is to solve the upper-half of the system in Eq. (A12) exactly (making allowances for the damped least-squares solution, i.e., $(A^TA + \delta J)^{-1}$ replaces $(A^TA)^{-1}$), and then seek the best solution of the lower-half system subject to the constraint equation on a. For the case of no damping, the solution for m is then

$$m = (A^TA)^{-1}A^TSa. \tag{A14}$$

This solution can be substituted for m in the lower-half system of Eq. (A12) to obtain

$$[S^TS - S^TA(A^TA)^{-1}A^TS]a =? 0 \tag{A15}$$

which can be rewritten as

$$La =? 0. \tag{A16}$$

The L matrix has some interesting properties. First and most important, L is a positive definite matrix, though it can have zero eigenvalues. We now try to find a that satisfies the constraint equation and minimizes the squared length of La, i.e., tries to achieve zero length as requested by Eq. (A16). If we adopt the constraint equation of $a^T a = 1$, where a^T is the transpose of the column vector a, then the solution to Eq. (A16) is given by a that is the eigenvector associated with the smallest eigenvalue of L. Other constraint equations require slightly more analysis. First, define a residual vector as: $k = L^{1/2}a$. Then, the scalar residual is $\rho = k^T k = a^T La$; and ρ is a quadratic surface as a function of a. The gradient vector of this surface is given by $v = La$; and v is orthogonal to the contours of constant ρ. These contours are a generalized elliptical shape in higher dimensional spaces. The existence of an exact solution corresponds to a singular L matrix and would be represented by an ellipse in which one of the axes extends to infinity. For this case, a solution is the eigenvector associated with the zero eigenvalue. Exact solutions are not encountered when inverting real data. We now seek to minimize ρ given various constraints on a. In geometric terms, we seek the inner-most elliptical contour that just touches the constraint curve for a, that is, the two surfaces share the same tangent direction at this point. In this geometric view, we seek the a vector that produces $v = La$ that is orthogonal to the constraint surface. If the constraint surface is linear—for example, $c^T a = 1$—then the vector normal to the constraint surface points in the same direction everywhere in the space. Let this normal vector be u. Then, the direction of the solution vector is α, where: $L\alpha = u$. With L a positive definite nonsingular matrix, $\alpha = L^{-1}u$; then a is the rescaled α: $a = (1/c^T \alpha)\, \alpha$.

The constraint surface for $\prod_i a_i = 1$ is "nonlinear" in the sense that the normal vector points in different directions as a function of a. For an a vector that satisfies this "nonlinear" constraint, the normal vector to this surface can be written as: $(1/a_1, 1/a_2, \ldots, 1/a_N)$. The solution is then the a vector that minimizes ρ and also produces $v = La$ such that $v_i = c(1/a_i)$, where c is some constant. Hence, we have a nonlinear equation for a

$$La = c[1/a_i]. \qquad (A17)$$

Although this equation is nonlinear as it cannot be solved in an explicit one-step operation, it is easily solved by iteration or geometric manipulation in the linear space of the a vector. The key is to notice that the solution vector will lie in the sector between the u and $\alpha = L^{-1}u$ vectors, as previously defined. Thus, provided with this good guess of the solution, we use Eq. (A17) to quickly converge on the solution.

We finish this discussion with a critical review of the constraint equations that might be used for omnilinear analysis. The constraint of $a^T a = 1$ is perhaps the simplest case to analyze because the solution is given by the eigenvector for the smallest eigenvalue of L. However, this constraint is difficult to justify from a statistical point of view. Furthermore, if two or more eigenvalues of L share the same minimum value, then there is no unique solution. Even for this situation, minimizing ρ with $\sum a_i = 1$ or $\prod a_i = 1$ will produce a unique solution. Given the symmetry of the error surface and the constraint surface of $\prod a_i = \pm 1$, an alternative solution to the "correct" a is to multiply all a_i's by -1; of course this would also reverse the sign of m in Eq. (A14). For the present application, we imagine that the polarities of most seismograms are correct, thus we choose the a vector that mostly points toward positive values of the individual components. Omnilinear analysis has the capability to find reversed polarity seismograms and automatically "flip" them to the correct polarity, but this option should be carefully monitored.

Once the scale factors are determined, the best model is calculated from Eq. (A14). We define the *a posteriori* data variance as: $\sigma^2 = e^T e / N$, where e is the error vector $Sa - Am$, and N is the total number of data vector elements. The model covariance matrix is then

$$[\text{Cov}] = A^*[\sigma^2 I]A^{*T} \qquad (A18)$$

where I is the identity matrix, and A^* is the generalized inverse. A single value of the overall data variance is used to be compatible with the construction of A^*. We can easily represent the diagonal elements of the model covariance matrix by plotting the plus and minus standard deviation about each time function point by dashed lines, as in Figure 2 in the main text. A clear graphical representation of the full covariance matrix is more diffcult. In the section below, we shall use the covariance matrix as part of the numerical calculations to extract the best time function and moment tensor.

A3. Extraction of Moment Tensor and Time Function from MTRFs

This lengthy section covers both theoretical and practical aspects of the quasi-nonlinear inversion problem to extract the best moment tensor and time function from the MTRFs. The first subsection sets up the forward problem; subsection A3.2 develops the inverse equations for the "bouncing" method; A3.3 shows the omnilinear formulation; A3.4 shows how to obtain a global solution; A3.5 generalizes A3.4 to include a data covariance matrix; A3.6 then returns to the statistical question of whether the earthquake can be represented by a single moment tensor and time function, or not; and A3.7 finishes with a practical discussion of several issues including error propagation to the faulting parameters.

A3.1 Statement of the problem

If an earthquake has a constant moment tensor during rupture, then the individual MTRFs will all have exactly the same shape. If we normalize the waveshape to have positive unit area, then the scale factors $(m_1, m_2, \ldots, m_5)$ that multiply this waveshape to produce the MTRFs will have units of seismic moment, and these scale factors are components of the moment tensor:

$$m_1 f(t) = \dot{M}_1(t), \quad m_2 f(t) = \dot{M}_2(t), \ldots, \quad m_5 f(t) = \dot{M}_5(t). \qquad (A19)$$

Since the MTRFs are organized into a column vector for the data inversion, the above relations can also be written as

$$\begin{bmatrix} m_1 f \\ \hline m_2 f \\ \hline \vdots \\ \hline m_5 f \end{bmatrix} = \begin{bmatrix} \dot{M}_1 \\ \hline \dot{M}_2 \\ \hline \vdots \\ \hline \dot{M}_5 \end{bmatrix} \qquad (A20)$$

where the horizontal lines within the column vectors separate the MTRFs. The moment tensor is then given by

$$M_{jk} = \begin{pmatrix} m_1 & m_3 & m_4 \\ m_3 & m_2 & m_5 \\ m_4 & m_5 & -m_1 - m_2 \end{pmatrix}. \qquad (A21)$$

The moment tensor can be further analyzed to see whether or not it represents a single double couple. Let the eigenvalues of M_{jk} be λ_1, the largest positive one, λ_2, the intermediate one, and λ_3 the minimum one. Also, designate the eigenvectors associated with the λ_1, λ_2, and λ_3 eigenvalues as: t, n, and p. Then, a pure double couple is represented by a moment tensor with $\lambda_2 = 0$, and consequently $\lambda_3 = -\lambda_1$. Then the t, n, and p unit vectors are the tension, nodal, and pressure axes of the focal mechanism. The seismic moment of the earthquake is $M_0 = \lambda_1 = -\lambda_3$. If $\lambda_2 \neq 0$, the moment tensor can be separated into either major or minor double couples, or a major double couple and a compensated linear vector dipole. The geometry of the major double couple is the same for either choice, only M_0 differs slightly. We us the former option, where the seismic moment is given by the larger of λ_1 or $|\lambda_3|$. The "size" of the secondary source is given by the ratio of $|\lambda_2|/M_0$, and is expressed as "percent second double couple."

When we invert real data to find the MTRFs, we do not expect to see the same identical shapes for each MTRF. The shapes will be different due to "noise", i.e., inadequate Green's functions, combined with low resolution, or possibly because the earthquake actually did rupture with time-varying moment tensor. The scientific challenge is: How can we assess if the variations between MTRFs are significant? We must make use of the estimates of data and model variance in answering this question. The procedure of RUFF and TICHELAAR (1990) is to first find the "best" single time function and moment tensor, and then see if the "synthetic" MTRFs obtained from recombining the time function and moment tensor fall within the error bounds of the MTRFs. This basic procedure is graphically displayed in several figures in the main text, Figure 2 for example. If the synthetic MTRFs fall within the error bounds of the MTRFs, then we conclude that the earthquake can be represented by a single time function and moment tensor. RUFF and TICHELAAR (1990) prefer to use two standard deviations as the error bounds. We would then expect about 95% of the samples of the synthetic MTRFs to fall within these error bounds, and a few points that lie outside are not cause for concern. If the synthetic MTRFs fall significantly outside the error bounds of the MTRFs, then we must retain some aspects of the more complicated description of the earthquake rupture process.

A3.2 How to find the best f(t) and M_{jk}

We must now find the best time function and moment tensor components given the MTRFs and their covariance. This is another problem in statistical estimation. This inverse problem requires some discussion because of the apparent nonlinearity of our model parameters, i.e., the product of the unknown $f(t)$ and m_i's. It seems intuitively clear that the "best" time function should be some average of the individual MTRFs. As one example of a quantitative application of this idea, VASCO (1988) assembled the MTRFs into individual column vectors of a matrix, and then used the "principal components" technique to find a single time function shape that best represents the MTRFs. Another example is the technique used by RUFF and TICHELAAR (1990), where a weighted average of the MTRFs produces the overall time function shape, and the weights are based on the average standard deviation of each MTRF, with positive and negative signs determined by whether an individual MTRF is mostly positive or negative. Here, we shall develop the details of the statistical inverse problem that allows for rigorous definition of the "best" time function and moment tensor with respect to the best-fit of the MTRFs; and we shall see how this formal development unites the above two *ad hoc* techniques as part of the overall best solution.

There is a special structure to the nonlinear inverse problem for $f(t)$ and m_i that allows a global analysis of this problem. Also, it is possible to pose this problem as one in omnilinear analysis, albeit an ill-posed problem that is not recommended. We first develop the inverse problem with no statistical information on the MTRFs, and then go back to add in the MTRF covariance matrix.

We now define the error vector, e, as the difference between the MTRFs and the "synthetic" MTRFs based on a single time function, $f(t)$, and moment tensor components, m_1 through m_5:

$$
e = \begin{bmatrix} \dot{M}_1 \\ \hline \dot{M}_2 \\ \hline \vdots \\ \hline \dot{M}_5 \end{bmatrix} - \begin{bmatrix} m_1 f \\ \hline m_2 f \\ \hline \vdots \\ \hline m_5 f \end{bmatrix} = d - (mf). \tag{A22}
$$

The discretized time function vector is f, and the five m_i's are collected into a vector, m. We seek the f and m that minimize the squared length of the error vector,

$\varepsilon = e^T e$. More generally, we can explicitly include the MTRF covariance matrix, Cov, by minimizing: $e^T[\text{Cov}^{-1}]e$.

To focus on the synthetic MTRF vector, note that the m and f vectors can be extracted in the following fashion:

$$(mf) = \begin{bmatrix} m_1 & & & 0 \\ & m_2 & & \\ & & \ddots & \\ 0 & & & m_5 \end{bmatrix} \begin{bmatrix} I \\ \overline{I} \\ \overline{\vdots} \\ \overline{I} \end{bmatrix} \quad f = A_m f \tag{A23a}$$

$$(mf) = \begin{bmatrix} [f] & & 0 \\ & [f] & \\ & & \ddots & \\ 0 & & [f] \end{bmatrix} \quad m = A_f m \tag{A23b}$$

where "I" is the identity matrix. Thus, we can extract either f or m as the unknown model parameters, but the resultant A matrices will then depend on m or f, respectively. Keeping f and m together in the (mf) vector, and with the MTRFs in the d vector, the error function is

$$\varepsilon = e^T e = d^T d + (mf)^T(mf) - 2(mf)^T d. \tag{A24}$$

Regardless of whether we use Eq. (A23a) or (A23b) for (mf), we find that

$$(mf)^T(mf) = (m^T m)(f^T f) \tag{A25}$$

that is, the product of the squared lengths of the m and f vectors. Define the D matrix as below:

$$D = [(\dot{M}_1)(\dot{M}_2)(\dot{M}_3)(\dot{M}_4)(\dot{M}_5)]. \tag{A26}$$

If we use Eq. (A23a) to extract the f vector from (mf), then Eq. (A24) becomes

$$\varepsilon = d^T d + (m^T m)(f^T f) - 2f^T Dm \tag{A27}$$

and a similar development based on Eq. (A23b) results in

$$\varepsilon = d^T d + (m^T m)(f^T f) - 2m^T D^T f. \tag{A28}$$

Equations (A27) and (A28) are virtually identical because: $m^T D^T f = f^T Dm$.

The least squares normal equations are obtained by finding the minimum in the error surface with respect to the model parameters. If the error surface is quadratic with respect to the model parameters, then setting the equations of the partial derivatives of ε (with respect to each model parameter) equal to zero results in linear equations for the model parameters. Equation (A28) represents a quadratic surface for m, keeping f fixed, and it is also a quadratic surface for f, keeping m fixed. However, if we form a model vector that combines both f and m, i.e.,

$h^T = (f^T, m^T)$, we encounter the problem that $(m^T m)(f^T f)$ term in the above equation displays a fourth-order dependence on components of the model vector. On the other hand, the last term in the above equations presents no problems. We shall return to analyze this nonlinear character, but first let us write down the separate normal equations for f and m:

$$(m^T m) I_f f = Dm \qquad \text{(A29a)}$$

$$(f^T f) I_m m = D^T f \qquad \text{(A29b)}$$

where I_f is the identity matrix with dimension that corresponds to the number of elements in f, and I_m is the same for m. The above two systems of equations cannot be solved simultaneously as a linear system due to the scalar factors on the LHS. However, if we consider (A29a) and (A29b) as two distinct systems, we have as solutions for f and m:

$$f = (1/m^T m) Dm \qquad \text{(A30a)}$$

$$m = (1/f^T f) D^T f. \qquad \text{(A30b)}$$

Thus, Eq. (A30a) gives the best f, for a given specified m; while Eq. (A30b) gives the best m, for a given specified f. This pair of equations can be used to find f and m by the "bouncing" method of alternating back and forth with the updated f and m vectors. To start this procedure, one must provide some initial estimate for m, or f, that is nonzero. We need to know whether this procedure converges, and if a unique solution exists.

A3.3 Omnilinear formulation

The special structure of this nonlinear inverse problem allows an omnilinear formulation of the problem, with certain restrictions. Using Eqs. (A20) and (A23a), we can write the statistical estimation problem

$$\begin{bmatrix} m_1 & & & 0 \\ & m_2 & & \\ & & \ddots & \\ 0 & & & m_5 \end{bmatrix} \begin{bmatrix} I \\ \overline{I} \\ \overline{\vdots} \\ \overline{I} \end{bmatrix} f = \begin{bmatrix} \dot{M}_1 \\ \overline{\dot{M}_2} \\ \overline{\vdots} \\ \overline{\dot{M}_5} \end{bmatrix}. \qquad \text{(A31)}$$

The purely diagonal matrix of m_i's can be inverted, but ONLY if none of the m_i's are equal to zero. If we proceed with this assumption, we can extract the $(1/m_i)$ components as a vector on the RHS as

$$\begin{bmatrix} I \\ \overline{I} \\ \overline{\vdots} \\ \overline{I} \end{bmatrix} f = \begin{bmatrix} [\dot{M}_1] & & & 0 \\ & [\dot{M}_2] & & \\ & & \ddots & \\ 0 & & & [\dot{M}_5] \end{bmatrix} \begin{bmatrix} 1/m_1 \\ 1/m_2 \\ \vdots \\ 1/m_5 \end{bmatrix}. \qquad \text{(A32)}$$

Thus, we have an omnilinear problem for the estimation of f and $(1/m_1, \ldots, 1/m_5)$, which could be solved for an unconstrained $(1/m_1, \ldots, 1/m_5)$ vector and a f vector with the unit area constraint of: $\sum_i f_i = 1$. However, examination of this solution shows that the moment tensor components $(m_1, \ldots, m_5)$ derived from the $(1/m_1, \ldots, 1/m_5)$ vector have the undesirable property that the m_i are larger for a smaller correlation between f and the individual MTRFs. Thus, the omnilinear formulation of this problem is NOT recommended for two reasons: (1) it presumes that none of the moment tensor components are zero—this condition is artificially restrictive; (2) the resultant solution for m has undesirable statistical properties.

A3.4 Global solution

The basic fact that we reparameterize the MTRFs into the product of two model components forces us to eventually prescribe additional information. The reason for this requirement is that: if we find an f and m that equals M, i.e. $(mf) = M$, we can always multiply all components of m by an arbitrary nonzero number c, and divide all components of f by that same number, and the resultant (mf) is exactly the same. Thus, the reparameterization of the model description from the MTRFs to f and m must be accompanied by some constraint equation on either f or m. Perhaps the most common choice for a constraint equation is to assign unit area to f: i.e. $\Delta t_f \sum_i f_i = 1$, where Δt_f is the sampling interval of the time function. An example of a constraint equation for m is: $m^T m = 1$.

Note that we must also assign units to f and m. Their product must have the units of moment rate, but there is no unique choice for dividing the units between f and m. A common choice is to split the units such that f has units of $(1/\text{time})$, and then m has units of (seismic moment). This choice motivates the above constraint equation of $\Delta t_f \sum_i f_i = 1$. In other applications, m is chosen to be dimensionless, and f carries the units of (moment rate). For the following applications, we let m have units of (moment).

Recalling Eq. (A28), we see that choosing the constraint equation to be either $(m^T m) = 1$ or $(f^T f) = 1$ is a judicious choice because it removes the nonliner coupling between f and m. For example, choose $m^T m = 1$, and then the least squares solution for f, Eq. (A30a), becomes

$$f = Dm. \tag{A33}$$

Equation (A30b) is no longer valid because it was derived for an unconstrained m. But we can substitute the above solution for f into the error function, Eq. (A28), to obtain

$$\xi = d^T d - m^T D^T D m. \tag{A34}$$

We can now minimize ξ with respect to m by analysis of the above scalar equation. To proceed, first recognize that $D^T D$ is a positive definite matrix with dimensions of 5×5, and the ith diagonal component is the squared length of the ith MTRF.

Now notice that $d^T d$ can be written as: $d^T d = |\dot{M}_1|^2 + \cdots + |\dot{M}_5|^2$. Thus, $d^T d$ equals the trace of the $D^T D$ matrix. Now note that the maximum value of the eigenvalues of $D^T D$ is trace($D^T D$) $= d^T d$, and this maximum value is only obtained if all other eigenvalues are zero. With the constraint of $m^T m = 1$, we thus have the upper bound on the second term of Eq. (A34):

$$m^T D^T D m \leq d^T d. \tag{A35}$$

Thus the overall minimum value of ξ is achieved by the maximum value of $(m^T D^T D m)$, which is achieved by choosing m as the eigenvector associated with the largest eigenvalue of $D^T D$. Zero error is only achieved when there is only one linearly-independent column vector of D, or in other words, when all MTRFs share the same shape.

The above analysis provides a unique solution for m and f with the following exceptions: (1) in all situations, all components of f and m can be multiplied by -1; (2) in special pathological circumstances, there may be two or more eigenvalues of $D^T D$ that have the same maximum value. The ambiguity presented by the first exception is easily resolved by choosing the sign of f to make it mostly positive. If the second exception is encountered in applications, then it probably means that the earthquake cannot be represented by a single f and m.

The basic idea of analyzing the D matrix to find a solution for m associated with the largest eigenvalue was advocated by VASCO (1988). We now see that there is a statistical basis for this approach in that it minimizes the least squares error between the synthetic and "observed" MTRFs, when combined with Eq. (A23) for the determination of f, plus the original constraint condition of $m^T m = 1$. Are there other solutions associated with other constraint conditions that further minimize ξ? Another option that can be easily analyzed would be to choose $(f^T f) = 1$. Then, we have $m = D^T f$ and substitution of this solution for m into Eq. (A28) produces

$$\xi = d^T d - f^T D D^T f. \tag{A36}$$

This scalar equation is complementary to Eq. (A34) for m. While the $(D D^T)$ matrix has the dimensions of f, it is still easily seen that Trace($D D^T$) $= d^T d$. Therefore, the minimum of the error function is the same number as above, and is found by choosing f as the eigenvector for the largest eigenvalue of $(D D^T)$. Does this complementary solution produce the same f and m as above, possibly scaled by a constant? We let the readers ponder this question.

Given the above assurances that a global solution can be found, does the "bouncing" method of Eqs. (A30a,b) also find this globally best solution? Let's write our initial guess of the m vector as: $m_{\text{init}} = \alpha \mu_1 + \beta \mu_2 + \ldots$, where μ_1 is the eigenvector associated with the largest eigenvalue of $D^T D$, λ_1, and μ_2 is the eigenvector for the next largest eigenvalue, λ_2, etc. Then Eq. (A30a) gives: $f_{\text{init}} = (1/m_{\text{init}}^T m_{\text{init}}) D m_{\text{init}}$. Substitution of this expression back into the least squares equation for m, Eq. (A30b), provides the "Ith" estimate for m

$$m_I = (1/f_{\text{init}}^T f_{\text{init}})[\lambda_1 \alpha \mu_1 + \lambda_2 \beta \mu_2 + \ldots]. \tag{A37}$$

Thus the first iteration for m changes the direction of the m_{init} vector by emphasizing the contribution of μ_1 since each eigenvector contribution is multiplied by its eigenvalue. It is easy to generalize the "bouncing" method to show that the Nth estimate for m will be

$$m_N = (1/f_{N-1}^T f_{N-1})[\lambda_1^N \alpha \mu_1 + \lambda_2^N \beta \mu_2 + \ldots]. \tag{A38}$$

We see that as N obtains a large value, the first eigenvector will dominate the m vector, though of course we must renormalize the length of the m vector in the end. Thus, as long as there is a distinct maximum eigenvalue of $D^T D$, then the "bouncing" method will converge to the overall globally best solution, unless our initial choice for m is exactly orthogonal to μ_1. If the only nonzero eigenvalue if λ_1, then the "bouncing" method will find the exact answer in one iteration. In practice, if λ_1 is about one hundred times larger than λ_2, then we find the solution in one iteration or so.

A3.5 Extraction analysis with covariance

The error function normalized by the inverse of the MTRF covariance matrix is

$$\xi = e^T[\text{Cov}^{-1}]e. \tag{A39}$$

Similar to above, we can derive two systems of equations by separately seeking the minimum of ξ with respect to f and m

$$f = [m_i m_j \, \text{Sub}_{ij}]^{-1}[I \mid I \mid \cdots \mid I]\hat{\mathcal{M}}\hat{d} \tag{A40a}$$

$$m = [\hat{F}^T \hat{F}]^{-1}\hat{F}^T \hat{d} \tag{A40b}$$

with the definitions,

$$\hat{d} = [\text{Cov}^{-1/2}]d$$

$$\hat{\mathcal{M}} = \begin{bmatrix} m_1 & & 0 \\ & m_2 & \\ & & \ddots \\ 0 & & m_s \end{bmatrix}[\text{Cov}^{-1/2}]$$

$$\tag{A41}$$

$$\hat{F} = [\text{Cov}^{-1/2}]\begin{bmatrix} [f] & & 0 \\ & [f] & \\ & & \ddots \\ 0 & & [f] \end{bmatrix}$$

and where Sub_{ij} is a submatrix of the $[\text{Cov}^{-1}]$, where $[\text{Cov}^{-1}]$ is divided into twenty-five submatrices, the divisions corresponding to the five MTRFs. The pair of equations (A40a,b) allows us to find f and m by the "bouncing" method. Can we also perform a global analysis of the error function, similar to above? If we substitute the least squares solution for f back into Eq. (A39), we find

$$e^T[\text{Cov}^{-1}]e = d^T[\text{Cov}^{-1}]d - m^T\hat{D}^T[m_i m_j \text{ Sub}_{ij}]^{-1}\hat{D}m \qquad (\text{A42})$$

where

$$\hat{D} = (I \,|\, I \,|\, \cdots \,|\, I)[\text{Cov}^{-1}]d. \qquad (\text{A43})$$

To find m from this equation, we must extract an m vector from the middle matrix of the second term:

$$(m_i m_j \text{ Sub}_{ij})^{-1}. \qquad (\text{A44})$$

Equation (A44) indicates that we must sum together the 25 submatrices of $[\text{Cov}^{-1}]$, each submatrix multiplied by $m_i m_j$, and then invert the resultant matrix. If we allow $[\text{Cov}^{-1}]$ to have nonzero off-diagonal values, then we cannot extract m from the above matrix.

If the $[\text{Cov}^{-1}]$ matrix is diagonal, then the problem is tractable, and it becomes quite straightforward if we assign a single value of covariance to each MTRF. Then all submatrices are zero except for $i = j$, where the five diagonal submatrices are: $\text{Sub}_{ii} = (1/sd_i^2)I_f$. We can then write Eq. (A44) as

$$(m_i m_j \text{ Sub}_{ij})^{-1} = (1/(m_1^2/sd_1^2 + m_2^2/sd_2^2 + \cdots + m_5^2/sd_5^2))I_f. \qquad (\text{A45})$$

Equations (A42) then becomes

$$e^T[\text{Cov}^{-1}]e = \hat{d}^T\hat{d} - \hat{n}^T\hat{D}^T\hat{D}\hat{n}$$

$$= \left[\frac{|\dot{M}_1|^2}{sd_1^2} + \cdots + \frac{|\dot{M}_5|^2}{sd_5^2}\right] - \hat{n}^T \begin{bmatrix} \frac{|\dot{M}_1|^2}{sd_1^2} & \cdots & \left(\frac{\dot{M}_i}{sd_i}\right)^T\left(\frac{\dot{M}_j}{sd_j}\right) \\ & \ddots & \\ & & \frac{|\dot{M}_5|^2}{sd_5^2} \end{bmatrix} \hat{n} \qquad (\text{A46a})$$

with the modified m vector,

$$\hat{n}^T = \overbrace{\frac{m_1}{sd_1} \,\Big|\, \frac{m_2}{sd_2} \,\Big|\, \cdots \,\Big|\, \frac{m_5}{sd_5}} \qquad (\text{A46b})$$

and explicit use of the constraint relation,

$$\hat{n}^T\hat{n} = 1. \qquad (\text{A46c})$$

We also have the reduced form of the covariance-weighted least squares solution for f

$$f = \hat{D}\hat{n}. \qquad (\text{A47})$$

The error function in Eq. (A46) has the same structure as the one analyzed in the problem with no covariance estimates. As a consequence, we are again guaranteed of finding the overall best solution with the following prescription:

(1) Normalize the MTRFs by dividing each MTRF by the average standard deviation for that MTRF.

(2) Assemble the MTRFs into the $\hat{D}$ matrix.

(3) Find the eigenvalues and eigenvectors of the $\hat{D}^T\hat{D}$ matrix.

(4) The best $\hat{n}$ is the eigenvector for the largest eigenvalue, and m is found by multiplying $\hat{n}$ by the standard deviations for each MTRF, i.e., $m_i = sd_i\hat{n}_i$.

(5) The best f is found from the relation: $f = \hat{D}\hat{n}$.

(6) Change the scaling for f and m such that $\Delta t_f \sum_i f_i = 1$, and then the m_i's will be in units of seismic moment.

As before, we also know that the "bouncing" method will arrive at the same globally best solution. RUFF and TICHELAAR (1990) used the above simplification of the covariance matrix and the "bouncing" method to find the best f and m for the 1989 Loma Prieta earthquake. The advantage of using the simplified covariance matrix is that we know a global solution exists and that we will find that solution. On the other hand, we would like to use the full covariance matrix as there might be significant trade-offs between components of the MTRFs. At this point, we can only use the full covariance matrix with the "bouncing" method. In the applications in the main text, we will show results for use of the full covariance matrix.

A3.6 Formal analysis of the adequacy of a single f and m with the full covariance matrix

The above analysis shows that the best f and m exist and can be found under certain conditions. We now return to the central scientific question as to whether the earthquake is adequately represented by the best f and m. For the case where we consider the covariance matrix to be diagonal, we can use the same arguments and graphical display as in Figure 2 of the main text. If we allow a covariance matrix with significant off-diagonal values, then it is possible that the synthetic MTRFs will fall outside the error bounds based on the diagonal elements, but in fact the synthetic MTRFs are within the higher-dimension error ellipsoid of the full covariance matrix. While a full graphical picture of the situation is not practical, it is possible to reduce this scientific test to a single number. Recall that the error function is given by

$$\xi = e^T[\text{Cov}^{-1}]e. \tag{A48}$$

We see that the above scalar function is nondimensional. We can attach significance to the numerical value of ξ by considering a diagonal covariance matrix, then let each component of e equal the standard deviation of that component

$$\xi = \sum_{i=1,N} (sd_i^2/sd_i^2) = N \tag{A49}$$

where N is the total number of components of the e vector. We could choose $\xi \leq N$ as our condition for an acceptable error vector, hence acceptance of a single f and m to represent the earthquake. Of course, an error vector that has all zeros except for one value that was a factor of N greater than its standard deviation would produce the same number for ξ; and this would also be considered as acceptable with the above criterion. To find the most suitable choice for the numerical value of ξ, write the components of the error vector as the ratio of the error to the standard deviation

$$e_i = r_i \, sd_i. \tag{A50}$$

Since the covariance matrix can be diagonalized, we can use a diagonal matrix for $[\text{Cov}^{-1}]$ in this discussion without loss of generality. Then for the case of a diagonal covariance matrix, but with different diagonal values, we have

$$\xi = e^T[\text{Cov}^{-1}]e = \sum_{i=1,N} r_i^2. \tag{A51}$$

If the e_i are in fact normally distributed, then the expected values of r_i should follow the normal distribution with zero mean and unit variance. If we assign an occurrence frequency to values of r_i based on the normal distribution, we find that for a sufficient number of components, the summation in Eq. (A51) equals N. This expectation is well-satisfied for $N = 100$ or more; clearly it may not be satisfied for small values of N less than 10. A typical size for N in our applications is about 50 to 100. Thus, we can characterize the overall consistency of a single time function and moment tensor with the MTRFs and their covariance by calculating the normalized scalar error, ζ (see Fig. A1 for simplified view):

$$\zeta = \xi/N = e^T[\text{Cov}^{-1}]e/N. \tag{A52}$$

If $\zeta \leq 1$, then the synthetic MTRF vector falls within the contour of the multi-dimensional error ellipse expected from the covariance matrix and normal distribution of errors. In applications, we will show the graphical representation of the

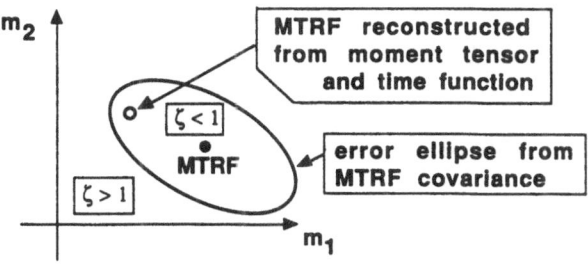

Figure A1
Pictorial representation of the MTRF parameter, ζ. Basic concept is illustrated for MTRFs with only two components, m_1 and m_2. Solid dot shows the MTRF solution, error ellipse is centered on MTRF. After finding the best moment tensor and time function, the reconstructed MTRF is shown as open dot. If this reconstructed MTRF is within the error ellipse, $\zeta < 1$; if it is outside, then $\zeta > 1$.

error bounds from the diagonal elements of the covariance matrix, i.e., Figure 2 in the main text, in addition to quoting a value of ζ, which is referred to as the "MTRF parameter" in the plots (see Fig. 3 in text).

A3.7 Error estimates of f_{best} and m_{best} and overall match to data

We are still not finished in our statistical analysis of the MTRF inverse problem. Let us suppose now we have f_{best} and m_{best}, and we find them to be an adequate representation of the earthquake from the above error analysis. We should provide error estimates for f_{best}, and especially for m_{best} since we must perform further analysis of the moment tensor to find M_0 and the faulting geometry of the major double couple. We use linear error estimation by choosing Eq. (A40b) for the basic least squares solution of m given f_{best}. Thus, the covariance matrix for m_{best} is then

$$[\text{Cov } m_{best}] = [\hat{F}^T \hat{F}]^{-1} \hat{F}^T [\text{Cov}^{-1}] \hat{F} [\hat{F}^T \hat{F}]^{-1}. \tag{A53}$$

Of course, we can simply quote individual standard deviations of each m_i by taking the square root of the diagonal elements of the above covariance matrix. This is easily accomplished, and previous investigators do give error estimates of the m_i, e.g., the Harvard CMT solutions. At this point it is not clear how to use these error estimates since the eigenvalue analysis of the moment tensor displays a nonlinear dependence on the original components of the moment tensor, and further nonlinear relationships are introduced when converting the t and p eigenvectors into fault strike, dip, and slip rake angles. Fortunately, new computer-intensive methods in statistics are now being used that circumvent these analytical difficulties (see TICHELAAR and RUFF, 1991). Resampling methods such as bootstrapping and jackknifing can find error estimates on final model parameters even if the connection is nonlinear and the error distribution is unknown. Our problem is a nonlinear mapping from the five components of m_i to five other parameters: (1) M_0; (2) % second double; (3) fault strike; (4) fault dip; and (5) slip vector rake angle. We use a somewhat primitive form of resampling where m_{test} is systematically varied about m_{best} by the standard deviations; the resultant determinations of the above five variables are then analyzed to determine the mean and standard deviation of each variable. It is more complicated to display the covariance between the five derived parameters. The best way to show covariance between the focal mechanism parameters is to simply plot all the focal mechanisms. Another simple approach is to plot four of the parameters versus the one parameter that shows the most variation. Since the resampling scheme is not exhaustive, we consider these error estimates of M_0 and faulting geometry to be lower bounds on the true errors. Indeed, for the Mexico earthquakes, errors in the focal mechanism geometry due to depth uncertainty are greater than the above formal errors.

The MTRF inversion method also tests the match between synthetic and observed seismograms at every step of the reduction from MTRFs to a single time function and double couple. For any estimate of the MTRFs in m, the discretized synthetic seismograms are: $d_{syn} = Am$. The match between d_{syn} and d is measured by the correlation coefficient

$$cc = (1/|d_{syn}| \, |d|)d_{syn}^T d \qquad (A54)$$

where cc is 1 for a perfect match, and 0 if d_{syn} is orthogonal to d. The correlation coefficient measures the match to data without regard to any slight scaling mismatches that might arise as the MTRFs are reduced to f and a focal mechanism. With the standard usage of MTRF inversion, there are three values of cc that are calculated:

(1) when m consists of the full MTRFs from the initial linear inversion;

(2) when m is composed of a single f and moment tensor extracted from the MTRFs; and

(3) when m is composed of a single f and the major double couple from the moment tensor.

Our experience with MTRF inversion shows that cc for above cases #2 and #3 is much more sensitive to variations in the key parameters that are used to construct the Green's functions. The one key seismological parameter is earthquake depth. For most earthquake studies, the best depth is determined by repeating the MTRF inversion at several trial depths, we then choose the depth that produces the best overall match to the seismograms, as measured by cc. The peak in cc *versus* depth is typically better defined by cc #2 and #3. It seems that omnilinear inversion for the MTRFs is quite good at matching seismograms, even if the MTRFs are "garbage" that vary widely. These "garbage" MTRFs present a "wild story" for the earthquake rupture, and hence the match to the data declines rapidly as we reduce the MTRFs to the simplest description of the earthquake with a single f and m. Given the sensitivity of body wave Green's functions to depth for shallow earthquakes, proper use of MTRF inversion should always include analysis for the best depth.

There is one further option of the MTRF inversion method. This option uses f_{best}, but then determines the moment tensor by reinverting the seismograms for just the five components of m. In other words, the seismograms are inverted twice: (1) full MTRF inversion, followed by extraction of f_{best}; and then (2) f_{best} is convolved with the Green's functions, and Eq. (7) in the main text is used to invert for the five independent components of the moment tensor. The reason for this complicated dual-inversion is to avoid the use of damping when inverting for the moment tensor. The A matrix for the full MTRF inversion is poorly conditioned and the damped least squares inversion must be used, i.e., Eq. (A7). One problem with this damped inversion is that it preferentially eliminates the poorly resolved components of the model. While this consequence of damping on time functions is well-understood and acceptable, it potentially has the undesirable effect of reducing the contribution of some moment tensor elements. This effect could bias the resultant focal mechanism. Thus, the complex dual-inversion performs the second inversion for the moment tensor with zero damping. Use of this special option is reserved for final tests of the robustness of the MTRF results. After several tests, we found no scientifically significant results from applying the reinversion scheme to large Mexico earthquakes, but it might be useful for other earthquakes.

The technical details of the MTRF inversion methodology have now been covered in sufficient detail such that any investigator should be able to reproduce the technique and results shown here. In this paper, we have applied this methodology to large Mexican earthquakes. We anticipate many other exciting applications in future studies.

References

AKI, K. (1979), *Characterization of Barriers on an Earthquake Fault*, J. Geophys. Res. *84*, 6140–6148.

AKI, K., and RICHARDS, P. G., *Quantitative Seismology* (Freeman, San Francisco 1980) 932 pp.

ASTIZ, L., KANAMORI, H., and EISSLER, H. (1987), *Source Characteristics of the Earthquakes in the Michoacan Seismic Gap in Mexico*, Bull. Seismol. Soc. Am. *77*, 1326–1346.

BARKER, J., and LANGSTON, C. (1981), *Inversion of Teleseismic Body Waves for the Moment Tensor of the 1978 Thessaloniki, Greece Earthquake*, Bull. Seismol. Soc. Am. *71*, 1423–1444.

BECK, S. L., and CHRISTENSEN, D. H. (1991), *Rupture Process of the February 4, 1965, Rat Islands Earthquake*, J. Geophys. Res. *96*, 2205–2221.

BECK, S., and RUFF, L. (1985), *The Rupture Process of the 1976 Mindanao Earthquake*, J. Geophys. Res. *90*, 6773–6782.

BECK, S., and RUFF, L. (1987), *Rupture Process of the Great 1963 Kurile Islands Earthquake Sequence: Asperity Interaction and Multiple Event Rupture*, J. Geophys. Res. *92*, 14123–14138.

BECK, S., and RUFF, L. (1989), *Great Earthquakes and Subduction along the Peru Trench*, Phys. Earth Planet. Int. *57*, 199–224.

BYRNE, D. E., DAVIES, D. M., and SYKES, L. R. (1988), *Loci and Maximum Size of Thrust Earthquakes and the Mechanics of the Shallow Region of Subduction Zones*, Tectonics *7*, 833–857.

CHAEL, E. P., and STEWART, G. S. (1982), *Recent Large Earthquakes along the Middle America Trench and their Implications for the Subduction Process*, J. Geophys. Res. *87*, 329–338.

DAS, S., and AKI, K. (1977), *Fault Planes with Barriers: A Versatile Earthquake Model*, J. Geophys. Res. *82*, 5658–5670.

DEMETS, C., GORDON, R. G., ARGUS, D. F., and STEIN, S. (1990), *Current Plate Motions*, Geophys. J. Int. *101*, 425–478.

DEWEY, J. W., and SPENCE, W. (1979), *Seismic Gaps and Source Zones of Recent Large Earthquakes in Coastal Peru*, Pure and Appl. Geophys. *117*, 1148–1171.

DMOWSKA, R., and LI, V. C. (1982), *A Mechanical Model of Precursory Source Processes for Some Large Earthquakes*, Geophys. Res. Lett. *9*, 393–396.

DZIEWONSKI, A. M., and WOODHOUSE, J. H. (1983), *An Experiment in Systematic Study of Global Seismicity: Centroid Moment Tensor Solutions for 201 Moderate and Large Earthquakes of 1981*, J. Geophys. Res. *88*, 3247–3271.

EISSLER, H., ASTIZ, L., and KANAMORI, H. (1986), *Tectonic Setting and Source Parameters of the September 19, 1985 Michoacan, Mexico Earthquake*, Geophys. Res. Lett. *13*, 569–572.

EKSTRÖM, G. (1989), *A Very Broad Band Inversion Method for the Recovery of Earthquake Source Parameters*, Tectonophysics *166*, 73–100.

EKSTRÖM, G., and DZIEWONSKI, A. M. (1986), *A Very Broad Band Analysis of the Michoacan, Mexico Earthquake of September 13, 1985*, Geophys. Res. Lett. *13*, 605–608.

GETTRUST, J. F., HSU, V., HELSLEY, C. E., HERRERO, E., and JORDAN, T. (1981), *Pattern of Seismicity Preceding the Petatlan Earthquake of 14 March 1979*, Bull. Seismol. Soc. Am. *71*, 761–770.

GUTENBERG, B., and RICHTER, C. F. (1956), *Magnitude and Energy of Earthquakes*, Ann. Geofis. Rome 9, 1–15.

HASKELL, N. A. (1964), *Total Energy and Energy Spectral Density of Elastic Wave Radiation from Propagating Faults*, Bull. Seismol. Soc. Am. 54, 1811–1842.

HASKOV, J., SINGH, S. K., NAVA, E., DOMINGUEZ, T., and RODRIGUEZ, M. (1983), *Playa Azul, Michoacan, Mexico Earthquake of 25 October 1981 (M_s = 7.3)*, Bull. Seismol. Soc. Am. 73, 449–457.

HOTELLING, H. (1936), *Relation between Two Sets of Variates*, Biometrika 28, 321–377.

HOUSTON, H., and KANAMORI, H. (1986), *Source Characteristics of the 1985 Michoacan, Mexico Earthquake at Periods of 1 to 30 Seconds*, Geophys. Res. Lett. 13, 597–600.

HSU, V., HELSLEY, C. E., BERG, E., and NOVELO-CASANOVA, D. A. (1985), *Correlation of Foreshocks and Aftershocks and Asperities*, Pure and Appl. Geophys. 122, 878–893.

JARRARD, R. D. (1986), *Relation among Subduction Parameters*, Rev. Geophysics 24, 217–284.

KANAMORI, H. (1977), *The Energy Release in Great Earthquakes*, J. Geophys. Res. 82, 2981–2987.

KANAMORI, H., *The nature of seismicity patterns before large earthquakes*. In *Earthquake Prediction — An International Review* (D. Simpson and P. Richards, eds.) (AGU, Washington D.C. 1981) pp. 1–19.

KANAMORI, H., MORI, J., HAUKSSON, E., HEATON, T., HUTTON, L., and JONES, L. (1993), *Determination of Earthquake Energy Release and M_L Using TERRAscope*, Bull. Seismol. Soc. Am. 83, 330–346.

KIKUCHI, M., *Strength and stickiness of earthquake source*. In *Proceedings of Internat. Sympos. Earthquake Source Physics and Earthquake Precursors* (University of Tokyo, 1990) pp. 163–166.

KIKUCHI, M., and FUKAO, Y. (1988), *Seismic Wave Energy Inferred from Long-period Body Wave Inversion*, Bull. Seismol. Soc. Am. 78, 1707–1724.

KOSTROV, B. V., and DAS, S., *Principles of Earthquake Source Mechanics* (Cambridge University Press, Cambridge 1988) 286 pp.

LAY, T., KANAMORI, H., and RUFF, L. (1982), *The Asperity Model and the Nature of Large Subduction Zone Earthquakes*, Earthquake Pred. Res. 1, 3–71.

MADARIAGA, R. (1977), *High-frequency Radiation from Crack (Stress Drop) Models of Earthquake Faulting*, Geophys. J. Roy. Astron. Soc. 51, 625–651.

MCCANN, W. R., and HABERMANN, R. E. (1989), *Morphologic and Geologic Effects of the Subduction of Bathymetric Highs*, Pure and Appl. Geophys. 129, 41–69.

MCNALLY, K. C., and MINSTER, J. B. (1981), *Non-uniform Seismicity Rates along the Middle America Trench*, J. Geophys. Res. 86, 4949–4959.

MENDEZ, A. J., and ANDERSON, J. G. (1991), *The Temporal and Spatial Evolution of the 19 September 1985 Michoacan Earthquake as Inferred from Near-source Ground Motion Records*, Bull. Seismol. Soc. Am. 81, 844–861.

MENDOZA, C. (1993), *Coseismic Slip of Two Large Mexican Earthquakes from Teleseismic Body Waveforms: Implications for Asperity Distribution in the Michoacan Plate-boundary Segment*, J. Geophys. Res. submitted.

MENDOZA, C., and HARTZELL, S. H. (1988), *Aftershock Patterns and Mainshock Faulting*, Bull. Seismol. Soc. Am. 78, 1438–1449.

MENDOZA, C., and HARTZELL, S. H. (1989), *Slip Distribution of the 19 September 1985 Michoacan, Mexico Earthquake: Near-source and Teleseismic Constraints*, Bull. Seismol. Soc. Am. 79, 655–669.

NABELEK, J., *Determination of Earthquake Source Parameters from Inversion of Body Waves*, Ph.D. Thesis (Mass. Inst. of Technol. Cambridge, 1984) 346 pp.

NISHENKO, S. P. (1991), *Circum-Pacific Seismic Potential: 1989–1999*, Pure and Appl. Geophys. 135, 169–259.

NUR, A., and BEN-AVRAHAM, Z. (1983), *Volcanic Gaps due to Oblique Consumption of Aseismic Ridges*, Tectonophysics 99, 355–362.

OHNAKA, M. (1992), *Earthquake Source Nucleation: A Physical Model for Short-term Precursors*, Tectonophysics 211, 149–178.

PRIESTLEY, K. F., and MASTERS, T. G. (1986), *Source Mechanism of the September 19, 1985 Michoacan Earthquake and its Implications*, Geophys. Res. Lett. 13, 601–604.

REYES, A., BRUNE, J. N., and LOMNITZ, C. (1979), *Source Mechanism and Aftershock Study of the Colima, Mexico Earthquake of January 20, 1973*, Bull. Seismol. Soc. Am. 69, 1819–1840.

RIEDESEL, M. A., JORDAN, T. H., SHEEHAN, A. F., and SILVER, P. G. (1986), *Moment-tensor Spectra of the 19 Sept. 85 and 21 Sept. 1985 Michoacan, Mexico Earthquakes*, Geophys. Res. Lett. *13*, 609–612.

RIKITAKE, T. (1976), *Recurrence of Great Earthquakes at Subduction Zones*, Tectonophysics *35*, 335–362.

RUFF, L., *Fault asperities inferred from seismic body waves*. In *Earthquakes: Observation, Theory, and Interpretation* (Kanamori, H. and Boschi, E., eds.) (North-Holland, Amsterdam 1983) pp. 251–276.

RUFF, L. (1984), *Tomographic Imaging of the Earthquake Rupture Process*, Geophys. Res. Lett. *11*, 629–632.

RUFF, L., *Tomographic imaging of seismic sources*. In *Seismic Tomography* (Nolet, G., ed.) (Reidel, Dordrecht, Holland 1987) 386 pp.

RUFF, L. (1989a), *Do trench sediments affect great earthquake occurrence in subduction zones?* In *Subduction Zones, Part II* (Ruff, L. and Kanamori, H., eds.), Pure and Appl. Geophys. *129*, 263–282.

RUFF, L. J. (1989b), *Multi-trace Deconvolution with Unknown Trace Scale Factors: Omnilinear Inversion of P and S Waves for Source Time Functions*, Geophys. Res. Lett. *16*, 1043–1046.

RUFF, L. J. (1992a), *Asperity Distributions and Large Earthquake Occurrence in Subduction Zones*, Tectonophysics *211*, 61–83.

RUFF, L. J., *Seismic energy release of great subduction earthquakes: Can irregular rupture models explain the "missing" energy?* Extended Abstract in *Wadati Conference on Great Subduction Earthquakes* (Christensen, D., Wyss, M., Habermann, R., and Davies, J., eds.) (University of Alaska, Fairbanks 1992b) pp. 12–14.

RUFF, L., and KANAMORI, H. (1983a), *Seismic Coupling and Uncoupling at Subduction Zones*, Tectonophysics *99*, 99–117.

RUFF, L., and KANAMORI, H. (1983b), *The Rupture Process and Asperity Distribution of Three Great Earthquakes from Long-period Diffracted P Waves*, Phys. Earth Planet. Int. *31*, 202–230.

RUFF, L., and TICHELAAR, B. (1990), *Moment Tensor Rate Functions for the 1989 Loma Prieta Earthquake*, Geophys. Res. Lett. *17*, 1187–1190.

SCHELL, M., and RUFF, L. (1989), *Rupture of a Seismic Gap in Southeastern Alaska: The 1972 Sitka Earthquake (M_s 7.6)*, Phys. Earth Planet. Int. *54*, 241–257.

SCHOLZ, C. H., *The Mechanics of Earthquakes and Faulting* (Cambridge University Press, New York 1990) 439 pp.

SCHWARTZ, S., and RUFF, L. (1987), *Asperity Distribution and Earthquake Occurrence in the Southern Kurile Islands Arc*, Phys. Earth Planet. Int. *49*, 54–77.

SINGH, S. K., and LERMO, J. (1985), *Mislocation of Mexico Earthquakes as Reported in International Bulletins*, Geof. Int. *24* (2), 333–351.

SINGH, S. K., and MORTERA, F. (1991), *Source Time Functions of Large Mexican Subduction Earthquakes, Morphology of the Benioff Zone, Age of the Plate, and their Tectonic Implications*, J. Geophys. Res. *96*, 21,487–21,502.

SINGH, S. K., and ORDAZ, M. (1993), *Seismic Energy Release in Mexican Subduction Zone Earthquakes*, preprint.

SINGH, S. K., RODRIGUEZ, M., and ESPINDOLA, J. M. (1984), *A Catalog of Shallow Earthquakes of Mexico from 1900–1981*, Bull. Seismol. Soc. Am. *74*, 267–279.

SIPKIN, S. (1986a), *Estimation of Earthquake Source Parameters by the Inversion of Waveform Data: Global Seismicity 1981–1983*, Bull. Seismol. Soc. Am. *76*, 1515–1541.

SIPKIN, S. (1986b), *Interpretation of Non-double-couple Earthquake Mechanisms Derived from Moment Tensor Inversion*, J. Geophys. Res. *91*, 531–549.

STOLTE, C., MCNALLY, K. C., GONZALEZ-RUIZ, J., SIMILA, G. W., REYES, A., REBOLLAR, C., MUNGUIA, L., and MENDOZA, L. (1986), *Fine Structure of a Postfailure Wadati-Benioff Zone*, Geophys. Res. Lett. *13*, 577–580.

STUMP, B. W., and JOHNSON, L. R. (1977), *The Determination of Source Properties by the Linear Inversion of Seismograms*, Bull. Seismol. Soc. Am. *67*, 1489–1502.

TAJIMA, F., RUFF, L., KANAMORI, H., ZHANG, Z., and MOGI, K. (1990), *Earthquake Source Processes and Subduction Regime in the Santa Cruz Islands Region*, Phys. Earth Planet. Int. *61*, 269–290.

TICHELAAR, B. W., CHRISTENSEN, D. H, and RUFF, L. J. (1992), *Depth Extent of Rupture of the 1981 Chilean Outer-rise Earthquake as Inferred from Long-period Body Waves*, Bull. Seismol. Soc. Am. 82, 1236–1252.

TICHELAAR, B. W., and RUFF, L. J. (1991), *Seismic Coupling along the Chilean Subduction Zone*, J. Geophys. Res. 96, 11,997–12,022.

TICHELAAR, B. W., and RUFF, L. J. (1993), *Depth of Seismic Coupling along Subduction Zones*, J. Geophys. Res. 98, 2017–2037.

UNAM SEISMOLOGY GROUP (1986), *The September 1985 Michoacan Earthquakes: Aftershock Distribution and History of Rupture*, Geophys. Res. Lett. 13, 573–576.

UYEDA, S., and KANAMORI, H. (1979), *Back-arc Opening and the Mode of Subduction*, J. Geophys. Res. 84, 1049–1061.

VALDES, C., MEYER, R. P., ZUNIGA, R., HASKOV, J., and SINGH, S. K. (1982), *Analysis of the Petatlan Aftershocks: Numbers, Energy Release, and Asperities*, J. Geophys. Res. 87, 8519–8527.

VASCO, D. W. (1989), *Deriving Source-time Functions Using Principal Component Analysis*, Bull. Seismol. Soc. Am. 79, 711–730.

YOMOGIDA, K. (1988), *Crack-like Rupture Processes Observed in Near-fault Strong Motion Data*, Geophys. Res. Lett. 15, 1223–1226.

(Received April 27, 1993, revised/accepted January 11, 1994)

PAGEOPH, Vol. 142, No. 1 (1994)

0033–4553/94/010173–52$1.50 + 0.20/0

Global Variability in Subduction Thrust Zone–Forearc Systems

Robert McCaffrey[1]

Abstract —Deviations of slip vector azimuths of interplate thrust earthquakes from expected plate convergence directions at oblique subduction zones provide kinematic information about the deformation of forearcs and indirect evidence on the dynamics of the plate boundary. A global survey of slip vectors at major trenches of the world reveals a large variability in the kinematic response of forearcs to shear produced by oblique convergence. The variability in forearc deformation inferred from slip vector deflections is suggested to be caused by variations in forearc rheology rather than in the stresses acting on subduction zone thrust faults. Estimated apparent macroscopic rheologies range from elastic to perfectly plastic (or viscous). Forearc rheologies inferred from slip vectors do not correlate with age of the subducting lithosphere, but continental forearcs or old arcs appear to deform less than oceanic or young arcs. The inferred absence of forearc deformation at continental arcs from this study is counter to inferences drawn from compiled geologic information on forearc faults. Correlations of the apparent forearc rheology with backarc spreading, convergence rate, slab dip, arc curvature, and downdip length of the thrust contact are poor. However, great subduction zone earthquakes occur where forearcs are apparently more elastic (i.e., less deformed by oblique convergence), which suggests that the mechanical properties of forearcs rather than stress magnitude on thrust faults control both the kinematic behavior of forearcs and where great subduction zone earthquakes occur.

Key words: Forearcs, subduction, rheology of lithosphere, earthquakes.

Introduction

Deflections of slip vectors of subduction zone thrust earthquakes away from the expected plate convergence vectors at oblique subduction zones are likely caused by rapid anelastic strain rates in the forearc of the upper plate (FITCH, 1972). The partitioning of the plate convergence vector into a thrust component more perpendicular to the strike of the subduction zone and a component of shear parallel to the trench provides both kinematic information about the deformation of the forearc and dynamic information about the ratio of the force acting on the upper plate to that on the thrust interface (BECK, 1991; McCAFFREY, 1992). The goal of this paper is to examine the global variability in this partitioning and to understand whether it is stress on the thrust interface or forearc rheology or both that control

[1] Department of Earth and Environmental Sciences, Rensselaer Polytechnic Institute, Troy NY 12180, U.S.A.

it. At the same time, estimates of the macroscopic rheologies of forearcs are made. Such estimates are important constraints in attempts to understand or model the long-term evolution of forearcs (e.g., ENGLAND et al., 1985; GEIST and SCHOLL, 1992; WDOWINSKI et al., 1989).

Where oblique convergence occurs, a component of relative plate motion parallel to the margin (Figure 1) produces a shear force parallel to the plate boundary. If the response of the upper plate to this force is elastic, so that there is no permanent deformation in the forearc, then slip vectors of interplate thrust earthquakes, that show relative motion between the subducting plate and the forearc, will be parallel to the relative plate convergence vector. Where earthquake

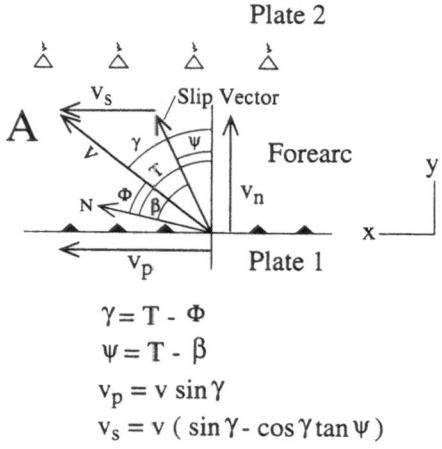

$$\gamma = T - \Phi$$
$$\psi = T - \beta$$
$$v_p = v \sin\gamma$$
$$v_s = v (\sin\gamma - \cos\gamma \tan\psi)$$

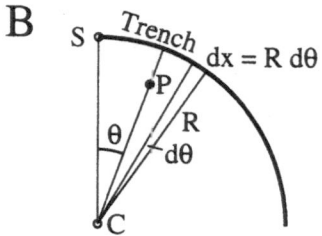

Figure 1(a, b)

(a) Geometrical relationship of the angles of plate convergence obliquity γ and slip vector obliquity ψ to the plate vector V at azimuth Φ, the trench normal azimuth (angle T), and the slip vector azimuth β. The plate vector V is partitioned into a thrust component (the slip vector) and an arc-parallel component of the forearc relative to the upper plate, V_s. N is the north arrow. (b) Description of the forearc binning procedure. The bin width dx in kilometers at the trench is specified and the angle $d\theta$ is determined; R is the radius of curvature of the arc, C is the center of curvature of the trench, and S is the start of the trench. For any point P (such as an earthquake epicenter), the angle θ between the lines CP and CS is found and the bin number is the integer part of $(\theta/d\theta + 1)$.

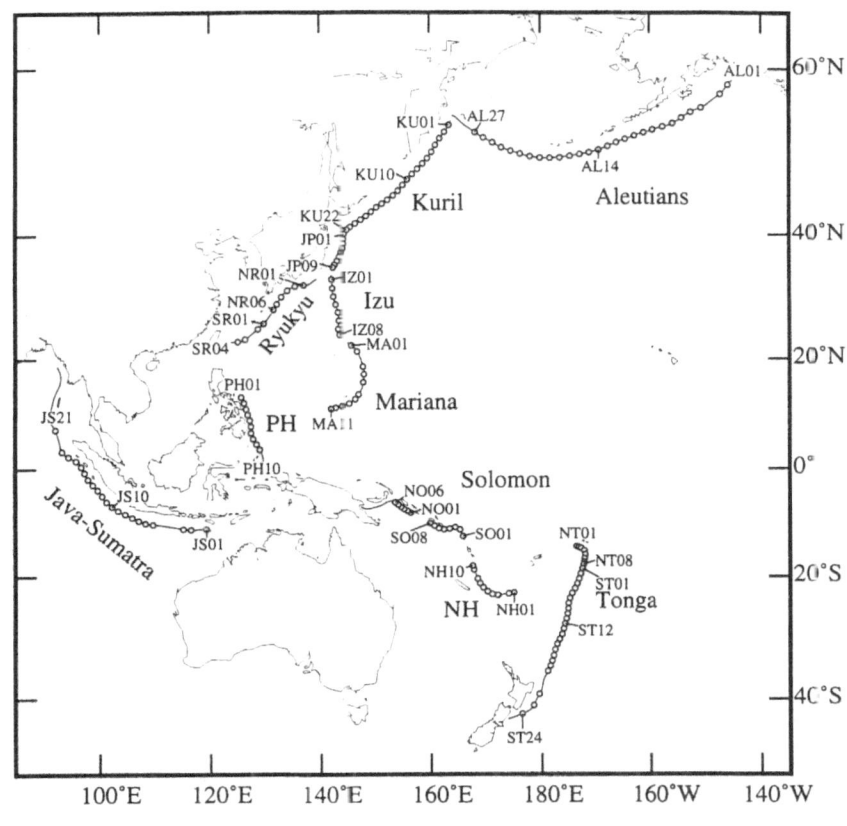

Figure 1(c)

(c) Map of western Pacific subduction zones included in this study. Centers of bins are shown with circles and are labeled to correspond to Tables 1 to 3. PH = Philippine trench; NH = New Hebrides trench.

slip vectors are systematically deflected away from the plate convergence vector and toward the trench-normal, the magnitudes of these deflections constrain the strain rates in the forearc. If the arc-parallel shear stress acting on the overriding plate increases in some manner with increasing convergence obliquity, the kinematic response of the upper plate can then be used to learn about the rheology of the forearc.

Since the kinematic response of the forearc depends not only on its rheology but also on the magnitude of and the variation in the stress acting on it, which are unknown, this approach reveals only the 'apparent' rheology; i.e., the rheology at some unknown level of stress. For this reason, I consider this estimated 'apparent' rheology to be a property of the coupled thrust zone–forearc system. Nevertheless, I argue that global variations in the true forearc rheologies and not stress levels on the thrust interface are largely responsible for the differences in the observed kinematic behaviors of forearcs.

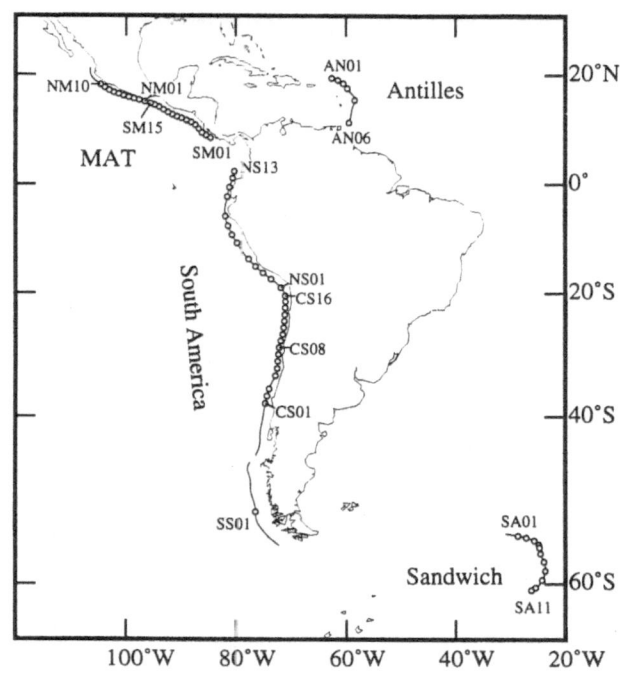

Figure 1(d)

(d) Map of eastern Pacific subduction zones included in this study. Centers of bins are shown with circles and are labeled to correspond to Tables 1 to 3. MAT = Middle America Trench.

List of Symbols and Abbreviations

γ	plate convergence obliquity
ψ	slip vector obliquity
β	slip vector azimuth (SVA)
T	trench-normal azimuth
Φ	azimuth of motion of subducting plate relative to upper plate (PVA)
$\beta - \Phi$	slip vector residual (SVR)
V	plate convergence rate
V_s	arc-parallel rate of motion of forearc relative to upper plate
V_p	arc-parallel rate of motion of subducting plate
V_n	arc-normal rate of motion of subducting plate
V_r	potential arc-parallel rate for motion of forearc
R_f	ratio of the shear forces resisting motion on the arc-parallel strike-slip and thrust faults
γ_{max}	maximum obliquity at trench
α	ratio of V_s to V_r
γ_c	critical obliquity above which forearc deforms plastically

Data

The goal is to compare the observed slip vectors from thrust earthquakes at subduction zones to the expected convergence directions predicted by rigid plate rotation poles and see how the difference varies with convergence obliquity (convergence obliquity is the angle between the direction normal to the trench and the plate convergence azimuth; Figure 1a). The three basic observations are slip vector azimuths from earthquakes, trench outlines and their normals, and plate convergence vectors. These and other subduction parameters (rakes and dip angles) are collected into bins (segments of trenches) in order to place statistical uncertainties on them. The binning procedure is shown in Figure 1b, binned data are listed in Tables 1 and 2, and bin locations are shown in Figures 1c and 1d.

Trench Outlines and their Normals

The traces of the trenches in map view were estimated by finding the deepest point in the seafloor using the ETOPO-5 database which includes 5 minute (approximately 8 km) averages of bathymetry (NATIONAL GEOPHYSICAL DATA CENTER, 1988). These points were smoothed by calculating values every 0.1° (approximately 11 km) with a least-squares spline using a correlation distance of 2° (≈ 220 km) to remove short wavelength signals that may be fictitious (see SMITH, 1993). The trench-normal is the azimuth perpendicular to the trench trend, points toward the hanging wall side, and is calculated with respect to this smoothed, discretized curve. The definition of the positive x direction for distance along the trench is from the seismological definition of strike of an earthquake fault plane; while facing the strike ($+x$) direction the fault dips to the observer's right (Figure 1). Trench-normal azimuths for each bin are shown in Table 1.

Slip Vectors

Slip vectors are taken from interplate earthquakes that are due to thrusting of oceanic lithosphere beneath the forearcs of the island arcs. These slip vectors strictly show the relative motion between the plates only along the thrust fault. The motion of the forearc as a whole may be different due to vertical and horizontal gradients in deformation. I average the slip vectors across the forearc so that arc-perpendicular horizontal gradients in deformation rates are averaged, but the vertical gradients are still unknown. Hence, what is inferred to be the response of the entire forearc may only be the response of a small part of it directly above the thrust plane, although forearc extension rates near the top of the Aleutian forearc estimated from geologic evidence (GEIST et al., 1988) are similar to strain rates estimated from slip vector deflections (EKSTRÖM and ENGDAHL, 1989; MCCAFFREY, 1992). Earthquakes from which the slip vectors are derived probably occur

Table 1

List of binned slip vector data by trench

Bin	Lat.	Lon.	Dist.	T	SD	PVA	SD	N	Rake	SD	Dip	SD	SVA	SD
						Aleutian								
					Pole: NAM-PAC (NUVEL-1) (48.7N -78.2E 0.78°/Ma) Start: (58.9N 214.8E)									
					Center: (66.0N 184.0E) Bin width: 125 km N:240 SVR:7.8									
AL01	58.43	214.04	68	-50.5	9.0	-16.5	1.5	5	51		11		-1.8	25.0
AL02	57.48	212.63	202	-44.3	6.7	-18.5	1.4	2					-13.5	20.0
AL03	56.04	209.29	464	-26.2	7.9	-22.1	1.4	3	98	29	8	1	-34.7	8.0
AL04	55.58	207.45	592	-28.8	5.4	-23.6	1.3	12	95	33	10	4	-33.9	7.0
AL05	54.94	205.82	718	-36.5	0.6	-25.2	1.3	1					-24.0	20.0
AL06	54.31	204.20	845	-27.8	5.8	-26.6	1.3	6	95	10	14	7	-29.0	6.3
AL07	53.93	202.39	971	-14.0	1.7	-28.0	1.2	4	108	22	20	5	-40.5	15.1
AL08	53.61	200.57	1097	-19.1	1.0	-29.2	1.2	6	90	21	24	6	-17.5	17.9
AL09	53.27	198.85	1218	-17.1	0.8	-30.4	1.2	6	100	12	21	9	-44.5	15.2
AL10	52.91	197.14	1340	-21.1	1.2	-31.6	1.2	4	72		19		-23.0	5.4
AL11	52.51	195.44	1462	-20.0	0.9	-32.8	1.1	6	111	26	30	4	-38.7	23.2
AL12	52.12	193.83	1580	-24.1	1.4	-33.9	1.1	7	84	22	26	6	-27.1	7.1
AL13	51.68	192.24	1700	-23.0	1.3	-35.0	1.1	19	85	22	25	6	-32.5	11.4
AL14	51.26	190.55	1828	-20.6	0.8	-36.1	1.1	16	92	30	25	6	-36.1	18.9
AL15	50.91	188.83	1955	-13.5	2.9	-37.2	1.1	8	98	29	27	4	-26.1	24.0
AL16	50.71	187.09	2080	-9.4	0.9	-38.2	1.1	9	81	30	20	4	-24.0	6.5
AL17	50.50	185.31	2210	-12.4	0.9	-39.2	1.0	35	98	22	22	5	-31.3	9.1
AL18	50.30	183.52	2341	-4.1	4.7	-40.3	1.0	19	106	14	25	6	-33.4	7.0
AL19	50.23	181.72	2472	-0.5	1.5	-41.2	1.0	9	114	14	25	7	-29.8	10.0
AL20	50.26	179.92	2603	4.2	3.4	-42.2	1.0	15	108	21	24	8	-24.8	8.7
AL21	50.42	178.13	2733	11.2	4.1	-43.0	1.0	10	109	22	16	5	-22.6	7.9
AL22	50.75	176.36	2864	18.3	1.3	-43.9	1.0	12	110	22	22	7	-14.8	15.4
AL23	51.10	174.60	2995	18.2	1.6	-44.7	1.0	12	115	21	18	5	-24.2	14.6
AL24	51.52	172.90	3123	25.6	3.2	-45.5	0.9	4	133	9	23	4	-25.5	3.9
AL25	52.10	171.30	3251	32.6	1.1	-46.2	0.9	5	148	11	28	5	-23.0	6.6
AL26	52.68	169.70	3377	30.1	2.2	-46.9	0.9	3	162	8	21	7	-32.7	2.1
AL27	53.30	168.18	3501	37.3	0.9	-47.5	0.9	1					-45.0	20.0
						Antilles								
					Pole: CAR-SAM (NUVEL-1) (50.0N -65.3E 0.19°/Ma) Start: (19.6N 296.7E)									
					Center: (14.0N 295.0E) Bin width: 125 km N:10 SVR:-20.6									
AN01	19.43	297.32	69	195.1	1.1	265.9	6.0	2	19	4	18	11	250.0	20.0
AN02	19.07	298.47	195	202.4	3.7	264.1	6.1	2	23	7	22	11	239.5	20.0
AN03	18.47	299.49	323	223.9	9.1	262.7	6.2	2	60	61	26	8	250.0	20.0
AN04	17.59	300.21	448	233.4	3.4	262.0	6.2	2	112	21	18	11	245.0	20.0
AN05	15.48	301.55	732	271.9	16.2	260.9	6.0	1	108		37		233.0	20.0
AN06	11.39	300.52	1200	290.0	0.4	263.0	5.1	1	124		27		257.0	20.0
						Izu								
					Pole: PAC-PHL (Seno et al., 1987) (-0.5N 133.3E 1.02°/Ma) Start: (34.2N 142.1E)									
					Center: (8.0N 84.0E) Bin width: 150 km N:39 SVR:-20.4									
IZ01	33.42	142.17	83	-94.0	2.5	-74.4	0.7	7	112	26	25	7	-91.1	12.7
IZ02	32.03	142.32	238	-94.9	0.7	-73.6	0.7	2	90	11	26	4	-87.0	20.0
IZ03	30.69	142.50	387	-100.5	3.0	-72.6	0.8	15	81	21	26	5	-92.1	6.0
IZ04	29.40	142.89	535	-107.6	0.8	-71.3	0.9	4	80	20	26	7	-89.3	6.9

Table 1 (*Contd*)

Bin	Lat.	Lon.	Dist.	T	SD	PVA	SD	N	Rake	SD	Dip	SD	SVA	SD
IZ05	28.07	143.33	689	-102.7	3.4	-69.7	0.9	4	100	16	20	4	-90.8	4.6
IZ06	26.69	143.47	843	-87.1	4.5	-68.5	0.9	1	71		25		-87.0	20.0
IZ07	25.30	143.35	997	-91.2	8.5	-67.8	1.0	1	129		28		-128.0	20.0
IZ08	24.27	143.62	1118	-120.1	6.2	-66.5	1.2	5	74	9	22	5	-103.0	10.2

Mariana
Pole: PAC-PHL (Seno et al., 1987) (-0.5N 133.3E 1.02°/Ma) Start: (23.0N 145.0E)
Center: (17.0N 141.0E) Bin width: 150 km N:34 SVR:-10.5

Bin	Lat.	Lon.	Dist.	T	SD	PVA	SD	N	Rake	SD	Dip	SD	SVA	SD
MA01	22.46	145.68	1310	-140.1	2.0	-60.6	1.7	1	86		29		-124.0	20.0
MA02	21.42	146.75	1469	-123.7	7.6	-57.3	1.8	1	41		15		-97.0	20.0
MA03	18.76	147.79	1787	-104.7	3.7	-51.9	2.1	7	85	30	23	10	-90.1	21.4
MA04	17.39	147.94	1941	-87.0	4.2	-49.5	2.2	3	111	11	27	6	-94.7	4.2
MA05	15.99	147.81	2095	-83.5	2.4	-47.6	2.4	1	110		37		-90.0	20.0
MA06	13.79	146.96	2359	-58.1	0.4	-45.3	2.7	1	58		37		-59.0	20.0
MA07	12.94	146.35	2475	-49.0	7.2	-45.0	2.9	6	110	26	27	6	-63.3	14.2
MA08	12.15	145.27	2623	-26.3	4.6	-45.8	3.1	3	96	32	23	5	-55.3	14.0
MA09	11.67	144.01	2770	-17.2	1.4	-48.0	3.2	6	89	47	24	8	-32.2	26.9
MA10	11.35	142.81	2907	-14.2	0.7	-50.6	3.3	2	78	13	25	4	-9.5	20.0
MA11	11.15	141.98	2999	-12.7	0.7	-52.7	3.2	3	58	14	24	11	7.7	12.7

Japan
Pole: NAM-PAC (NUVEL-1) (48.7N -78.2E 0.78°/Ma) Start: (40.6N 144.7E)
Center: (40.0N 133.0E) Bin width: 75 km N:119 SVR:-0.3

Bin	Lat.	Lon.	Dist.	T	SD	PVA	SD	N	Rake	SD	Dip	SD	SVA	SD
JP01	40.19	144.44	2177	287.5	8.8	296.7	0.8	19	78	25	18	7	293.7	8.5
JP02	39.51	144.32	2254	270.0	1.8	296.6	0.8	14	73	23	19	7	293.7	7.9
JP03	38.87	144.31	2325	273.9	3.4	296.6	0.8	7	70	20	23	6	295.0	4.4
JP04	38.18	144.17	2402	284.3	2.4	296.5	0.8	20	53	20	22	6	304.4	15.1
JP05	37.54	143.91	2478	290.8	1.1	296.3	0.8	5	96	42	28	9	284.2	23.0
JP06	36.92	143.58	2553	295.2	1.7	296.1	0.8	9	88	19	24	7	293.4	13.2
JP07	36.29	143.17	2631	300.7	1.2	295.9	0.8	13	88	11	23	5	293.2	3.9
JP08	35.70	142.71	2709	301.0	1.8	295.6	0.8	26	97	18	23	6	293.8	7.9
JP09	35.28	142.44	2762	292.8	2.9	295.5	0.8	6	69	42	22	8	293.0	10.2

Java-Sumatra
Pole: AUS-SEA (0.0N 30.0E 0.60°/Ma) Start: (-11.6N 120.1E)
Center: (12.0N 116.0E) Bin width: 150 km N:80 SVR:23.5

Bin	Lat.	Lon.	Dist.	T	SD	PVA	SD	N	Rake	SD	Dip	SD	SVA	SD
JS01	-11.29	119.29	93	10.7	13.1	0.1	4.1	7	82	19	17	6	-0.1	10.6
JS02	-11.41	116.52	400	2.8	2.5	0.7	4.1	1	32		25		13.0	20.0
JS03	-11.27	115.13	553	8.8	2.0	1.0	4.1	1	163		8		-3.0	20.0
JS04	-10.46	109.70	1155	3.9	2.3	1.9	4.1	1	106		35		24.0	20.0
JS05	-10.24	108.32	1308	17.0	4.7	2.1	4.2	3	97	24	27	8	7.3	7.4
JS06	-9.72	107.08	1457	26.6	1.5	2.2	4.2	4	94	8	17	8	5.8	5.1
JS07	-9.11	105.88	1605	26.9	1.0	2.3	4.2	2	128	9	20	8	14.5	20.0
JS08	-8.44	104.65	1759	29.4	2.8	2.3	4.2	6	107	33	26	6	3.8	15.6
JS09	-7.78	103.41	1913	29.1	1.8	2.3	4.3	2	111	26	28	1	25.5	20.0
JS10	-7.04	102.29	2062	39.1	3.2	2.2	4.3	11	114	36	19	8	23.9	10.3
JS11	-6.09	101.33	2210	49.7	2.9	2.0	4.3	14	105	25	20	10	23.2	13.9
JS12	-4.97	100.50	2365	55.1	0.5	1.7	4.4	5	100	24	12	8	33.0	9.0

Table 1 (*Contd*)

Bin	Lat.	Lon.	Dist.	T	SD	PVA	SD	N	Rake	SD	Dip	SD	SVA	SD
JS13	-3.88	99.70	2513	51.0	2.7	1.4	4.4	3	89	34	24	7	29.7	14.6
JS14	-2.92	98.83	2657	45.9	1.5	1.1	4.4	3	104	15	11	8	26.0	12.1
JS15	-1.88	97.98	2807	57.1	3.8	0.8	4.5	4	83	24	13	5	54.0	4.5
JS16	-0.69	97.34	2956	64.0	0.6	0.3	4.5	4	90	34	14	10	39.8	3.9
JS17	0.47	96.76	3099	60.9	2.9	-0.2	4.5	2	60	42	16	16	56.5	20.0
JS18	1.52	95.87	3254	36.0	9.3	-0.7	4.5	3	92	44	16	11	38.3	10.4
JS19	2.29	94.53	3426	29.9	2.4	-1.1	4.6	1	90		25		34.0	20.0
JS20	3.25	93.36	3598	56.7	15.4	-1.6	4.6	2	90	1	22	8	32.5	20.0
JS21	7.29	92.18	4066	64.6	5.6	-3.8	4.6	1	38		35		90.0	20.0

Kuriles
Pole: NAM-PAC (NUVEL-1) (48.7N -78.2E 0.78°/Ma) Start: (54.6N 163.9E)
Center: (59.0N 122.0E) Bin width: 100 km N:233 SVR:2.9

Bin	Lat.	Lon.	Dist.	T	SD	PVA	SD	N	Rake	SD	Dip	SD	SVA	SD
KU01	54.11	163.51	58	297.0	4.3	310.0	0.9	5	70	13	27	5	309.6	6.8
KU02	53.34	162.72	158	304.5	0.9	309.3	0.9	5	87	10	33	2	305.0	2.2
KU03	52.59	161.86	260	303.6	1.3	308.5	0.9	14	92	17	34	3	304.1	10.2
KU04	51.81	161.08	362	298.9	1.3	307.9	0.9	12	85	13	26	4	307.8	5.2
KU05	51.01	160.40	463	299.5	1.6	307.3	0.9	4	89	19	30	6	306.0	7.7
KU06	50.24	159.62	565	306.7	2.4	306.7	0.9	5	96	12	29	8	299.8	17.7
KU07	49.57	158.75	663	312.7	1.2	306.0	0.9	3					309.0	5.2
KU08	48.92	157.80	764	314.7	0.3	305.3	0.9	8	85	9	30	3	307.8	4.2
KU09	48.27	156.85	864	313.4	0.8	304.7	0.8	15	88	16	29	6	305.7	8.1
KU10	47.61	155.97	963	309.8	1.2	304.1	0.8	8	84	6	30	6	303.8	4.8
KU11	46.89	155.14	1065	307.1	0.5	303.5	0.8	8	94	13	31	4	296.8	8.1
KU12	46.16	154.31	1167	311.0	2.4	302.9	0.8	18	91	22	31	4	303.2	9.4
KU13	45.53	153.41	1266	318.7	2.1	302.3	0.8	3	78	41	27	12	327.0	42.7
KU14	44.97	152.41	1367	323.8	1.1	301.7	0.8	23	101	26	20	7	309.2	12.0
KU15	44.45	151.39	1467	324.9	0.3	301.0	0.8	17	110	16	26	10	299.5	7.2
KU16	43.94	150.41	1564	323.4	1.0	300.4	0.8	26	92	15	26	4	305.5	6.8
KU17	43.37	149.41	1666	320.7	0.6	299.8	0.8	20	88	14	21	6	309.1	5.9
KU18	42.78	148.42	1770	322.2	1.0	299.2	0.8	16	110	30	24	7	304.6	8.3
KU19	42.26	147.46	1869	325.2	0.7	298.6	0.8	13	104	16	22	6	303.5	6.5
KU20	41.76	146.47	1967	325.7	0.5	298.0	0.8	4	124	16	34	1	302.5	12.4
KU21	41.23	145.51	2067	321.2	2.5	297.4	0.8	1	146		22		301.0	20.0
KU22	40.84	144.93	2132	312.6	2.0	297.1	0.8	5	120	18	17	7	294.8	6.2

S. Mid America
Pole: COC-CAR (NUVEL-1) (24.1N -119.4E 1.37°/Ma) Start: (8.3N 275.9E)
Center: (-19.0N 252.0E) Bin width: 100 km N:101 SVR:5.6

Bin	Lat.	Lon.	Dist.	T	SD	PVA	SD	N	Rake	SD	Dip	SD	SVA	SD
SM01	8.59	275.42	60	27.6	0.3	29.2	1.8	9	84	34	25	10	23.9	17.0
SM02	9.03	274.59	165	30.2	1.7	29.1	1.8	3	71	31	16	5	32.3	11.9
SM03	9.54	273.85	264	41.8	5.6	28.8	1.9	3	90	13	13	6	31.3	2.3
SM04	10.27	273.24	368	55.0	2.0	28.2	1.9	12	90	27	18	5	31.2	5.4
SM05	10.98	272.62	473	39.8	5.6	27.6	2.0	9	86	28	23	12	33.0	8.9
SM06	11.46	271.87	572	27.3	2.2	27.3	2.0	8	57	26	23	12	37.1	8.2
SM07	11.84	271.00	676	22.5	0.8	27.2	2.1	2	26		10		40.5	20.0
SM08	12.19	270.11	779	21.1	0.2	27.2	2.1	3	76	35	24	13	46.3	16.0
SM09	12.51	269.27	877	21.7	0.4	27.2	2.2	6	56	23	26	10	36.2	12.7
SM10	12.85	268.44	975	24.4	1.2	27.1	2.2	8	78	13	19	10	30.8	3.7
SM11	13.25	267.63	1073	29.1	1.7	26.9	2.3	14	72	16	25	5	32.6	10.2

Table 1 (*Contd*)

Bin	Lat.	Lon.	Dist.	T	SD	PVA	SD	N	Rake	SD	Dip	SD	SVA	SD
SM12	13.74	266.82	1176	32.5	0.5	26.6	2.4	11	69	25	22	9	29.2	9.2
SM13	14.22	266.00	1280	28.3	2.0	26.2	2.5	3					23.3	7.6
SM14	14.59	265.18	1377	22.4	1.5	26.0	2.6	7	75	16	21	2	30.9	15.7
SM15	14.87	264.43	1463	19.2	0.9	26.0	2.6	3					21.3	6.7

N. Mid America
Pole: COC-NAM (NUVEL-1) (27.9N -120.7E 1.42°/Ma) Start: (15.0N 264.0E)
Center: (25.0N 262.0E) Bin width: 100 km N:59 SVR:-4.7

Bin	Lat.	Lon.	Dist.	T	SD	PVA	SD	N	Rake	SD	Dip	SD	SVA	SD
NM01	15.19	263.38	1555	16.6	0.4	33.7	1.3	8	46	10	17	7	35.0	5.8
NM02	15.47	262.37	1668	15.5	0.5	34.0	1.4	6	96		36		26.3	15.0
NM03	15.73	261.40	1776	15.5	0.0	34.5	1.4	9	45	17	21	9	30.8	10.3
NM04	15.98	260.49	1878	15.9	0.5	34.9	1.5	6	64	29	23	15	27.2	17.4
NM05	16.25	259.57	1980	17.5	0.6	35.4	1.5	3	68		18		38.3	9.8
NM06	16.54	258.67	2083	19.8	0.9	35.8	1.6	10	87	27	27	8	21.6	7.5
NM07	16.86	257.83	2179	23.7	1.4	36.1	1.7	4	91	17	18	4	23.0	6.8
NM08	17.24	257.02	2276	28.8	1.7	36.3	1.7	8	84	15	20	4	26.6	9.8
NM09	17.70	256.24	2373	34.0	1.4	36.3	1.8	3	87		18		28.3	4.7
NM10	18.23	255.46	2476	36.5	0.8	36.0	1.9	2	78		29		31.0	20.0

New Hebrides
Pole: PAC-AUS (NUVEL-1) (-60.1N -178.3E 1.12°/Ma) Start: (-22.7N 175.5E)
Center: (-18.0N 173.0E) Bin width: 100 km N:81 SVR:-32.8

Bin	Lat.	Lon.	Dist.	T	SD	PVA	SD	N	Rake	SD	Dip	SD	SVA	SD
NH01	-22.77	174.98	57	-10.8	1.4	84.5	0.7	2	9	1	30	6	45.5	20.0
NH02	-22.92	173.99	160	-8.6	0.6	83.7	0.7	3	59	59	29	8	20.0	31.2
NH03	-23.21	172.01	366	-3.0	5.8	82.1	0.7	3	88	24	28	6	-10.0	1.7
NH04	-23.07	171.04	469	21.4	7.2	81.3	0.7	10	111	22	24	8	8.7	8.0
NH05	-22.58	170.17	573	39.0	3.8	80.8	0.7	14	108	14	29	7	29.4	13.3
NH06	-21.90	169.44	680	50.3	3.2	80.3	0.7	8	85	14	25	10	43.9	26.1
NH07	-21.15	168.87	783	59.3	2.3	80.1	0.7	8	88	36	32	5	74.9	29.3
NH08	-20.36	168.44	881	65.4	1.5	79.9	0.7	6	94	7	31	6	69.3	4.0
NH09	-18.79	167.79	1068	71.0	0.4	79.8	0.6	6	99	10	29	6	69.3	6.7
NH10	-18.08	167.55	1151	72.0	0.1	79.7	0.6	21	115	19	22	6	68.7	16.7

Philippine
Pole: EUR-PHL (Seno et al., 1987) (47.3N 154.4E 1.06°/Ma) Start: (13.9N 125.4E)
Center: (-19.0N 5.0E) Bin width: 125 km N:74 SVR:-25.4

Bin	Lat.	Lon.	Dist.	T	SD	PVA	SD	N	Rake	SD	Dip	SD	SVA	SD
PH01	13.28	125.61	72	249.9	7.0	299.5	2.5	9	72	26	17	5	256.2	7.6
PH02	12.20	126.12	203	245.9	3.3	298.5	2.4	4	72	24	28	4	273.0	14.2
PH03	11.11	126.49	330	256.2	1.7	297.7	2.3	2	72	10	18	1	284.0	20.0
PH04	10.05	126.78	452	251.6	2.7	297.0	2.3	13	78	28	27	8	269.2	12.2
PH05	9.02	127.16	573	251.8	3.6	296.2	2.2	11	76	21	29	8	274.5	11.2
PH06	7.90	127.35	699	268.9	4.7	295.6	2.1	8	86	39	27	8	277.1	25.7
PH07	6.70	127.37	831	262.9	7.4	295.1	2.1	3	102	32	28	7	276.0	12.5
PH08	5.58	127.75	964	241.4	4.0	294.4	2.0	15	66	26	29	7	277.3	15.4
PH09	4.60	128.36	1090	236.9	0.4	293.6	2.0	8	64	34	22	11	264.5	19.0
PH10	3.62	128.95	1217	241.7	2.5	292.8	1.9	1	66		29		262.0	20.0

Table 1 (*Contd*)

Bin	Lat.	Lon.	Dist.	T	SD	PVA	SD	N	Rake	SD	Dip	SD	SVA	SD

N. Ryukyu
Pole: EUR-PHL (Seno et al., 1987) (47.3N 154.4E 1.06°/Ma) Start: (33.4N 139.5E)
Center: (26.0N 139.0E) Bin width: 150 km N:17 SVR:3.8

Bin	Lat.	Lon.	Dist.	T	SD	PVA	SD	N	Rake	SD	Dip	SD	SVA	SD
NR01	32.48	137.06	256	-2.4	8.2	-53.2	5.0	2					-55.5	20.0
NR02	32.32	135.50	407	-20.6	8.5	-51.5	4.8	1					-50.0	20.0
NR03	31.63	134.12	559	-38.2	5.1	-51.0	4.6	1					-54.0	20.0
NR04	30.63	133.02	713	-54.2	3.6	-51.3	4.4	4	87		7		-45.3	5.5
NR05	29.45	132.17	868	-60.1	0.6	-52.1	4.1	4	61	20	22	4	-50.0	8.9
NR06	28.49	131.55	991	-59.9	0.5	-52.7	4.0	5	56	19	20	4	-37.2	8.5

S. Ryukyu
Pole: EUR-PHL (Seno et al., 1987) (47.3N 154.4E 1.06°/Ma) Start: (27.9N 131.2E)
Center: (32.0N 121.0E) Bin width: 150 km N:7 SVR:7.2

Bin	Lat.	Lon.	Dist.	T	SD	PVA	SD	N	Rake	SD	Dip	SD	SVA	SD
SR01	26.18	129.74	1259	-47.4	2.9	-53.8	3.6	2	74	16	24	3	-37.5	20.0
SR02	25.29	128.67	1406	-39.9	1.0	-53.8	3.5	2	76	35	19	13	-44.5	20.0
SR03	23.53	126.37	1711	-31.2	7.2	-53.7	3.2	2	108	6	20	1	-52.0	20.0
SR04	23.11	125.01	1861	-7.7	4.8	-53.1	3.2	1	158		12		-46.0	20.0

S. South America
Pole: ANT-SAM (NUVEL-1) (86.4N -40.7E 0.27°/Ma) Start: (-56.2N 287.7E)
Center: (-46.0N 303.0E) Bin width: 200 km N:1 SVR:-10.2

Bin	Lat.	Lon.	Dist.	T	SD	PVA	SD	N	Rake	SD	Dip	SD	SVA	SD
SS01	-52.40	283.44	519	83.3	4.0	93.2	2.0	1	104		8		83.0	20.0

Central S. America
Pole: NAZ-SAM (NUVEL-1) (56.0N -94.0E 0.76°/Ma) Start: (-45.4N 283.5E)
Center: (0.0N 215.0E) Bin width: 125 km N:107 SVR:5.9

Bin	Lat.	Lon.	Dist.	T	SD	PVA	SD	N	Rake	SD	Dip	SD	SVA	SD
CS01	-38.10	285.19	2051	100.9	1.1	79.3	0.9	9	86	29	19	4	87.9	14.5
CS02	-37.00	285.50	2177	104.3	1.1	79.2	0.9	2	88	11	16	1	87.0	20.0
CS03	-35.92	285.89	2302	109.5	2.5	79.0	0.9	1	131		16		72.0	20.0
CS04	-33.86	287.01	2552	111.6	4.1	78.4	0.9	24	107	23	21	7	85.5	7.4
CS05	-32.72	287.35	2684	96.9	3.5	78.2	0.9	24	104	19	25	7	85.4	9.8
CS06	-31.57	287.45	2810	93.6	0.6	78.2	1.0	5	116	18	21	10	72.2	17.5
CS07	-30.48	287.55	2932	95.0	0.3	78.1	1.0	6	93	4	34	2	82.0	5.3
CS08	-29.34	287.69	3060	99.4	3.1	78.0	1.0	2	75		37		103.0	20.0
CS09	-28.23	288.00	3186	106.8	1.1	77.9	1.0	5	112	2	31	3	82.8	6.3
CS10	-27.13	288.34	3312	102.2	3.1	77.7	1.0	5	130	38	14	6	89.0	30.6
CS11	-26.00	288.53	3439	95.8	1.1	77.5	1.0	6	109	16	21	7	81.5	9.5
CS12	-24.90	288.63	3560	94.6	0.7	77.5	1.1	9	90	9	18	4	82.1	9.3
CS13	-23.76	288.73	3687	94.0	1.0	77.4	1.1	3	119	37	32	1	82.7	8.6
CS14	-22.61	288.81	3814	93.6	0.2	77.3	1.1	3	101	6	24	11	71.7	10.7
CS15	-21.51	288.87	3935	90.8	2.3	77.2	1.1	1					86.0	20.0
CS16	-20.52	288.80	4045	79.7	4.5	77.2	1.1	2	122	11	20	13	53.0	20.0

Table 1 (*Contd*)

Bin	Lat.	Lon.	Dist.	T	SD	PVA	SD	N	Rake	SD	Dip	SD	SVA	SD

N. South America
Pole: NAZ-SAM (NUVEL-1) (56.0N -94.0E 0.76°/Ma) Start: (-20.0N 288.7E)
Center: (-4.0N 293 0E) Bin width: 200 km N:40 SVR:-0.8

Bin	Lat.	Lon.	Dist.	T	SD	PVA	SD	N	Rake	SD	Dip	SD	SVA	SD
NS01	-19.07	288.03	4224	52.5	8.4	77.6	1.1	4	79	10	16	8	56.3	18.2
NS02	-17.50	286.29	4478	38.3	0.8	78.4	1.1	2	42	19	32	6	72.5	20.0
NS03	-16.39	284.85	4674	40.0	1.6	79.2	1.0	2	41		21	4	81.5	20.0
NS04	-15.18	283.51	4870	46.1	1.8	79.8	1.0	7	61	16	28	8	81.9	6.1
NS05	-13.86	282.30	5067	50.2	1.3	80.4	1.0	2	74	11	26	4	73.0	20.0
NS06	-10.88	280.28	5462	58.7	1.2	81.4	0.9	1					60.0	20.0
NS07	-9.33	279.37	5660	63.1	3.1	81.9	0.9	2	82	13	16		81.0	20.0
NS08	-7.69	278.65	5858	68.4	0.8	82.2	0.9	2	82		20		76.0	20.0
NS09	-5.92	278.13	6062	81.6	6.2	82.4	0.9	5	97	14	18	3	76.8	6.1
NS10	-2.33	278.52	6464	100.1	2.7	81.9	1.0	1	81		17		78.0	20.0
NS11	-0.57	278.89	6663	108.2	5.0	81.5	1.0	3	105	38	20	6	85.3	8.6
NS12	1.13	279.47	6861	104.3	4.6	80.9	1.1	5	122	8	22	4	86.0	4.4
NS13	2.39	279.77	7004	109.3	4.8	80.6	1.1	4	112	28	21	1	99.8	19.9

Sandwich
Pole: SAN-SAM(Pelayo) (-36.6N -29.6E 1.88°/Ma) Start: (-54.9N 329.1E)
Center: (-58.0N 330.0E) Bin width: 100 km N:53 SVR:1.1

Bin	Lat.	Lon.	Dist.	T	SD	PVA	SD	N	Rake	SD	Dip	SD	SVA	SD
SA01	-55.10	331.39	152	192.2	1.4	267.5	1.1	1	37		22		230.0	20.0
SA02	-55.30	332.88	251	195.8	2.1	263.8	1.1	5	70	17	29	7	238.6	12.2
SA03	-55.62	334.34	351	210.5	9.2	260.4	0.9	1	104		26		238.0	20.0
SA04	-56.05	335.17	423	246.2	8.9	258.6	0.1	4	73	20	28	6	257.0	7.8
SA05	-56.36	335.30	460	263.8	1.9	258.5	0.0	5	63	20	28	4	260.6	14.0
SA06	-56.98	335.49	529	254.6	6.8	258.3	0.1	5	96	18	24	5	251.4	8.5
SA07	-57.82	336.10	631	250.0	5.5	257.5	0.3	14	74	12	28	9	261.6	11.4
SA08	-58.74	336.34	736	276.2	7.3	257.5	0.2	8	78	12	25	5	273.1	15.7
SA09	-59.65	335.77	843	298.0	7.4	259.0	0.7	1	133		29		269.0	20.0
SA10	-60.36	334.59	948	321.3	5.5	261.7	0.9	8	141	18	26	6	279.8	33.4
SA11	-60.62	333.83	999	328.7	0.0	263.2	0.0	1	136		36		302.0	20.0

S. Solomon
Pole: PAC-AUS (NUVEL-1) (-60.1N -178.3E 1.12°/Ma) Start: (-13.5N 166.1E)
Center: (-39.0N 154 0E) Bin width: 100 km N:58 SVR:-17.1

Bin	Lat.	Lon.	Dist.	T	SD	PVA	SD	N	Rake	SD	Dip	SD	SVA	SD
SO01	-12.70	165.85	1762	72.1	2.2	79.6	0.6	17	96	18	32	7	75.1	8.5
SO02	-11.47	165.31	1911	54.6	11.2	79.4	0.6	10	88	31	17	5	67.2	17.6
SO03	-11.01	164.43	2031	-6.0	16.1	79.0	0.6	2	39	38	24	14	61.0	20.0
SO04	-11.31	163.44	2146	-16.0	5.0	78.3	0.6	1	8		22		53.0	20.0
SO05	-11.44	162.45	2256	2.3	5.7	77.7	0.6	8	68	50	23	10	44.6	37.0
SO06	-11.24	161.54	2360	23.4	6.3	77.2	0.6	7	71	25	21	5	57.7	20.1
SO07	-10.73	160.68	2469	34.9	1.0	76.8	0.6	9	84	34	19	11	55.1	19.0
SO08	-10.23	159.88	2573	25.9	5.7	76.5	0.6	4	68	45	26	4	43.8	20.4

Table 1 (*Contd*)

Bin	Lat.	Lon.	Dist.	T	SD	PVA	SD	N	Rake	SD	Dip	SD	SVA	SD

N. Solomon
Pole: PAC-AUS (NUVEL-1) (-60.1N -178.3E 1.12°/Ma) Start: (-8.5N 156.7E)
Center: (7.7N 165.8E) Bin width: 75 km N:29 SVR:-30.3

Bin	Lat.	Lon.	Dist.	T	SD	PVA	SD	N	Rake	SD	Dip	SD	SVA	SD
NO01	-8.24	156.38	3099	36.2	2.9	75.0	0.6	3	105	14	30	1	47.0	7.9
NO02	-7.86	155.78	3177	29.3	1.1	74.7	0.6	5	79	12	28	4	39.8	5.2
NO03	-7.52	155.18	3253	32.2	2.5	74.5	0.6	1	41		14		54.0	20.0
NO04	-7.12	154.63	3328	39.7	1.8	74.3	0.6	12	88	14	33	6	45.3	4.3
NO05	-6.68	154.13	3402	42.1	0.6	74.1	0.6	5	95	21	31	6	46.2	6.1
NO06	-6.23	153.58	3481	33.8	5.3	73.9	0.6	3	107	3	36	1	38.3	4.0

N. Tonga
Pole: PAC-AUS (NUVEL-1) (-60.1N -178.3E 1.12°/Ma) Start: (-14.5N 186.0E)
Center: (-17.0N 185.0E) Bin width: 75 km N:59 SVR:-2.3

Bin	Lat.	Lon.	Dist.	T	SD	PVA	SD	N	Rake	SD	Dip	SD	SVA	SD
NT01	-14.65	186.34	43	198.7	1.3	273.2	0.6	4	38	58	30	6	265.8	40.0
NT02	-14.80	186.77	92	201.5	0.6	273.5	0.6	5	81	72	17	10	246.6	44.7
NT03	-14.98	187.18	140	207.8	3.8	273.8	0.6	12	57	39	24	10	259.9	33.9
NT04	-15.43	187.76	221	234.2	14.4	274.3	0.6	6	78	17	19	5	273.0	10.9
NT05	-16.17	187.93	309	275.3	6.3	274.5	0.6	12	97	20	18	4	279.8	13.8
NT06	-16.86	187.82	386	279.1	1.1	274.4	0.6	11	108	27	18	7	274.6	25.2
NT07	-17.51	187.72	458	278.0	0.7	274.4	0.6	8	110	24	24	6	289.6	28.8
NT08	-17.90	187.66	502	280.3	0.0	274.4	0.8	1	140		28		271.0	20.0

S. Tonga
Pole: PAC-AUS (NUVEL-1) (-60.1N -178.3E 1.12°/Ma) Start: (-18.1N 187.6E)
Center: (3.0N 123.0E) Bin width: 100 km N:306 SVR:13.1

Bin	Lat.	Lon.	Dist.	T	SD	PVA	SD	N	Rake	SD	Dip	SD	SVA	SD
ST01	-18.66	187.44	562	288.6	2.1	274.3	0.6	5	113	22	26	6	277.0	13.6
ST02	-19.54	187.11	667	288.9	1.1	274.1	0.6	8	85	20	28	6	287.8	8.3
ST03	-20.40	186.82	766	287.6	0.7	274.0	0.7	15	96	26	22	6	290.7	8.2
ST04	-21.24	186.50	865	292.6	2.2	273.8	0.7	28	89	21	25	5	288.9	11.8
ST05	-22.08	186.06	969	298.3	1.0	273.5	0.7	17	83	17	22	4	296.8	8.6
ST06	-22.90	185.57	1072	298.5	1.4	273.2	0.7	11	98	22	28	7	292.4	11.8
ST07	-23.74	185.14	1175	290.5	5.4	272.9	0.7	25	97	22	24	5	284.9	9.3
ST08	-24.65	184.91	1280	276.4	2.6	272.8	0.7	14	91	20	23	6	279.9	5.4
ST09	-25.55	184.84	1379	273.6	0.9	272.8	0.7	3	99	15	22	6	268.3	3.2
ST10	-26.44	184.73	1478	279.7	2.2	272.7	0.7	2	80	26	33	8	286.0	20.0
ST11	-27.32	184.52	1577	283.4	0.7	272.6	0.8	20	93	24	22	6	279.5	9.9
ST12	-28.18	184.28	1676	283.6	0.3	272.4	0.8	20	101	20	22	6	282.4	8.0
ST13	-29.05	184.04	1775	284.8	0.8	272.3	0.8	12	94	17	28	6	282.5	6.9
ST14	-29.94	183.72	1879	291.2	2.7	272.0	0.8	26	98	16	26	6	280.5	8.4
ST15	-30.78	183.27	1981	296.9	0.7	271.6	0.8	35	100	19	28	6	281.1	14.0
ST16	-31.56	182.82	2078	294.5	2.0	271.2	0.9	12	99	15	23	7	282.4	7.0
ST17	-32.43	182.45	2181	285.8	2.6	270.8	0.9	15	95	17	19	8	288.8	5.7
ST18	-33.34	182.18	2285	284.2	3.1	270.5	0.9	11	101	38	24	10	281.5	11.5
ST19	-34.19	181.90	2383	286.5	1.0	270.2	0.9	8	114	19	28	8	276.5	11.4

Table 1 (*Contd*)

Bin	Lat.	Lon.	Dist.	T	SD	PVA	SD	N	Rake	SD	Dip	SD	SVA	SD
ST20	-35.02	181.57	2481	290.6	1.9	269.9	1.0	8	119	20	25	11	280.8	10.3
ST21	-35.85	181.11	2582	297.3	1.1	269.3	1.0	5	125	24	23	7	279.8	5.9
ST22	-39.29	179.58	2990	298.1	2.1	267.0	1.2	1	65		25		315.0	20.0
ST23	-40.88	178.55	3188	298.0	6.0	265.2	1.3	4	84	73	15	9	291.5	28.8
ST24	-42.12	176.45	3417	336.4	2.0	261.6	1.5	1	145		38		285.0	20.0

Pole: the pole of rotation used and the latitude, longitude and angular velocity are given in parentheses. Start: the starting point of the trench (i.e., x=0). Center: the center of curvature of the trench, latitude and longitude in parentheses. Bin width: the width of each trench segment measured at the trench. N: the number of slip vectors. SVR: the average slip vector residual. For each bin, Lat. and Lon. are the latitude and longitude at the trench of the mid-point of the bin, Dist. is the distance in km from the start of the trench, T is the average trench-normal azimuth (in each case, SD is the standard deviation for the parameter that it follows), PVA is the average plate vector azimuth, N is the number of slip vectors in the bin, Rake is the average Rake angle on the fault plane, Dip is average dip of the fault plane, SVA is the average slip vector azimuth. When the entry is blank, not enough information is available.

downdip of the deepest point of contact between the subducting plate and the accretionary prism; hence this study pertains to deformation of the forearc lithosphere and not the accretionary prism.

The major source of earthquake data is the Harvard centroid-moment tensor (HCMT) catalog from 1977 through June, 1992 (e.g., DZIEWONSKI *et al*, 1981). Earthquakes are selected by the following criteria: focal depth is less than 60 km, seismic moment is greater than 10^{17} Nm, rake angle is between $0°$ and $180°$, strike of the nodal plane with the smaller dip angle is within $45°$ of the local strike of the trench, fault plane dip is less than $45°$ (PACHECO *et al.*, 1993) and is in the same direction as the subduction thrust fault, and the earthquake occurs landward of a trench. The slip vector azimuth is calculated by rotating it about the fault plane strike into the horizontal. Additional non-HCMT slip vectors were taken from DEMETS *et al.* (1990), MCCAFFREY (1988), RANKEN *et al.* (1984) and from C. DEMETS and D. ARGUS (personal communication, 1992). The total number of slip vectors is 1747 in 249 bins; binned slip vector azimuths are shown in Table 1. This set of slip vectors differs from that used by MCCAFFREY (1993), owing to the seismic moment cutoff used here.

Plate Convergence Vectors

Expected convergence vectors between plates meeting at a particular trench were calculated from NUVEL-1 (DEMETS *et al.*, 1990) for most trenches, from SENO *et al.* (1987) for trenches bordering the Philippine Sea plate, from MCCAFFREY (1990) for the Java-Sumatra trench, and from PELAYO and WIENS (1989) for the Sandwich trench. Statistical uncertainties in the plate vector directions for each bin

Table 2

List of additional binned data by trench

Bin	SVR	SD	γ	SD	ψ	SD	V	Vp	Vn	Vs	SD	Mo/yr
					Aleutian							
AL01	13.0	24.7	-34.1	8.6	-48.7	26.6	55.3	-30.9	45.8	21.2	65.5	20.90
AL02	0.2	20.1	-25.9	7.2	-30.8	21.1	56.4	-24.6	50.8	5.7	25.5	15.43
AL03	-12.6	8.4	-4.2	8.4	8.4	11.3	58.9	-4.3	58.8	-13.0	14.3	18.57
AL04	-10.5	6.9	-5.2	5.0	5.2	8.8	60.2	-5.4	59.9	-10.8	10.2	18.07
AL05	1.4	20.0	-11.4	0.8	-12.5	20.0	61.4	-12.1	60.2	1.3	16.8	17.67
AL06	-2.6	6.2	-1.2	6.2	1.2	8.6	62.6	-1.3	62.5	-2.6	11.3	18.28
AL07	-12.6	15.4	13.9	1.7	26.5	15.2	63.7	15.3	61.8	-15.5	16.4	19.19
AL08	11.2	17.9	10.1	1.0	-1.6	18.0	64.8	11.3	63.8	13.2	15.3	18.16
AL09	-14.4	14.9	13.3	0.9	27.4	15.2	65.9	15.2	64.1	-18.1	16.5	17.62
AL10	8.2	5.2	10.5	1.1	1.9	5.5	66.9	12.2	65.8	10.0	5.3	19.07
AL11	-6.3	23.0	12.8	0.9	18.7	23.2	67.9	15.0	66.3	-7.4	23.4	17.69
AL12	6.8	7.0	9.8	1.2	3.0	7.2	68.9	11.7	67.9	8.1	7.0	16.92
AL13	2.2	11.5	12.0	1.6	9.5	11.5	69.9	14.5	68.3	3.1	11.4	17.94
AL14	-0.5	18.8	15.6	1.1	15.5	19.0	70.8	19.0	68.2	0.1	18.8	18.91
AL15	10.5	24.1	23.7	3.2	12.7	24.2	71.8	28.9	65.7	14.1	24.8	17.71
AL16	14.0	6.5	28.8	0.9	14.6	6.5	72.6	34.9	63.6	18.4	6.3	16.79
AL17	7.6	9.1	26.8	1.0	18.9	9.2	73.4	33.1	65.5	10.8	9.3	20.07
AL18	6.2	7.1	36.1	4.9	29.3	8.4	74.2	43.7	60.0	10.1	13.3	18.18
AL19	11.0	10.2	40.7	1.7	29.2	10.1	74.9	48.9	56.8	17.0	11.2	18.36
AL20	16.7	8.8	46.4	3.6	29.0	9.4	75.6	54.7	52.1	25.8	11.6	19.10
AL21	19.9	7.9	54.2	4.4	33.8	8.9	76.1	61.7	44.5	32.0	11.8	20.19
AL22	28.8	15.5	62.2	1.1	33.1	15.5	76.5	67.7	35.7	44.4	11.3	17.69
AL23	20.1	14.6	62.8	1.9	42.3	14.7	76.9	68.4	35.1	36.5	14.9	16.62
AL24	19.3	3.6	71.1	3.4	51.1	5.0	77.2	73.1	25.0	42.0	8.9	17.75
AL25	22.7	6.3	78.7	1.0	55.6	6.7	77.4	75.9	15.1	53.9	5.6	18.70
AL26	13.8	2.1	77.0	2.3	62.8	3.0	77.6	75.6	17.4	41.7	7.8	18.45
AL27	2.2	20.0	84.9	0.9	82.3	20.0	77.7	77.4	6.9	25.9	45.4	15.13
					Antilles							
AN01	-17.0	20.9	-71.6	1.5	-54.9	20.0	11.0	-10.4	3.5	-5.5	4.0	15.54
AN02	-29.6	20.9	-62.9	4.1	-37.1	20.3	11.1	-9.9	5.1	-6.1	2.2	15.99
AN03	-15.6	20.9	-40.2	9.3	-26.1	22.0	11.4	-7.3	8.7	-3.1	4.1	17.43
AN04	-20.4	20.9	-30.0	3.2	-11.6	20.3	11.6	-5.8	10.1	-3.8	2.3	16.23
AN05	-31.6	20.9	9.6	16.1	38.9	25.7	12.3	2.1	12.1	-7.8	7.5	15.06
AN06	-8.9	20.7	25.9	0.4	33.0	20.0	13.4	5.9	12.1	-2.0	4.0	15.38
					Izu							
IZ01	-15.5	12.4	-19.7	2.8	-2.9	13.0	65.0	-21.9	61.2	-18.8	12.0	18.65
IZ02	-11.5	20.0	-21.3	0.7	-7.9	20.0	62.9	-22.8	58.6	-14.8	15.8	15.96
IZ03	-18.0	6.0	-27.8	3.3	-8.3	6.7	60.8	-28.4	53.8	-20.5	6.8	16.91
IZ04	-16.6	7.2	-36.3	1.2	-18.3	7.0	58.9	-34.9	47.5	-19.2	5.4	16.00
IZ05	-19.2	5.3	-33.0	3.0	-12.0	5.8	57.0	-31.1	47.8	-20.9	5.5	16.27
IZ06	-16.7	20.0	-18.5	4.3	-0.1	20.5	54.9	-17.5	52.0	-17.3	16.3	15.87
IZ07	-58.3	20.0	-23.4	8.8	36.8	21.7	52.6	-20.8	48.3	-56.9	25.8	15.15
IZ08	-35.2	10.6	-53.6	6.6	-17.1	11.9	51.1	-41.1	30.3	-31.8	8.4	17.07

Table 2 (*Contd*)

Bin	SVR	SD	γ	SD	ψ	SD	V	Vp	Vn	Vs	SD	Mo/yr
					Mariana							
MA01	-59.6	20.1	-79.4	1.0	-16.1	20.1	49.7	-48.8	9.1	-46.2	2.9	17.82
MA02	-38.0	20.1	-66.3	6.7	-26.7	21.4	49.0	-44.9	19.7	-35.0	10.8	15.09
MA03	-36.7	21.4	-52.7	3.0	-14.5	21.7	46.1	-36.7	27.9	-29.5	9.8	18.28
MA04	-43.4	5.0	-37.4	3.7	7.6	5.9	44.4	-27.0	35.2	-31.7	4.0	15.72
MA05	-40.1	20.1	-35.8	1.9	6.5	20.1	42.3	-24.7	34.3	-28.7	9.7	15.55
MA06	-12.4	20.2	-12.7	0.3	0.9	20.0	38.3	-8.4	37.4	-9.0	9.7	15.82
MA07	-20.7	13.7	-3.9	7.2	19.4	15.9	36.4	-2.5	36.3	-15.2	11.2	16.09
MA08	-8.4	14.0	19.6	5.1	30.0	14.8	33.9	11.4	32.0	-7.1	10.5	15.80
MA09	17.2	27.2	30.9	2.2	14.9	27.0	31.7	16.3	27.2	9.0	11.4	16.47
MA10	41.2	20.3	36.7	1.5	-4.7	20.0	29.8	17.8	23.9	19.7	6.5	16.49
MA11	48.0	11.4	40.2	1.1	-5.1	12.7	28.5	18.4	21.8	20.3	3.7	16.01
					Japan							
JP01	-2.4	8.6	-9.3	8.7	-6.2	12.2	86.2	-13.9	85.0	-4.6	22.1	19.64
JP02	-2.3	8.0	-26.6	1.7	-23.7	8.1	86.3	-38.6	77.1	-4.8	11.2	19.74
JP03	0.1	4.3	-22.7	3.4	-22.1	5.5	86.3	-33.3	79.7	-1.0	10.2	18.28
JP04	8.7	15.1	-12.2	2.4	-20.1	15.3	86.4	-18.3	84.5	12.6	21.3	18.35
JP05	-11.1	23.0	-5.5	1.1	6.6	23.0	86.5	-8.3	86.1	-18.3	27.2	18.63
JP06	-1.7	13.2	-1.0	1.7	1.7	13.3	86.6	-1.4	86.6	-4.1	16.2	19.00
JP07	-1.8	4.0	4.8	1.2	7.5	4.1	86.6	7.3	86.3	-4.1	5.7	18.89
JP08	-1.0	7.9	5.4	1.7	7.3	8.1	86.7	8.2	86.3	-2.8	10.6	18.63
JP09	-1.7	10.2	-2.7	2.9	-0.2	10.6	86.7	-4.1	86.6	-3.7	14.4	17.69
					Java-Sumatra							
JS01	-0.3	10.6	10.5	13.2	10.8	16.8	66.6	12.2	65.5	-0.3	24.9	17.25
JS02	17.5	20.4	2.1	2.4	-15.2	20.2	66.4	2.4	66.4	20.5	20.4	15.05
JS03	-3.9	20.4	7.9	1.9	11.8	20.1	66.3	9.1	65.7	-4.7	19.3	16.25
JS04	22.5	20.4	2.0	2.3	-20.1	20.1	65.6	2.3	65.5	26.3	21.2	15.44
JS05	5.7	7.4	14.9	4.7	9.7	8.8	65.3	16.8	63.1	6.0	10.7	15.73
JS06	4.9	5.2	24.4	1.5	19.9	5.3	65.0	26.9	59.2	5.5	5.6	16.59
JS07	12.6	20.4	24.6	1.0	12.4	20.0	64.7	27.0	58.8	14.0	16.6	18.12
JS08	7.1	15.7	27.1	2.8	20.5	15.9	64.4	29.3	57.3	7.8	15.5	17.13
JS09	24.8	20.4	26.8	1.8	2.6	20.1	64.0	28.8	57.1	26.2	15.9	18.12
JS10	19.2	10.3	36.8	3.2	18.2	10.8	63.6	38.1	50.9	21.4	10.0	17.84
JS11	26.7	13.9	47.6	3.0	21.5	14.2	63.3	46.7	42.6	30.0	11.1	18.61
JS12	28.6	9.0	53.4	0.6	25.1	9.0	62.9	50.5	37.5	32.9	5.6	17.86
JS13	28.6	14.6	49.6	2.6	21.4	14.8	62.6	47.7	40.6	31.8	10.7	17.97
JS14	25.3	12.0	44.8	1.6	19.9	12.2	62.3	43.9	44.2	27.9	8.8	18.28
JS15	53.5	4.5	56.4	3.9	3.1	5.9	61.9	51.5	34.3	49.7	4.2	16.68
JS16	39.9	3.7	63.7	0.7	24.2	3.9	61.6	55.2	27.3	42.9	2.0	18.84
JS17	56.6	20.5	61.1	2.7	4.4	20.2	61.4	53.7	29.6	51.4	8.8	16.30
JS18	39.2	10.5	36.7	9.2	-2.4	13.9	61.0	36.4	48.9	38.4	13.7	16.30
JS19	35.5	20.5	31.0	2.6	-4.1	20.1	60.3	31.1	51.7	34.8	14.9	18.71
JS20	34.3	20.5	58.3	15.6	24.2	25.2	59.8	50.9	31.4	36.8	23.9	16.50
JS21	93.5	20.5	68.4	5.4	-25.4	20.8	59.3	55.1	21.8	65.5	6.7	15.68

Table 2 (*Contd*)

Bin	SVR	SD	γ	SD	ψ	SD	V	Vp	Vn	Vs	SD	Mo/yr
Kuriles												
KU01	0.0	6.8	-12.9	4.5	-12.6	8.1	78.5	-17.5	76.5	-0.5	0.6	16.46
KU02	-3.2	2.1	-4.7	1.1	-0.5	2.4	79.1	-6.5	78.8	-5.9	3.4	19.91
KU03	-3.4	10.3	-5.0	1.0	-0.6	10.3	79.6	-6.9	79.3	-6.1	11.1	19.14
KU04	0.4	5.1	-9.0	1.1	-8.9	5.4	80.1	-12.5	79.2	-0.1	6.4	20.57
KU05	-0.8	7.7	-7.8	1.7	-6.5	7.9	80.7	-10.9	79.9	-1.8	9.5	18.77
KU06	-6.0	17.9	0.0	2.6	6.9	17.8	81.2	0.0	81.2	-9.8	21.1	17.97
KU07	3.9	5.1	6.7	1.4	3.7	5.3	81.6	9.6	81.0	4.3	6.7	17.68
KU08	3.3	4.2	9.3	0.4	6.9	4.2	82.1	13.3	81.0	3.5	4.6	16.66
KU09	1.4	8.2	8.7	0.6	7.7	8.1	82.5	12.5	81.5	1.5	8.9	17.82
KU10	0.3	4.7	5.8	1.1	6.1	4.9	82.9	8.3	82.5	-0.4	6.1	16.91
KU11	-6.3	8.2	3.6	0.4	10.4	8.1	83.3	5.3	83.1	-9.9	9.0	19.00
KU12	0.7	9.4	8.1	2.5	7.8	9.7	83.7	11.9	82.8	0.5	12.8	17.12
KU13	25.0	42.7	16.4	2.3	-8.3	42.7	84.0	23.7	80.6	35.5	56.2	15.74
KU14	8.0	12.0	22.2	1.3	14.6	12.1	84.3	31.8	78.1	11.5	13.9	19.43
KU15	-1.1	7.2	23.9	0.4	25.5	7.2	84.6	34.3	77.3	-2.5	8.8	19.04
KU16	5.5	6.7	23.0	0.8	17.9	6.8	84.8	33.1	78.1	7.9	8.1	19.92
KU17	9.7	5.9	20.9	0.5	11.6	5.9	85.1	30.4	79.5	14.1	6.6	19.75
KU18	5.8	8.2	23.0	1.2	17.6	8.3	85.3	33.3	78.6	8.5	10.1	19.37
KU19	5.2	6.4	26.6	0.8	21.6	6.5	85.5	38.3	76.5	7.9	8.0	18.65
KU20	5.0	12.2	27.7	0.4	23.2	12.4	85.7	39.8	75.9	7.3	14.4	18.92
KU21	4.1	20.0	23.8	2.4	20.2	20.2	85.9	34.6	78.6	5.7	26.0	15.08
KU22	-1.8	6.2	15.6	2.0	17.8	6.5	86.0	23.1	82.9	-3.6	9.5	16.85
S. Mid America												
SM01	-4.6	16.9	-1.7	0.3	3.7	17.0	91.1	-2.6	91.0	-8.5	0.5	18.60
SM02	4.3	11.8	1.0	1.8	-2.2	12.1	89.1	1.6	89.0	5.0	15.2	18.27
SM03	3.7	2.3	12.9	5.7	10.4	6.1	87.1	19.4	85.0	3.8	12.7	18.83
SM04	4.5	5.5	26.8	2.0	23.9	5.7	85.2	38.4	76.1	4.7	8.7	18.20
SM05	6.7	9.1	12.2	5.4	6.8	10.5	83.3	17.6	81.4	7.8	16.2	18.23
SM06	11.6	8.3	-0.1	2.2	-9.8	8.5	81.3	-0.1	81.3	14.0	10.9	18.19
SM07	15.0	20.1	-4.8	0.8	-18.0	20.0	79.2	-6.6	79.0	19.1	23.3	15.45
SM08	20.7	16.0	-6.1	0.2	-25.2	16.0	77.1	-8.2	76.7	27.9	19.4	17.70
SM09	10.5	12.7	-5.5	0.4	-14.5	12.7	75.1	-7.2	74.8	12.1	13.0	16.70
SM10	5.4	3.8	-2.8	1.2	-6.4	3.9	73.1	-3.5	73.0	4.6	4.6	18.83
SM11	7.3	10.5	2.2	1.8	-3.5	10.3	71.0	2.7	71.0	7.0	10.6	18.47
SM12	4.1	9.2	5.8	0.5	3.3	9.2	68.8	7.0	68.5	3.1	8.2	18.57
SM13	-1.8	7.6	2.0	1.9	4.9	7.9	66.6	2.3	66.6	-3.4	8.0	17.68
SM14	6.4	15.5	-3.7	1.5	-8.4	15.7	64.5	-4.1	64.4	5.4	14.2	16.60
SM15	-3.3	6.7	-6.8	0.9	-2.1	6.7	62.7	-7.5	62.2	-5.2	5.8	15.61
N. Mid America												
NM01	3.3	5.7	-17.0	0.5	-18.4	5.8	68.5	-20.0	65.5	1.7	0.3	19.25
NM02	-5.8	14.7	-18.5	0.6	-10.8	15.0	66.1	-21.0	62.6	-9.0	12.6	18.47
NM03	-1.8	10.5	-19.0	0.1	-15.3	10.3	63.8	-20.8	60.3	-4.3	8.3	18.52
NM04	-5.8	17.5	-19.0	0.4	-11.2	17.4	61.6	-20.0	58.2	-8.5	13.6	19.17
NM05	5.1	10.5	-17.9	0.4	-20.9	9.8	59.4	-18.2	56.5	3.3	8.2	18.56
NM06	-11.7	7.6	-15.9	0.8	-1.8	7.6	57.1	-15.7	54.9	-14.0	5.7	18.65
NM07	-10.2	6.4	-12.4	1.3	0.7	6.9	55.0	-11.8	53.7	-12.5	5.4	18.65

Table 2 (*Contd*)

Bin	SVR	SD	γ	SD	ψ	SD	V	Vp	Vn	Vs	SD	Mo/yr
NM08	-6.7	10.1	-7.5	1.6	2.2	9.9	52.7	-6.9	52.3	-8.8	7.5	19.19
NM09	-5.0	4.9	-2.3	1.5	5.7	4.9	50.5	-2.0	50.4	-7.0	4.0	18.82
NM10	-2.6	20.1	0.5	0.9	5.5	20.0	48.1	0.4	48.1	-4.2	12.9	18.89

New Hebrides

Bin	SVR	SD	γ	SD	ψ	SD	V	Vp	Vn	Vs	SD	Mo/yr
NH01	-39.3	20.0	84.7	1.7	-56.3	20.1	76.0	75.7	7.0	86.3	4.8	17.28
NH02	-63.8	31.0	87.7	0.5	-28.6	31.2	75.9	75.9	3.0	77.5	1.5	16.35
NH03	-92.1	2.1	-85.1	6.1	7.0	6.1	75.8	-75.6	6.5	-76.3	0.9	16.19
NH04	-72.7	8.1	-59.9	7.4	12.7	10.7	76.3	-66.0	38.2	-74.7	7.3	18.27
NH05	-51.6	13.4	-41.8	3.9	9.6	13.9	77.3	-51.5	57.7	-61.3	12.6	19.02
NH06	-31.7	26.1	-30.1	3.3	1.4	26.3	78.7	-39.4	68.1	-41.1	26.2	18.36
NH07	-5.4	29.2	-20.7	2.4	-15.5	29.4	80.1	-28.4	74.9	-7.5	35.1	16.96
NH08	-11.0	4.0	-14.5	1.5	-3.9	4.3	81.5	-20.4	78.9	-15.0	5.6	18.02
NH09	-10.7	6.6	-8.8	0.5	1.7	6.7	84.1	-12.8	83.2	-15.3	7.3	18.32
NH10	-11.2	16.7	-7.7	0.1	3.3	16.7	85.3	-11.5	84.5	-16.4	18.0	17.30

Philippine

Bin	SVR	SD	γ	SD	ψ	SD	V	Vp	Vn	Vs	SD	Mo/yr
PH01	-44.2	7.6	-49.6	6.7	-6.4	10.3	78.4	-59.7	50.8	-54.0	11.3	18.36
PH02	-25.6	14.2	-52.6	3.6	-27.1	14.5	79.5	-63.1	48.2	-38.4	14.9	18.50
PH03	-13.8	20.1	-41.5	1.9	-27.8	20.1	80.6	-53.4	60.4	-21.5	22.6	16.79
PH04	-28.0	12.2	-45.4	2.4	-17.6	12.5	81.8	-58.2	57.5	-40.0	12.2	17.56
PH05	-22.1	11.2	-44.4	3.8	-22.7	11.8	83.0	-58.0	59.3	-33.3	14.1	18.82
PH06	-18.7	25.7	-26.6	4.8	-8.2	26.1	84.3	-37.8	75.3	-26.9	31.3	18.58
PH07	-19.6	12.8	-32.2	7.2	-13.1	14.5	85.8	-45.7	72.6	-28.8	21.2	19.53
PH08	-17.7	15.2	-53.0	3.7	-36.0	15.9	87.0	-69.5	52.4	-31.4	21.6	18.60
PH09	-29.7	18.9	-56.6	0.6	-27.6	19.0	87.9	-73.4	48.3	-48.2	15.9	19.17
PH10	-31.1	20.1	-51.1	2.7	-20.3	20.2	88.8	-69.1	55.8	-48.4	19.4	17.68

N. Ryukyu

Bin	SVR	SD	γ	SD	ψ	SD	V	Vp	Vn	Vs	SD	Mo/yr
NR01	-4.6	20.6	51.0	7.7	53.1	21.6	40.1	31.2	25.2	-2.4	35.2	19 21
NR02	0.7	20.6	31.1	8.8	29.4	21.7	41.9	21.6	35.9	1.4	19.2	19 22
NR03	-6.2	20.5	13.0	5.2	15.8	20.6	44.3	10.0	43.2	-2.3	15.3	18.28
NR04	2.7	5.6	-2.7	3.4	-8.9	6.6	46.9	-2.2	46.9	5.2	5.8	18.99
NR05	0.1	9.2	-7.9	0.4	-10.1	8.9	49.5	-6.8	49.0	2.0	5.8	17.69
NR06	13.6	8.7	-7.0	0.7	-22.7	8.5	51.5	-6.3	51.1	15.1	6.7	17.69

S. Ryukyu

Bin	SVR	SD	γ	SD	ψ	SD	V	Vp	Vn	Vs	SD	Mo/yr
SR01	13.9	20.3	6.5	3.0	-9.9	20.2	56.6	6.4	56.3	16.2	17.0	16.23
SR02	7.9	20.3	14.0	1.0	4.6	20.0	59.0	14.2	57.2	9.7	15.4	17.68
SR03	0.3	20.3	22.6	7.1	20.8	21.3	63.6	24.4	58.7	2.1	24.6	18.42
SR04	6.4	20.3	45.5	4.6	38.3	20.6	65.5	46.7	45.9	10.4	26.6	15.70

S. South America

Bin	SVR	SD	γ	SD	ψ	SD	V	Vp	Vn	Vs	SD	Mo/yr
SS01	-10.2	20.1	-9.9	4.0	0.3	20.4	19.5	-3.4	19.2	-3.5	5.9	15.46

Table 2 (*Contd*)

Bin	SVR	SD	γ	SD	ψ	SD	V	Vp	Vn	Vs	SD	Mo/yr
					Central South America							
CS01	9.3	14.5	21.6	1.1	13.0	14.6	84.0	30.9	78.1	12.8	16.2	18.62
CS02	8.6	20.0	25.1	1.1	17.3	20.0	84.1	35.7	76.2	11.9	22.8	17.29
CS03	-6.5	20.0	30.5	2.6	37.5	20.2	84.2	42.7	72.6	-13.0	35.7	18.75
CS04	7.6	7.4	33.1	4.1	26.1	8.5	84.4	46.1	70.7	11.6	13.8	17.87
CS05	7.6	9.8	18.7	3.4	11.5	10.4	84.4	27.0	80.0	10.7	14.3	19.94
CS06	-5.5	17.5	15.4	0.6	21.4	17.5	84.4	22.4	81.4	-9.5	21.7	16.39
CS07	4.4	5.3	16.9	0.3	13.0	5.3	84.4	24.5	80.7	5.8	5.9	19.43
CS08	25.4	20.0	21.3	3.1	-3.6	20.2	84.3	30.7	78.5	35.6	23.3	15.46
CS09	5.5	6.3	28.9	1.1	24.0	6.4	84.2	40.7	73.7	7.9	8.2	20.18
CS10	11.7	30.6	24.5	3.1	13.2	30.8	84.1	34.9	76.5	16.9	37.5	16.70
CS11	4.3	9.4	18.2	1.1	14.3	9.6	84.0	26.3	79.8	5.9	11.2	18.81
CS12	5.1	9.3	17.1	0.7	12.5	9.3	83.8	24.6	80.1	6.9	10.4	19.00
CS13	5.8	8.5	16.6	0.9	11.4	8.7	83.6	23.9	80.1	7.8	9.9	17.83
CS14	-5.2	10.5	16.3	0.2	22.0	10.7	83.4	23.4	80.0	-8.9	12.6	15.85
CS15	9.3	20.0	13.6	2.3	4.8	20.1	83.1	19.5	80.8	12.7	23.2	18.18
CS16	-23.3	20.0	2.5	4.6	26.7	20.5	82.8	3.6	82.8	-38.1	32.8	18.21
					N. South America							
NS01	-21.3	18.4	-25.1	8.6	-3.8	20.0	82.4	-35.0	74.6	-30.0	27.0	18.24
NS02	-5.6	20.0	-40.2	1.0	-34.2	20.0	81.7	-52.7	62.4	-10.2	26.1	18.81
NS03	2.4	20.0	-39.2	1.4	-41.5	20.1	81.2	-51.3	63.0	4.4	34.0	18.66
NS04	2.5	6.1	-33.7	1.6	-35.7	6.4	80.6	-44.7	67.0	3.5	9.9	16.47
NS05	-6.9	20.0	-30.3	1.1	-22.8	20.0	79.9	-40.3	69.0	-11.3	22.5	17.68
NS06	-20.8	20.0	-22.7	1.3	-1.3	20.0	78.2	-30.3	72.2	-28.6	19.6	19.33
NS07	-0.4	20.0	-18.8	3.0	-17.9	20.2	77.3	-24.9	73.2	-1.2	24.1	19.04
NS08	-5.6	20.0	-13.8	0.8	-7.6	20.0	76.2	-18.2	74.0	-8.3	20.0	18.75
NS09	-5.1	6.2	-0.8	6.1	4.8	8.7	75.0	-1.0	75.0	-7.4	13.7	16.44
NS10	-3.5	20.0	18.2	2.6	22.1	20.2	72.5	22.7	68.9	-5.3	23.6	17.86
NS11	4.0	8.6	26.7	5.2	22.9	10.0	71.3	32.0	63.7	5.2	14.1	18.72
NS12	5.5	4.6	23.3	4.4	18.3	6.3	70.0	27.7	64.2	6.5	9.5	20.04
NS13	19.9	19.8	28.7	4.9	9.6	20.5	69.0	33.1	60.5	22.9	20.3	16.71
					Sandwich							
SA01	-36.5	20.0	-75.2	2.1	-37.8	20.0	66.4	-64.2	16.9	-51.1	9.5	15.49
SA02	-24.8	12.0	-68.0	3.1	-42.8	12.4	67.3	-62.4	25.2	-39.1	10.8	16.14
SA03	-24.2	20.0	-49.8	10.1	-27.5	22.0	68.8	-52.6	44.4	-29.5	26.0	15.53
SA04	-3.6	7.8	-12.4	9.0	-10.8	11.8	70.5	-15.2	68.9	-2.0	18.2	16.72
SA05	0.2	14.0	5.3	1.8	3.2	14.1	71.6	6.6	71.3	2.7	14.2	16.89
SA06	-8.1	8.3	-3.7	6.6	3.2	10.9	73.7	-4.8	73.6	-8.9	15.8	17.85
SA07	1.4	11.4	-7.4	5.8	-11.5	12.7	76.8	-9.9	76.2	5.6	18.0	17.01
SA08	12.9	15.8	18.7	7.1	3.0	17.4	79.9	25.6	75.7	21.6	23.1	16.79
SA09	6.5	20.0	39.0	6.7	29.0	21.3	82.7	52.0	64.3	16.4	32.0	15.47
SA10	16.3	33.5	59.6	4.6	41.5	33.8	84.7	73.1	42.8	35.1	65.7	17.93
SA11	37.8	20.0	65.4	0.0	26.7	20.0	85.4	77.7	35.5	59.9	11.7	15.18

Table 2 (*Contd*)

Bin	SVR	SD	γ	SD	ψ	SD	V	Vp	Vn	Vs	SD	Mo/yr
					S. Solomon							
SO01	-4.7	8.4	-7.4	2.2	-3.0	8.7	93.7	-12.1	93.0	-7.3	12.5	19.34
SO02	-12.4	17.6	-24.9	11.2	-12.6	20.9	95.6	-40.2	86.7	-20.8	37.2	18.80
SO03	-18.0	20.0	-85.0	16.0	-67.0	25.7	96.4	-96.1	8.4	-76.3	99.9	18.50
SO04	-25.2	20.0	85.7	5.2	-69.0	20.6	96.3	96.0	7.3	115.0	80.2	17.83
SO05	-33.0	37.0	-75.4	5.9	-42.3	37.5	96.4	-93.3	24.3	-71.2	66.6	18.33
SO06	-19.7	20.1	-53.9	6.4	-34.4	21.0	96.9	-78.3	57.1	-39.3	33.7	18.89
SO07	-21.8	19.0	-42.0	1.0	-20.2	19.0	97.9	-65.5	72.7	-38.6	21.5	19.19
SO08	-32.9	20.4	-50.6	5.5	-17.8	21.2	98.7	-76.3	62.7	-56.1	25.1	18.46
					N. Solomon							
NO01	-28.0	8.0	-38.8	2.9	-10.8	8.5	102.3	-64.1	79.8	-48.8	11.8	17.69
NO02	-35.1	5.2	-45.4	1.0	-10.5	5.3	102.9	-73.3	72.2	-59.9	6.0	18.55
NO03	-20.7	20.0	-42.2	2.6	-21.8	20.2	103.5	-69.6	76.7	-39.0	26.7	18.11
NO04	-29.2	4.3	-34.6	1.8	-5.6	4.6	104.2	-59.1	85.8	-50.8	6.9	18.87
NO05	-28.0	6.2	-32.0	0.5	-4.1	6.2	104.8	-55.6	88.9	-49.2	7.4	16.63
NO06	-35.8	4.0	-40.1	5.2	-4.5	6.6	105.5	-67.9	80.8	-61.6	12.0	19.22
					N. Tonga							
NT01	-7.6	39.9	-74.6	1.2	-67.1	40.0	88.9	-85.7	23.6	-29.8	64.2	18.28
NT02	-26.8	44.7	-72.1	0.5	-45.1	44.7	88.8	-84.4	27.3	-57.0	99.9	17.19
NT03	-13.7	33.9	-66.1	3.6	-52.2	34.2	88.5	-80.9	35.9	-34.7	99.9	19.88
NT04	-0.8	10.9	-40.0	14.3	-38.8	18.1	87.9	-56.6	67.3	-2.5	47.4	16.83
NT05	5.9	13.7	0.9	6.3	-4.5	15.1	86.8	1.3	86.8	8.2	23.1	18.62
NT06	0.5	24.9	4.6	1.1	4.4	25.2	85.7	6.9	85.4	0.3	29.5	16.36
NT07	15.8	28.9	3.6	0.7	-11.6	28.8	84.6	5.3	84.5	22.7	35.0	16.26
NT08	-2.8	20.0	5.9	0.0	9.3	20.0	84.0	8.6	83.6	-5.0	21.9	15.31
					S. Tonga							
ST01	3.4	13.7	14.3	2.1	11.6	13.8	82.8	20.5	80.2	4.0	16.6	19.47
ST02	14.2	8.1	14.8	1.1	1.2	8.4	81.3	20.8	78.6	19.1	9.1	16.61
ST03	17.4	8.0	13.6	0.8	-3.2	8.2	79.8	18.8	77.6	23.1	8.6	18.72
ST04	15.7	11.7	18.8	2.3	3.7	12.0	78.4	25.3	74.2	20.4	13.2	19.17
ST05	23.8	8.5	24.7	1.1	1.5	8.7	76.9	32.2	69.9	30.4	8.4	16.98
ST06	19.8	11.9	25.4	1.3	6.2	11.9	75.5	32.3	68.2	24.9	11.3	16.60
ST07	12.8	9.4	17.6	5.4	5.6	10.8	74.0	22.4	70.5	15.4	14.2	17.18
ST08	7.9	5.4	3.6	2.6	-3.5	6.0	72.4	4.6	72.2	9.0	7.6	18.57
ST09	-3.2	3.0	0.9	0.9	5.3	3.4	70.8	1.1	70.8	-5.5	3.7	18.65
ST10	14.5	20.0	7.0	2.2	-6.3	20.1	69.2	8.4	68.6	16.0	19.6	15.37
ST11	7.7	9.9	10.9	0.7	4.0	9.9	67.5	12.7	66.3	8.1	8.8	16.99
ST12	10.8	8.0	11.2	0.2	1.2	8.0	65.9	12.8	64.7	11.4	6.6	17.38
ST13	11.3	6.9	12.6	0.9	2.3	7.0	64.3	14.0	62.8	11.4	6.1	19.48
ST14	9.6	8.2	19.2	2.8	10.7	8.8	62.6	20.6	59.2	9.4	8.7	18.81
ST15	10.7	14.0	25.3	0.7	15.8	14.0	61.0	26.1	55.2	10.5	11.1	18.58
ST16	12.1	6.9	23.3	1.9	12.0	7.3	59.5	23.5	54.7	11.9	6.5	17.23
ST17	18.9	5.7	15.0	2.5	-3.0	6.2	57.9	15.0	55.9	17.9	5.9	16.94
ST18	12.1	11.7	13.6	3.2	2.7	12.0	56.1	13.2	54.5	10.6	10.2	17.19
ST19	7.8	11.2	16.2	1.1	10.0	11.4	54.4	15.2	52.3	6.0	8.3	16.82

Table 2 (*Contd*)

Bin	SVR	SD	γ	SD	ψ	SD	V	Vp	Vn	Vs	SD	Mo/yr
ST20	12.3	10.0	20.7	2.0	9.8	10.4	52.8	18.7	49.4	10.1	7.9	16.98
ST21	11.7	6.0	28.0	1.3	17.5	6.0	51.2	24.0	45.2	9.8	4.5	16.57
ST22	51.2	20.0	31.1	2.4	-17.9	20.1	44.4	22.9	38.0	35.2	11.8	15.72
ST23	29.5	28.8	32.7	6.3	6.5	29.4	41.2	22.3	34.7	18.4	16.7	16.74
ST24	24.8	20.1	74.8	2.6	51.4	20.1	39.1	37.7	10.2	24.9	11.1	15.25

Bin labels are as listed in Table 1. SVR is the average slip vector residual, in each case SD is the standard deviation, γ is the average convergence obliquity, ψ is the average slip vector obliquity, V is the plate convergence rate (mm/yr), Vp is the trench-parallel component of plate rate, Vn is the trench-normal component of plate rate, Vs is the trench-parallel motion rate of the forearc, and Mo/yr is the moment rate (log Nm/year) from the HCMT and Pacheco and Sykes [1992] catalogs.

include the variation of the mean vector in the bin and the variation due to uncertainties in the rotation poles (Table 1).

Slip vector azimuths and convergence obliquity are not independent in the cases where the slip vectors are important in estimating the poles of rotations. In NUVEL-1 (DEMETS *et al.*, 1990), subduction zone slip vectors were used selectively, but in some cases they are the only data on a plate boundary. For Pacific-N. America (Aleutian, Kuril, and Japan trenches) Cocos-N. America (N. Mid America trench), Nazca-S. America (Peru-Chile trench), Antarctica-S. America (S. Chile trench), and Australia-Pacific (Tonga, New Hebrides, and Solomon trenches) exclusion of subduction zone slip vectors from the NUVEL-1 data set makes little difference in the locations of the rotation poles and typically less than 2° difference in the predicted convergence direction at the trenches (DEMETS, 1993). Dependence of the plate vectors on slip vectors cannot be ruled out for the motions of the Philippine Sea, Caribbean, and Southeast Asian plates relative to surrounding plates. Furthermore, the NUVEL-1 predictions may be inaccurate at the Tonga, New Hebrides, Solomon, and Antilles subduction zones due to non-rigid plate behavior (backarc spreading in particular) near these margins. Therefore, for these and the Izu-Mariana, Java-Sumatra, southern Mid America, Philippine, and Ryukyu trenches I estimate poles of rotation that provide the best fit by least squares to the slip vectors (see DEMETS *et al.*, 1990 for the fitting procedure). I then estimate the rheology of these forearcs using convergence directions predicted by the new poles as well as the published poles. While the new poles probably do not correspond to the actual poles of rotation for the plate pairs, they do allow an unbiased estimate of how much the slip vectors deviate from rigid plate motions (new pole locations are given in Table 3).

Binning

To examine the variations in subduction parameters along the trenches, I divide the trenches into segments (bins) along strike (the binning procedure is shown in

Table 3

Results using new poles of rotation

Bin	PVA	SD	N	SVA	SD	SVR	SD	γ	SD	ψ	SD
						Antilles					
				Pole: (54.0N 273.0E) N:10 SVR:-2.3							
AN01	247.2	5.0	2	250.0	20.0	2.0	20.6	-52.4	1.3	-54.9	20.0
AN02	246.6	5.0	2	239.5	20.0	-9.6	20.6	-44.5	3.9	-37.1	20.3
AN03	246.2	4.9	2	250.0	20.0	2.5	20.6	-22.5	9.1	-26.1	22.0
AN04	246.1	4.8	2	245.0	20.0	-2.8	20.6	-12.9	3.4	-11.6	20.3
AN05	246.2	4.6	1	233.0	20.0	-14.9	20.5	25.6	16.0	38.9	25.7
AN06	248.3	4.4	1	257.0	20.0	7.4	20.5	41.5	0.4	33.0	20.0
						Izu					
				Pole: (1.0N 143.0E) N:39 SVR:-1.3							
IZ01	-91.6	7.7	7	-91.1	12.7	1.8	12.5	-2.5	2.6	-2.9	13.0
IZ02	-91.3	8.0	2	-87.0	20.0	6.6	21.6	-3.5	0.7	-7.9	20.0
IZ03	-91.0	8.4	15	-92.1	6.0	0.5	6.1	-9.5	3.1	-8.3	6.7
IZ04	-90.2	8.7	4	-89.3	6.9	2.6	7.2	-17.4	1.0	-18.3	7.0
IZ05	-89.3	9.1	4	-90.8	4.6	0.7	5.4	-13.4	3.2	-12.0	5.8
IZ06	-88.9	9.6	1	-87.0	20.0	4.0	22.2	1.8	4.6	-0.1	20.5
IZ07	-89.2	10.1	1	-128.0	20.0	-36.8	22.4	-2.1	8.6	36.8	21.7
IZ08	-88.4	10.6	5	-103.0	10.2	-12.6	10.4	-31.7	6.5	-17.1	11.9
						Mariana					
				Pole: (1.0N 143.0E) N:33 SVR:17.7							
MA01	-82.7	11.3	1	-124.0	20.0	-36.7	22.9	-57.4	1.0	-16.1	20.1
MA02	-79.4	11.7	1	-97.0	20.0	-15.3	23.2	-44.3	6.7	-26.7	21.4
MA03	-74.6	13.2	7	-90.1	21.4	-12.8	21.2	-30.1	3.1	-14.5	21.7
MA04	-72.9	14.1	3	-94.7	4.2	-19.3	4.9	-14.1	3.9	7.6	5.9
MA05	-71.9	15.4	1	-90.0	20.0	-13.9	25.2	-11.5	2.1	6.5	20.1
MA06	-72.7	18.3	1	-59.0	20.0	15.9	27.1	14.5	0.6	0.9	20.0
MA07	-74.3	19.7	6	-68.3	14.2	10.2	13.6	25.3	8.0	19.4	15.9
MA08	-78.5	21.6	3	-56.3	14.0	23.4	14.7	52.1	6.3	30.0	14.8
MA09	-84.5	23.0	6	-32.2	26.9	53.0	28.2	67.3	3.3	14.9	27.0
MA10	-91.1	24.1	2	-9.5	20.0	79.7	31.3	76.9	2.6	-4.7	20.0
MA11	-95.8	24.8	2	-12.5	20.0	85.8	31.8	83.0	1.6	-0.2	20.0
						Java- Sumatra					
				Pole: (-11.3N 122.0E) N:80 SVR:3.8							
JS01	119.6	87.5	7	-0.1	10.6	-24.5	14.9	10.5	15.2	10.8	16.8
JS02	-0.7	58.1	1	18.0	20.0	5.3	61.4	3.5	2.2	-15.2	20.2
JS03	0.7	39.1	1	-3.0	20.0	-16.4	43.9	8.0	1.5	11.8	20.1
JS04	5.0	20.2	1	24.0	20.0	11.5	28.4	-1.3	2.3	-20.1	20.1
JS05	5.7	18.1	3	7.3	7.4	-6.0	7.2	11.3	4.3	9.7	8.8
JS06	7.5	16.5	4	6.8	5.1	-6.6	3.8	19.2	1.0	19.9	5.3
JS07	9.3	15.1	2	14.5	20.0	1.9	25.1	17.7	1.2	12.4	20.0
JS08	10.9	14.0	6	8.8	15.6	-7.5	14.5	18.5	3.2	20.5	15.9
JS09	12.3	13.0	2	26.5	20.0	8.7	23.9	16.7	1.4	2.6	20.1
JS10	13.8	12.2	11	20.9	10.3	2.7	10.0	25.3	2.7	18.2	10.8
JS11	15.7	11.6	14	28.2	13.9	8.2	14.2	34.0	2.3	21.5	14.2

Table 3 (*Contd*)

Bin	PVA	SD	N	SVA	SD	SVR	SD	γ	SD	ψ	SD
JS12	17.8	11.0	5	30.0	9.0	9.6	9.1	37.2	0.7	25.1	9.0
JS13	19.8	10.4	3	29.7	14.6	7.1	14.3	31.2	3.1	21.4	14.8
JS14	21.2	9.9	3	26.0	12.1	2.1	13.0	24.7	1.3	19.9	12.2
JS15	22.6	9.5	4	54.0	4.5	29.5	5.3	34.5	3.3	3.1	5.9
JS16	24.3	9.1	4	39.8	3.9	12.7	4.7	39.6	0.3	24.2	3.9
JS17	25.9	8.8	2	56.5	20.0	30.1	21.9	35.0	3.3	4.4	20.2
JS18	26.9	8.5	3	38.3	10.4	9.0	10.1	9.1	9.4	-2.4	13.9
JS19	27.0	8.1	1	34.0	20.0	3.8	21.6	2.9	2.4	-4.1	20.1
JS20	27.4	7.8	2	32.5	20.0	1.9	21.5	29.2	15.1	24.2	25.2
JS21	31.8	7.2	1	90.0	20.0	58.6	21.2	32.8	5.9	-25.4	20.8

S. Mid America
Pole: (30.0N 172.0E) N:101 SVR:0.5

Bin	PVA	SD	N	SVA	SD	SVR	SD	γ	SD	ψ	SD
SM01	31.9	4.1	9	23.9	17.0	-8.3	17.0	-4.3	0.3	3.7	17.0
SM02	31.8	4.1	3	32.3	11.9	0.4	11.9	-1.6	1.7	-2.2	12.1
SM03	31.7	4.1	3	31.3	2.3	-0.6	2.3	10.1	5.7	10.4	6.1
SM04	31.6	4.1	12	31.2	5.4	-0.6	5.3	23.5	2.0	23.9	5.7
SM05	31.5	4.1	9	33.0	8.9	1.3	8.9	8.4	5.5	6.8	10.5
SM06	31.3	4.1	8	37.1	8.2	5.6	8.2	-4.0	2.2	-9.8	8.5
SM07	31.2	4.1	2	40.5	20.0	9.2	20.4	-8.7	0.7	-18.0	20.0
SM08	31.0	4.1	3	46.3	16.0	15.3	16.1	-9.9	0.2	-25.2	16.0
SM09	30.8	4.1	6	36.2	12.7	5.2	12.7	-9.1	0.5	-14.5	12.7
SM10	30.6	4.1	8	30.8	3.7	0.0	3.7	-6.2	1.2	-6.4	3.9
SM11	30.4	4.1	14	32.6	10.2	2.2	10.2	-1.3	1.7	-3.5	10.3
SM12	30.2	4.1	11	29.2	9.2	-1.1	9.2	2.2	0.5	3.3	9.2
SM13	30.0	4.1	3	23.3	7.6	-6.8	7.6	-1.8	1.9	4.9	7.9
SM14	29.8	4.1	7	30.9	15.7	1.0	15.6	-7.4	1.5	-8.4	15.7
SM15	29.7	4.1	3	21.3	6.7	-8.3	6.6	-10.4	0.9	-2.1	6.7

New Hebrides
Pole: (-65.0N 260.0E) N:81 SVR:-15.6

Bin	PVA	SD	N	SVA	SD	SVR	SD	γ	SD	ψ	SD
NH01	62.9	4.5	2	45.5	20.0	-17.5	20.5	-73.7	1.4	-56.3	20.1
NH02	62.9	4.5	3	20.0	31.2	-43.0	31.3	-71.5	0.6	-28.6	31.2
NH03	63.0	4.4	3	-10.0	1.7	-73.0	1.7	-66.0	5.8	7.0	6.1
NH04	63.1	4.4	10	8.7	8.0	-54.4	8.0	-41.6	7.1	12.7	10.7
NH05	63.2	4.4	14	29.4	13.3	-33.9	13.3	-24.2	3.7	9.6	13.9
NH06	63.4	4.3	8	48.9	26.1	-14.5	26.1	-13.1	3.1	1.4	26.3
NH07	63.5	4.3	8	74.9	29.3	11.4	29.3	-4.2	2.3	-15.5	29.4
NH08	63.7	4.3	6	69.3	4.0	5.7	4.0	1.8	1.4	-3.9	4.3
NH09	63.9	4.3	6	69.3	6.7	5.4	6.7	7.1	0.4	1.7	6.7
NH10	64.1	4.3	21	68.7	16.7	4.6	16.7	8.0	0.1	3.3	16.7

Philippine
Pole: (85.0N 166.0E) N:74 SVR:-2.1

Bin	PVA	SD	N	SVA	SD	SVR	SD	γ	SD	ψ	SD
PH01	273.4	2.9	9	256.2	7.6	-17.2	7.6	-23.5	7.0	-6.4	10.3
PH02	273.4	2.8	4	273.0	14.2	-0.4	14.2	-27.5	3.3	-27.1	14.5
PH03	273.3	2.8	2	284.0	20.0	10.7	20.2	-17.1	1.7	-27.8	20.1
PH04	273.3	2.8	13	269.2	12.2	-4.1	12.2	-21.7	2.7	-17.6	12.5
PH05	273.3	2.7	11	274.5	11.2	1.2	11.2	-21.4	3.6	-22.7	11.8

Table 3 (*Contd*)

Bin	PVA	SD	N	SVA	SD	SVR	SD	γ	SD	ψ	SD
PH06	273.2	2.7	8	277.1	25.7	3.9	25.7	-4.3	4.7	-8.2	26.1
PH07	273.2	2.7	3	276.0	12.5	2.8	12.5	-10.3	7.4	-13.1	14.5
PH08	273.2	2.6	15	277.3	15.4	4.1	15.4	-31.8	4.0	-36.0	15.9
PH09	273.1	2.6	8	264.5	19.0	-8.7	19.0	-36.2	0.4	-27.6	19.0
PH10	273.1	2.6	1	262.0	20.0	-11.1	20.2	-31.4	2.5	-20.3	20.2

N. Ryukyu
Pole: (48.0N 166.0E) N:7 SVR: -3.0

Bin	PVA	SD	N	SVA	SD	SVR	SD	γ	SD	ψ	SD
NR01	-43.7	8.6	2	-55.5	20.0	-13.5	21.8	41.6	8.0	53.1	21.6
NR02	-43.1	8.4	1	-50.0	20.0	-7.7	21.7	22.7	8.6	29.4	21.7
NR03	-43.2	8.0	1	-54.0	20.0	-12.6	21.6	5.3	5.0	15.8	20.6
NR04	-44.0	7.7	4	-45.3	5.5	-3.6	5.6	-10.0	3.3	-8.9	6.6
NR05	-44.9	7.4	4	-50.0	8.9	-6.3	9.3	-15.0	0.3	-10.1	8.9
NR06	-45.7	7.2	5	-37.2	8.5	7.3	8.7	-14.0	0.7	-22.7	8.5

S. Ryukyu
Pole: (48.0N 166.0E) N:7 SVR: 1.6

Bin	PVA	SD	N	SVA	SD	SVR	SD	γ	SD	ψ	SD
SR01	-47.1	6.7	2	-37.5	20.0	8.0	21.1	-0.1	3.1	-9.9	20.2
SR02	-47.5	6.5	2	-44.5	20.0	1.9	21.0	7.7	0.9	4.6	20.0
SR03	-48.0	6.1	2	-52.0	20.0	-4.9	20.9	16.9	7.2	20.8	21.3
SR04	-47.8	6.0	1	-46.0	20.0	1.2	20.9	40.2	4.7	38.3	20.6

S. Solomon
Pole: (-22.0N 170.0E) N:58 SVR: -0.7

Bin	PVA	SD	N	SVA	SD	SVR	SD	γ	SD	ψ	SD
SO01	67.0	23.3	17	75.1	8.5	5.8	8.0	4.6	2.4	-3.0	8.7
SO02	67.0	20.4	10	67.2	17.6	-1.6	17.8	-13.0	10.9	-12.6	20.9
SO03	64.4	19.1	2	61.0	20.0	-4.4	27.7	-70.9	15.0	-67.0	25.7
SO04	60.0	18.9	1	53.0	20.0	-6.9	27.5	-76.4	6.3	-69.0	20.6
SO05	56.3	18.3	8	44.6	37.0	-12.0	36.8	-54.3	6.6	-42.3	37.5
SO06	53.7	17.2	7	57.7	20.1	2.3	20.2	-30.8	6.8	-34.4	21.0
SO07	52.4	16.1	9	55.1	19.0	1.0	19.4	-17.9	1.1	-20.2	19.0
SO08	51.4	15.3	4	43.8	20.4	-9.3	20.5	-25.8	5.3	-17.8	21.2

N. Solomon
Pole: (-22.0N 170.0E) N:30 SVR:-6.2

Bin	PVA	SD	N	SVA	SD	SVR	SD	γ	SD	ψ	SD
NO01	47.7	12.5	3	47.0	7.9	-1.4	8.0	-11.7	2.8	-10.8	8.5
NO02	47.3	12.1	5	39.8	5.2	-8.9	5.1	-18.2	1.0	-10.5	5.3
NO03	46.8	11.7	1	54.0	20.0	5.4	23.2	-14.7	2.6	-21.8	20.2
NO04	46.5	11.4	12	45.3	4.3	-2.8	4.4	-7.0	1.8	-5.6	4.6
NO05	46.5	11.0	5	46.2	6.1	-1.1	6.6	-4.5	0.5	-4.1	6.2
NO06	46.3	10.6	4	22.0	32.8	-26.1	32.9	-12.6	5.2	11.8	33.3

N. Tonga
Pole: (-54.0N 173.0E) N:58 SVR:-9.9

Bin	PVA	SD	N	SVA	SD	SVR	SD	γ	SD	ψ	SD
NT01	282.2	4.2	4	265.8	40.0	-16.4	40.0	-83.3	1.2	-67.1	40.0
NT02	282.6	4.5	4	263.0	29.6	-19.3	29.5	-80.9	0.5	-61.5	29.6
NT03	283.0	4.2	12	259.9	33.9	-22.6	33.9	-75.0	3.6	-52.2	34.2

Table 3 (Contd)

Bin	PVA	SD	N	SVA	SD	SVR	SD	γ	SD	ψ	SD
NT04	283.6	4.3	6	273.0	10.9	-9.9	10.9	-49.2	14.3	-38.8	18.1
NT05	284.0	4.4	12	279.8	13.8	-3.4	13.7	-8.4	6.2	-4.5	15.1
NT06	284.1	4.5	11	274.6	25.2	-8.8	24.9	-4.8	1.1	4.4	25.2
NT07	284.2	4.5	8	289.6	28.8	6.3	28.9	-5.9	0.7	-11.6	28.8
NT08	284.3	5.5	1	271.0	20.0	-12.3	20.7	-3.8	0.0	9.3	20.0

S. Tonga
Pole: (-54.0N 173.0E) N:306 SVR:1.4

Bin	PVA	SD	N	SVA	SD	SVR	SD	γ	SD	ψ	SD
ST01	284.4	4.6	5	277.0	13.6	-6.1	13.7	4.5	2.1	11.6	13.8
ST02	284.4	4.7	8	287.8	8.3	4.4	8.1	4.9	1.1	1.2	8.4
ST03	284.4	4.8	15	290.7	8.2	7.5	8.0	3.5	0.7	-3.2	8.2
ST04	284.4	4.9	28	288.9	11.8	5.6	11.7	8.5	2.2	3.7	12.0
ST05	284.3	5.0	17	296.8	8.6	13.5	8.5	14.3	1.0	1.5	8.7
ST06	284.1	5.1	11	292.4	11.8	9.4	12.0	14.8	1.3	6.2	11.9
ST07	284.0	5.2	25	284.9	9.3	2.2	9.4	6.9	5.4	5.6	10.8
ST08	284.1	5.4	14	279.9	5.4	-2.8	5.4	-7.4	2.7	-3.5	6.0
ST09	284.4	5.6	3	268.3	3.2	-14.2	2.9	-10.4	0.9	5.3	3.4
ST10	284.7	5.7	2	286.0	20.0	3.3	20.8	-4.6	2.1	-6.3	20.1
ST11	284.9	5.9	20	279.5	9.9	-3.8	10.0	-1.0	0.6	4.0	9.9
ST12	285.1	6.1	20	282.4	8.0	-1.1	8.1	-1.0	0.3	1.2	8.0
ST13	285.3	6.3	12	282.5	6.9	-0.8	6.9	0.1	0.8	2.3	7.0
ST14	285.4	6.5	26	280.5	8.4	-2.8	8.1	6.3	2.7	10.7	8.8
ST15	285.3	6.7	35	281.1	14.0	-1.9	14.0	12.2	0.7	15.8	14.0
ST16	285.1	7.0	12	282.4	7.0	-0.9	6.9	9.9	2.0	12.0	7.3
ST17	285.2	7.2	15	288.8	5.7	5.6	5.8	1.2	2.6	-3.0	6.2
ST18	285.4	7.6	11	281.5	11.5	-1.5	11.8	-0.5	3.1	2.7	12.0
ST19	285.6	7.9	8	276.5	11.4	-6.0	11.2	1.6	1.0	10.0	11.4
ST20	285.7	8.3	8	280.8	10.3	-2.1	9.7	5.6	1.9	9.8	10.4
ST21	285.5	8.6	5	279.8	5.9	-2.8	6.0	12.5	1.2	17.5	6.0
ST22	285.8	10.7	1	316.0	20.0	35.8	22.7	13.4	2.2	-17.9	20.1
ST23	285.3	12.0	4	291.5	28.8	13.7	29.0	14.0	6.1	6.5	29.4
ST24	280.8	13.3	1	285.0	20.0	8.8	24.0	56.7	2.9	51.4	20.1

Pole: latitude and longitude of the pole of rotation used. N: the number of slip vectors. SVR: the average slip vector residual. Angular velocities were not estimated so no rate data are shown here. Bin labels are as listed in Table 1 and other information about the bins can be found there. PVA is the average plate vector azimuth predicted by the pole of rotation (in each case, SD is the standard deviation for the parameter that it follows), N is the number of slip vectors in the bin, SVA is the average slip vector azimuth, SVR is the average slip vector residual, γ is the average convergence obliquity, and ψ is the average slip vector obliquity.

Figure 1b). For each bin, the means and standard deviations of slip vector azimuths, trench normals, plate vectors, fault plane dip and rake angles, and convergence rates are calculated (Tables 1 and 2). Because the expected plate vector with azimuth Φ can be calculated at the same point that the trench-normal azimuth T is estimated, the mean convergence obliquity γ is $1/n \sum_{i=1,n} (T_i - \Phi_i)$, where n is the number of estimates of T and Φ in the bin. However, because the trench-normal at the epicenter of the earthquake (from which the slip vector azimuth β is derived) is not clearly defined, the mean slip vector obliquity ψ is the mean trench-normal azimuth minus the mean slip vector azimuth within the bin and its variance is $\sigma_\psi^2 = \sigma_T^2 + \sigma_\beta^2$, where σ represents standard deviation.

Bins were made large enough to maximize the number of slip vectors within them but small enough to keep the standard deviations of the other binned parameters small. Final bin widths are 75 to 200 km at the trench and vary from trench to trench (Table 1). Binned slip vectors from the Aleutian trench display similar variations along the trench for four different bin widths (Figure 2). Standard errors in slip vector azimuths in bins that have 10 or more slip vectors are typically less than 20° (Figure 3). Because the standard deviation of misfit to trench slip vectors in the NUVEL-1 solution is 12° (C. DeMets, personal communication, 1992), 20° is probably a conservative estimate for the measuring error of an individual slip vector and is used as the slip vector standard deviation for bins with fewer than 3 slip vectors.

Weighting

In calculating the means and standard deviations of the earthquake parameters in bins, a choice must be made on the weighting of individual values. Figure 2 (c, e, f, and g) shows that for four different weighting schemes, the differences in calculated means of ψ are small.

One method of weighting is to use some function of the seismic moment of the earthquake from which the parameter is derived (DeMets, 1992). Weighting directly with the seismic moment (Figure 2g) clearly produces smaller standard deviations. This happens because the statistics are dominated by a few events in the bin that might have one or two orders of magnitude more weight than the rest. Is this appropriate? The goal of the statistics is to produce an unbiased estimate of the true slip direction on the fault plane. Suppose the scatter in the slip vectors is due to intrinsic variation in the slip directions on the thrust plane with no measurement error. In this case weighting by seismic moment is appropriate because the mean slip vector is the vector sum of all the slip vectors and the moment, proportional to the slip during the earthquake, is a measure of the length of the slip vector averaged over the fault plane. However, if the measurement error is comparable to the intrinsic scatter in slip direction, then weighting by seismic moment may give a biased slip vector mean; this happens when the slip vector of the largest (and most weighted) event is not in the true slip direction. It is difficult to separate these two sources of error, but, based on my experience with estimating focal mechanism parameters, I judge that measurement error for slip vectors is at least 10° and is therefore a large part of the observed scatter (for example, Figure 3). Only when enough time has elapsed so that each bin contains several of the largest events will moment-weighting provide a good estimate of the mean and variance. Because I think that measurement error is a large part of the observed scatter in slip vector azimuths, in calculating means and standard deviations within the bins I give each slip vector

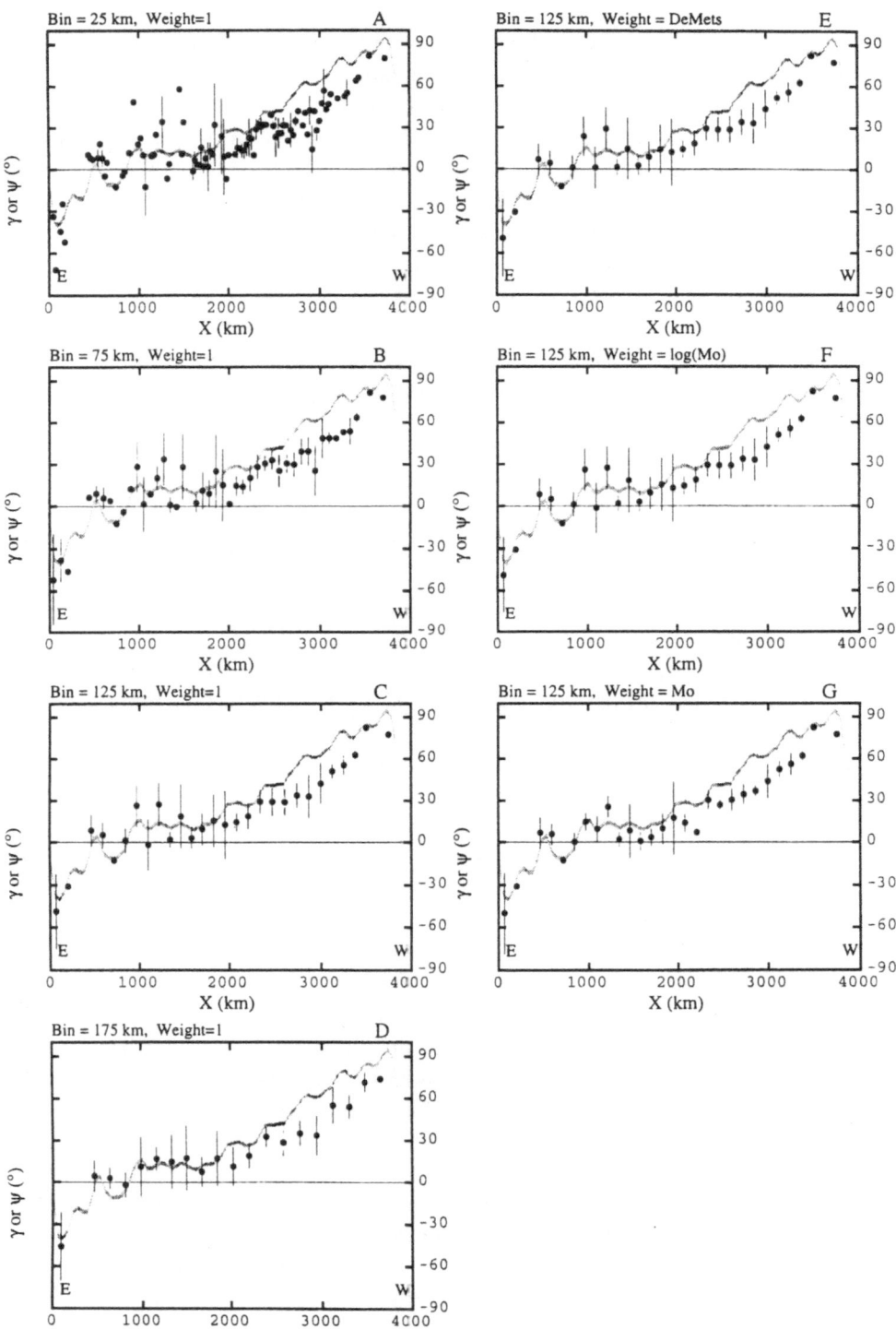

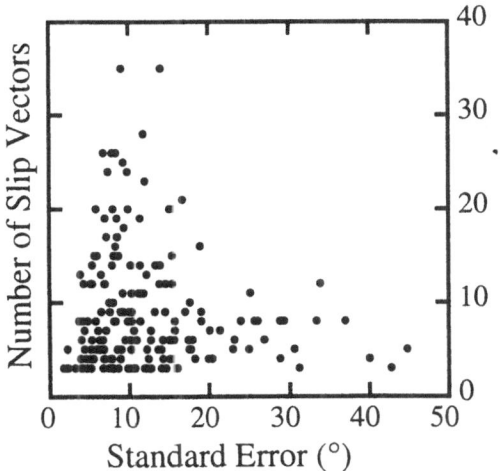

Figure 3
Relationship of slip vector standard errors .n each bin (in degrees) to the number of slip vectors in the
bin.

equal weight. However, in the estimation of rheology parameters by comparison of
weighted χ^2 values, each bin is weighted by the inverse of the variance of the
uniformly-weighted slip vectors within it.

Relationship of Slip Vector Obliquity to Plate Obliquity for the World's Trenches

Figure 4 shows earthquake slip vector obliquity, plate convergence obliquity,
dip and rake angles, and convergence rates with distance along the world's large
subduction zones. Maps showing many of these slip vectors are presented by YU *et
al.* (1993). Slip vectors along some of the subduction zones have been studied in

---

Figure 2
Variations of earthquake slip vector obliquity ψ and plate convergence obliquity γ with distance along
the Aleutian trench for different binning and weighting schemes. The shaded curve shows the range of
plate obliquity allowed by uncertainties in the pole of rotation. Dots with error bars show the mean
angle ψ for each bin with its standard deviation. (Those without error bars are for bins with fewer than
3 slip vectors.) Where the dots follow the shaded line, slip vectors parallel the plate vector (i.e., $\psi = \gamma$);
where the dots stay near the zero line, the slip vectors are perpendicular to the trench (i.e. $\psi = 0$). (a)
through (d) show different widths with uniform weighting while (c) and (e) through (g) show the slip
vector data for bin width of 125 km when individual slip vector azimuths are weighted in different ways.
In (c) all slip vectors are given equal weight, in (e) weight is according to expected measurement
uncertainties based on seismic moment (weight = σ^{-2}; for $M_0 < 10^{17}$ Nm, $\sigma = 25°$, for $10^{17} < M_0 <
10^{18}$ Nm, $\sigma = 20°$, and for $M_0 > 10^{18}$ Nm, $\sigma = 15°$, DEMETS, 1992), in (f) they are weighted by the log
of the seismic moment, and in (g) slip vectors are weighted by the seismic moment.

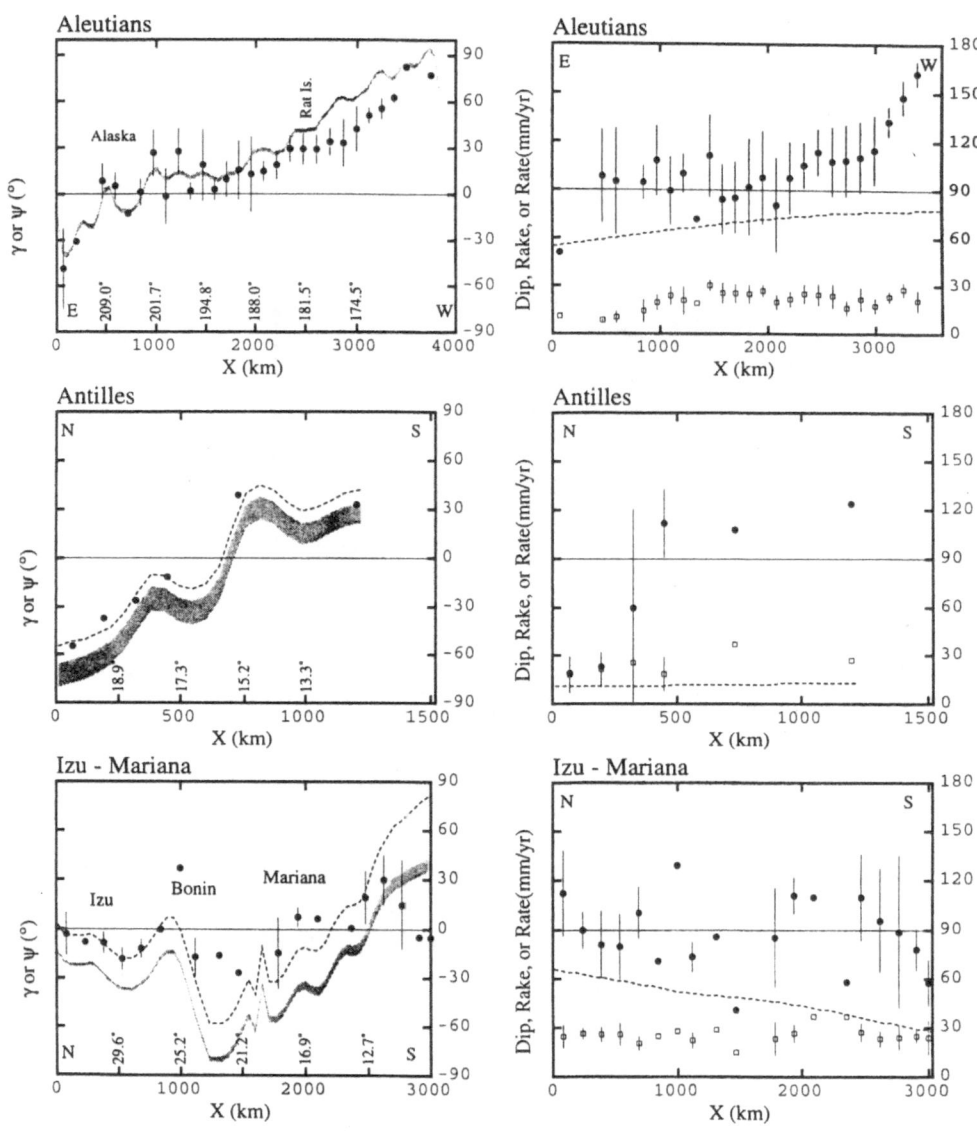

Figure 4

detail for evidence of variations in deformation (e.g., EKSTRÖM and ENGDAHL, 1989; DEMETS, 1992; MCCAFFREY, 1991) and more work of this type remains. In this study I am concerned with variations on the scale of the entire length of the trenches and therefore will focus on the general trends of the slip vectors. Table 2 gives the estimated arc-parallel slip rate of the forearc (V_s) for each bin.

Visual examination reveals a large difference from one trench to another in how the slip vectors respond to convergence obliquity. In cases where the slip vector

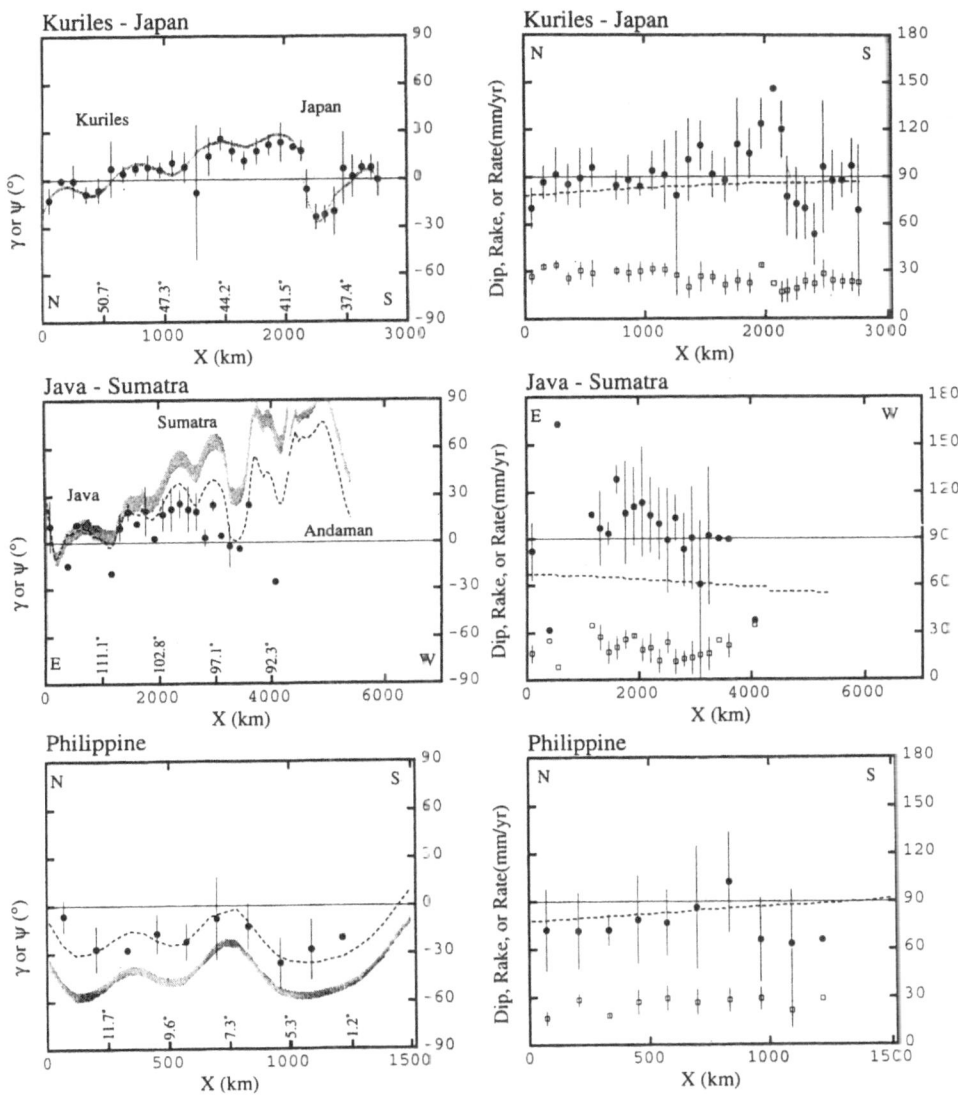

Figure 4

obliquity falls along the shaded line (or the dashed line for the new estimate of the pole of rotation) in Figure 4, slip vectors are parallel to the plate vector and suggest that no significant deformation of the forearc occurs. This behavior is observed in the Antilles (relative to the new pole), Izu (relative to the new pole), Kuriles-Japan, Philippine (relative to the new pole), Solomons (relative to new pole), Ryukyu, South America, N. Tonga, and Tonga (relative to new pole) forearcs. The other extreme is where the forearc deforms enough to prevent significant oblique slip and

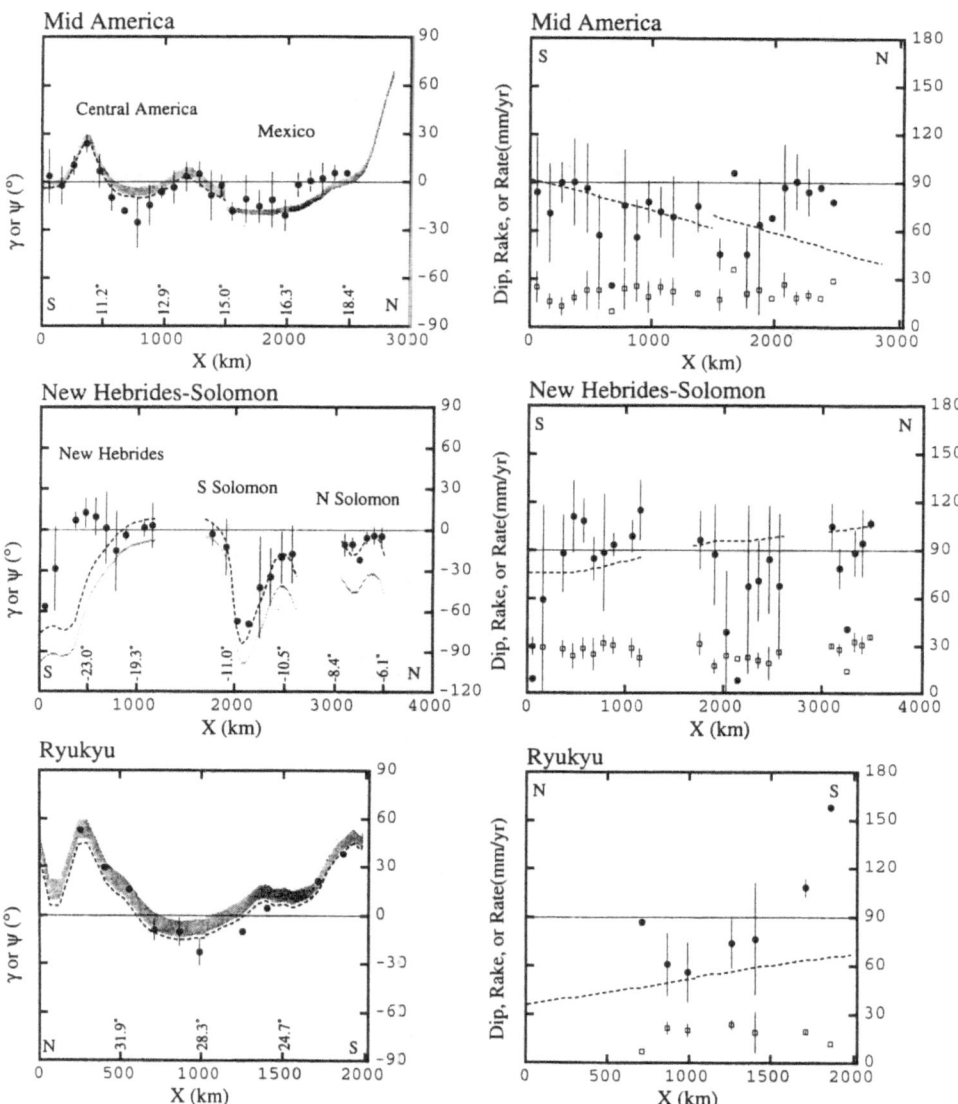

Figure 4

the slip vectors remain nearly normal to the trench, in which case the slip vector obliquity is near zero (Figure 4). Examples of this behavior are the Mariana and New Hebrides forearcs.

Intermediate behavior is characterized by slip vectors that fall between the plate convergence vector and the trench-normal, indicative of partial partitioning of slip. This suggests that the forearc has some strength and also exhibits anelastic

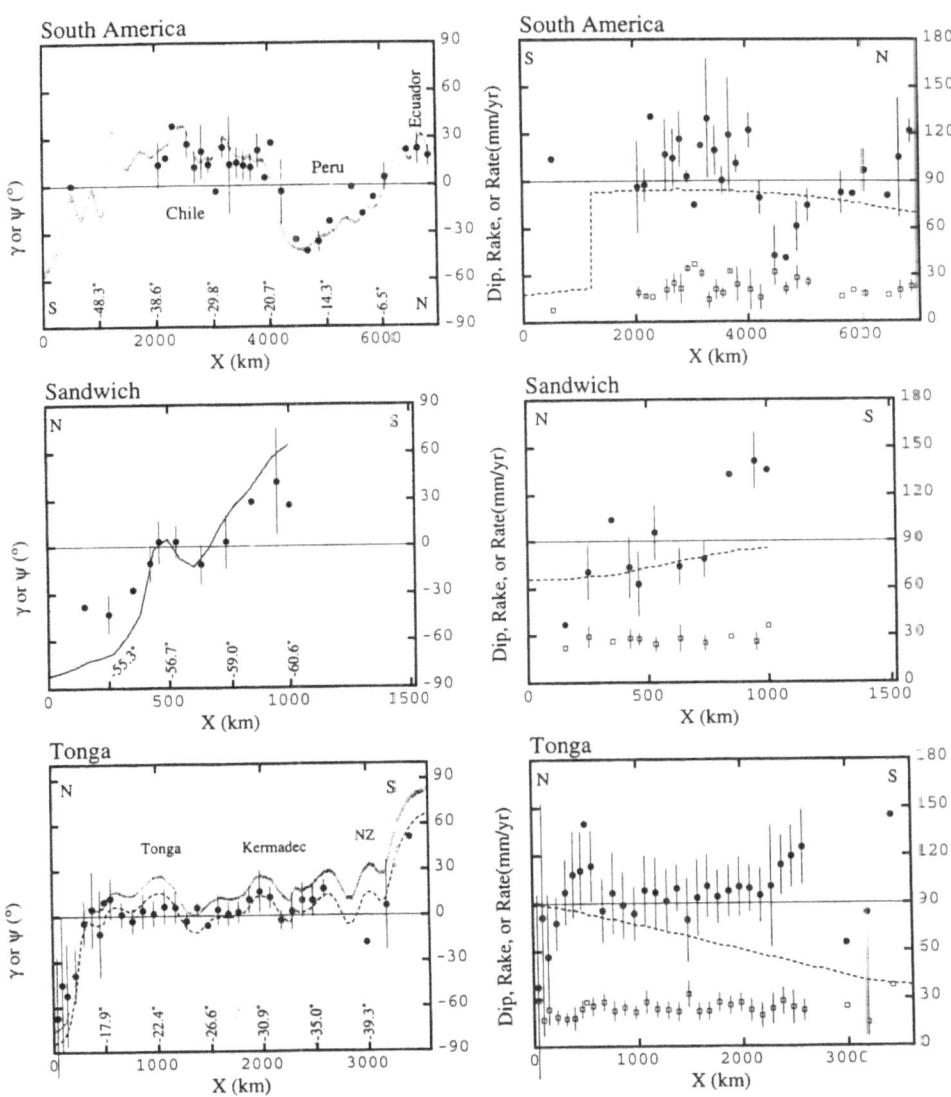

Figure 4

(Left column) Distributions of earthquake slip vector obliquity and plate obliquity with distance along the world's trenches. The positive x direction is in the direction of trench strike and distance is measured along the trench axis. The shaded curves show the range of plate obliquity allowed by the uncertainties in the published poles of rotation. Positive slope in the obliquity curve indicates that the trench is convex toward the subducting plate and negative slope indicates it is concave. Dashed lines on some plots show the plate obliquity for recalculated poles of rotation as discussed in the text. Dots with error bars show the mean angle ψ for each bin with its standard deviation. (Those without error bars are for bins with fewer than 3 slip vectors.) Where the dots follow the shaded line the slip vectors are parallel to the plate vector (i.e., $\psi = \gamma$); where the dots stay near the dashed line ($\psi = 0$), slip vectors are perpendicular to the trench. Horizontal axes are also marked with appropriate latitude (for N-S plots) or longitude (for E-W plots) for points along the trench. (Right column) Plots along each trench of binned rake angles (dots) and dip angles (boxes) with their standard deviations. The dashed line shows the convergence rate between the two plates in mm/yr (numbers on vertical axis are both degrees and mm/yr).

deformation. Examples of this behavior are the Aleutian, Java-Sumatra, Mexico, Solomon, Sandwich, and Tonga forearcs.

Slip partitioning characteristics are also revealed by rake angles of the earthquake mechanisms (Figure 4). First, ψ and rake angles show similar trends along the trenches (note that the rake angle is offset by $90°$ relative to ψ). The angle ψ is a measure of the amount of oblique slip derived from the slip vector and the trench outline. The rake angle also reveals oblique slip but is derived solely from the earthquake mechanism (the rake angle is the angle the slip vector makes with the strike direction in the plane of the fault). Hence, the agreement in the trends of ψ and rake angles indicates that the bathymetric expressions of the trenches (that I use to reveal the strike of the thrust fault) are good indicators of the strikes of the thrust faults derived from fault plane solutions of the subduction zone earthquakes (see also McCaffrey, 1992).

Where the rake angles deviate from $90°$, oblique slip is occurring and where they are near $90°$ and plate convergence is oblique, slip is partitioned. Significant oblique slip occurs in the following places: the western Aleutians, Antilles, Japan, Sumatra, southern Mexico, Chile; Peru, Sandwich, and the northern and southern Tonga forearcs. The rake angles provide important information because they reveal the forearc behavior in the absence of knowledge of the actual convergence direction. For example, Izu-Mariana and New Hebrides show rake angles near $90°$ (pure thrust; i.e., the slip vector is perpendicular to the strike of the fault plane) throughout their length. The great curvatures of these arcs preclude poles of rotation that predict convergence normal to them everywhere (Figure 4). Therefore, whatever the pole of rotation, these forearcs show nearly complete slip partitioning. Likewise it is evident that some forearcs display less partitioning because rake angles deviate significantly from $90°$; for example, the western Aleutians, Sumatra, northern Tonga, and others.

Rheological Description of the Thrust Zone–Forearc System

The strategy is to use slip vector constraints on forearc deformation rates within forearcs under oblique convergence to estimate the apparent rheology of the thrust zone–forearc system. Because the observed forearc deformation is influenced by both the rheology of the forearc and the magnitude of the stress acting on it, the 'true' rheology of the forearc cannot be estimated independently of the stress on the subduction thrust fault. The 'apparent' rheology of the forearc is therefore the macroscopic rheology of the forearc under some unknown stress at the thrust interface. Care must be used in comparing one forearc to another; the apparently stiff (elastic) forearc may simply have less stress acting on it and may in fact be less stiff. It is in this sense that the characterizations of forearc rheologies inferred from slip vectors apply to the thrust zone–forearc system as a whole.

Moreover, because of spatial averaging, this approach reveals an effective forearc rheology rather than the actual rheology. For example, deformation may be controlled by friction on faults but a system of distributed faults may, averaged in space and time, appear to have an elastic-plastic or viscous rheology. This effective rheology, while not applicable locally, is useful in predicting the long-term geologic evolution of large regions.

To constrain the rheology of the thrust-forearc system, I first estimate the forearc deformation and the resulting slip vector deflections for a simple class of rheologies. I assume that the horizontal shear stress that acts on the forearc sliver across the thrust interface increases in some manner with increasing obliquity so that gradients in oblique convergence act to stretch or compress the forearc parallel to strike. I further assume that the downgoing plate is fairly stiff and elastic so that strain rates within it are small and do not contribute to the deflection of slip vectors. Permanent, time-dependent deformation of the forearc will cause the angle ψ to differ from γ and this observed difference as a function of γ will be used to characterize rheology.

Consider the class of elastic-plastic rheologies that include elastic (E), elastic-perfectly plastic (EPP), elastic-plastic with slip hardening (EPSH), and elastic-plastic with slip softening (EPSS). Idealized stress-slip curves for these rheologies are shown in Figure 5. To see how these forearc rheologies relate slip vectors to obliquity, first suppose that the motion of the forearc relative to the upper plate occurs along an arc-parallel, vertical, strike-slip fault with a vertically-averaged frictional yield stress of τ_s'. The case in which τ_s' is nonzero and does not vary along the strike-slip fault is the elastic-perfectly plastic case (BECK, 1991; McCAFFREY,

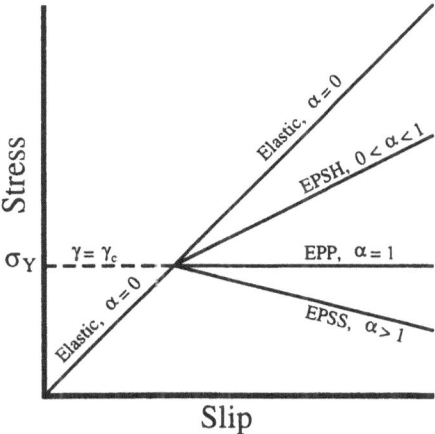

Figure 5

Idealized stress-slip relationship for elastic-plastic system. σ_Y represents the plastic yield stress. γ_c and α are parameters estimated from slip vector data that describe the rheology of the forearc. EPSH = elastic-plastic with slip hardening; EPP = elastic-perfectly plastic; EPSS = Elastic-plastic with slip softening.

1992) and will be described here as a starting point. The other cases are derived from this.

The shear stress acting on a vertical plane parallel to the strike of the subduction zone is derived from the horizontal component of traction on the subduction thrust fault, which is assumed to vary as the sine of the convergence obliquity. At a point along the subduction zone where the convergence obliquity is below a critical value (called γ_c here), the arc-parallel strike-slip fault will not move because, by definition of γ_c, the averaged shear stress on it is less than its yield stress τ'_s. At another point where the obliquity exceeds γ_c, the strike-slip fault will be active because its yield stress has been reached. If τ'_s is invariant along strike, ψ will increase with γ until it reaches the angle γ_c and no more, even if obliquity continues to increase along the margin. This relation arises because the arc-parallel, horizontal component of shear force on the thrust fault produced by oblique convergence cannot exceed the maximum shear force that the strike-slip fault can support (McCaffrey, 1992). The slip rate on the strike-slip fault is simply the amount necessary to deflect the slip vector from the angle γ back toward the trench-normal until $\psi = \gamma_c$.

At sections of the forearc where the strike-slip fault is active, the vertically and time averaged shear stress acting on it is close to its yield stress, τ'_s. In this case ψ is given by

$$\sin \psi = Z_s \tau'_s \sin \delta / Z_t \tau'_t \qquad (1)$$

where Z_t is the depth of the thrust fault, τ'_t is the vertically averaged shear stress on the thrust fault, δ is the average dip angle of the thrust fault from the surface to Z_t, Z_s is the depth of the strike-slip fault, and $0 \leq \psi \leq \gamma$ (McCaffrey, 1992). The right-hand side of (1) is simply the ratio, called R_f, of the shear forces resisting motion on the strike-slip and thrust faults. If the strike-slip fault sustains little force compared to the thrust fault, so that R_f is small, then $\psi \approx 0$ and the slip is completely partitioned between thrust and strike-slip. If the strike-slip fault sustains a relatively large force, then $\psi \approx \gamma$ and there is little partitioning of the slip. If the strike-slip fault is absent then $\psi = \gamma$, slip is not partitioned and the observed value of ψ provides a lower limit for R_f. Clearly, if R_f changes along strike, due to changes in any of the five parameters on the right side of (1), ψ will change as well. Steeper dips will increase ψ, resulting in an apparently stronger forearc. Dip angles vary little along the trenches (Figure 4) and trench averages range only from 20° (Java) to 28° (New Hebrides).

The slip rate on the strike-slip fault needed to deflect the slip vector from the plate vector obliquity angle γ to the angle ψ is

$$V_s = V(\sin \gamma - \cos \gamma \tan \psi) \qquad (2)$$

(Figure 1; Table 2) which increases with obliquity γ for any value of ψ except $\psi = \gamma$ (recall that $0 \leq \psi \leq \gamma$). Assuming that the total slip on the strike-slip fault is proportional to the slip rate and, for now, that the shear force on the thrust fault is uniform along strike, the relation between γ and ψ can be predicted for various rheological descriptions of the strike-slip fault. First, the elastic case is characterized by an average yield stress τ_s' that is larger than the average shear stress acting on the strike-slip fault at the largest obliquity so that the fault does not slip (i.e., there is no fault) and slip vectors on the thrust fault are not deflected ($\psi = \gamma$). For the elastic-perfectly plastic rheology (EPP), the forearc will be elastic until the yield stress is reached at the critical obliquity, γ_c. For convergence obliquity greater than γ_c, because stress on the strike-slip fault is uniform for EPP (independent of total slip) and therefore independent of convergence obliquity, ψ will equal γ_c for all convergence obliquities above γ_c. For elastic-plastic with slip hardening rheology (EPSH), ψ will again equal γ until $\gamma = \gamma_c$ but for $\gamma > \gamma_c$, because fault slip increases with γ, ψ will increase with increasing γ due to increase in stress on the strike-slip fault with increasing slip (1; R_f increases with γ). For the elastic-plastic with slip softening rheology (EPSS), ψ will equal γ until $\gamma = \gamma_c$ but for $\gamma > \gamma_c$, ψ will decrease with increasing γ due to decrease in stress on the strike-slip fault with increasing slip (R_f decreases with γ).

Because few forearcs have a single, arc-parallel strike-slip fault that takes up the motion of a rigid forearc relative to the upper plate, the representation of the forearc deformation given above must be modified. Instead of allowing slip on only a single strike-slip fault, suppose that arc-parallel normal and shear strains occur throughout the forearc block. The deviation of (1) is based on a two-dimensional model with constant obliquity and a rigid forearc block, and ignores the force needed to deform the forearc. A rigorous three-dimensional examination of stress within a curved forearc with gradients in convergence obliquity requires information about the geometry of deformation within it and the forces acting on its boundary. To apply the results above to three dimensions, I will simply assume that the force that stretches (or contracts) the forearc along strike increases with increasing obliquity. At this stage I restrict the discussion to elastic-plastic type rheologies and later show how this representation may also include viscous rheology.

Assume that the forearc deforms rapidly enough to deflect slip vectors only after its yield strength has been exceeded, which occurs at an obliquity of γ_c. For obliquities less than γ_c the arc parallel forces from oblique subduction are sustained elastically in the forearc and the earthquake slip vectors are not deflected away from the plate vector, in which case $\psi = \gamma$ and $V_s = 0$. Once the obliquity exceeds γ_c, the forearc will deform. If it deforms at uniform stress, the forearc slip rate relative to the upper plate will equal

$$V_r = V(\sin \gamma - \cos \gamma \tan \gamma_c),$$

which increases with obliquity γ. V_r is used here as a 'potential' slip rate in the sense that it represents the possible perfectly-plastic component of slip at constant stress; i.e., it is the full arc-parallel slip rate V_p minus the arc-parallel component that is taken up by oblique slip at $\gamma = \gamma_c$. If the resisting force in the forearc increases with slip (slip hardening) then V_s will increase less rapidly than V_r as γ increases ($V_s < V_r$). If the resisting force in the forearc decreases with slip (slip softening) then V_s will increase more rapidly than V_r as γ increases ($V_s > V_r$). With this in mind, the forearc slip rate V_s is described by

$$V_s = \alpha V_r = \alpha V(\sin \gamma - \cos \gamma \tan \gamma_c) \tag{3}$$

where α is a coefficient that determines how V_s changes with plate obliquity relative to V_r. This definition of α allows it to discriminate the behavior of the forearc during plastic deformation. Combining (3) and (2),

$$\tan \psi = (1 - \alpha) \tan \gamma + \alpha \tan \gamma_c \quad \text{for } \gamma > \gamma_c \tag{4a}$$

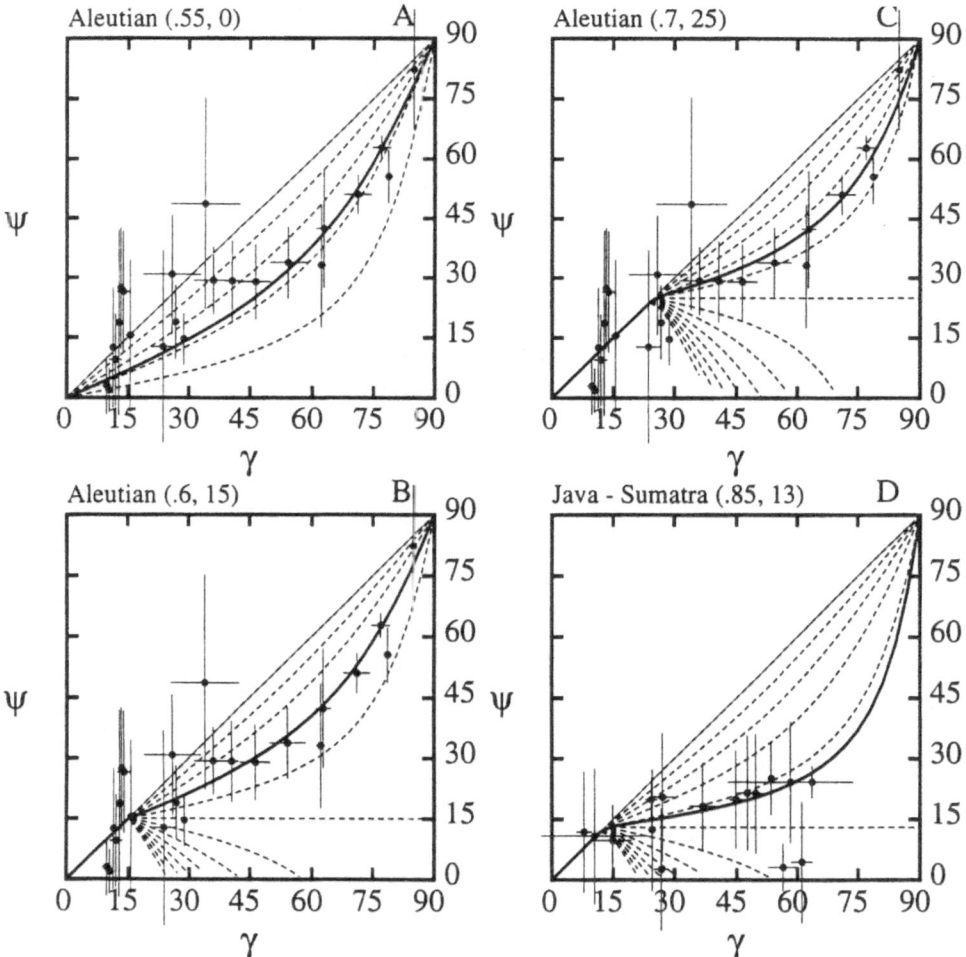

Figure 6

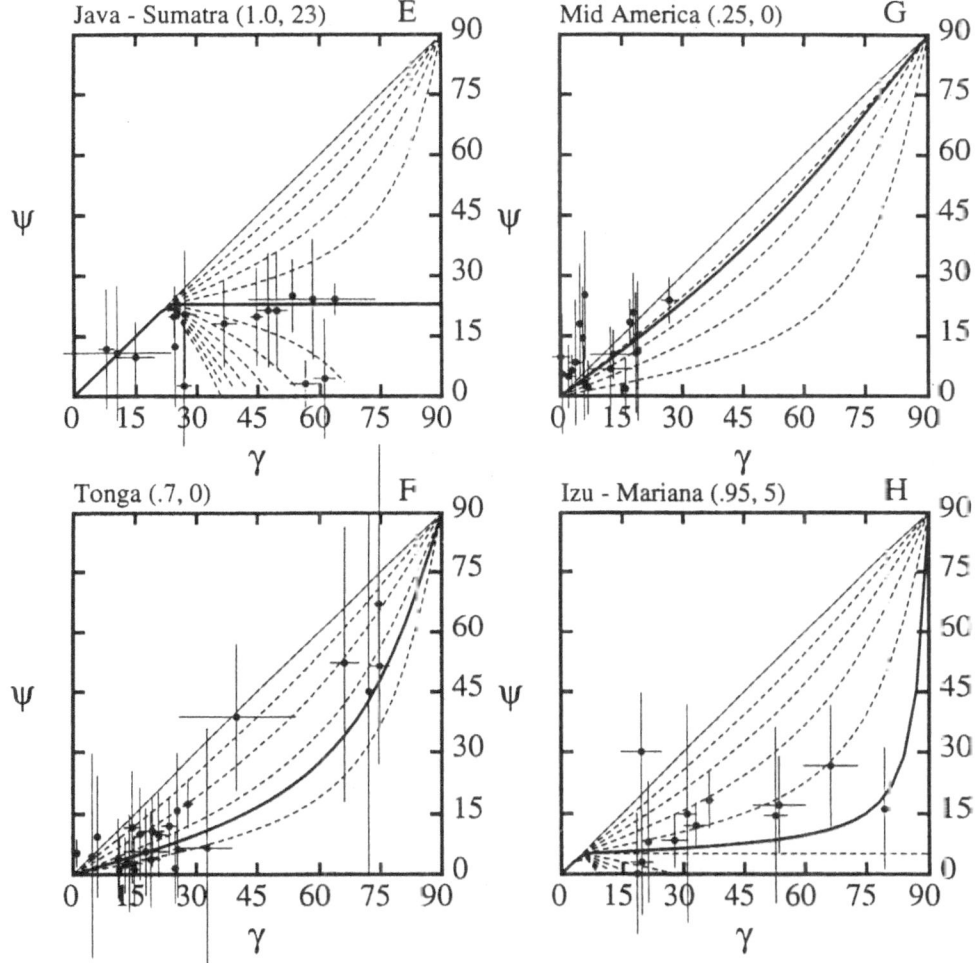

Figure 6

Theoretical relationships between plate obliquity γ and slip vector obliquity ψ for different values of α and γ_c (Equation (4)). Diagonal lines show $\psi(\gamma)$ for $\alpha = 0$ ($\psi = \gamma$). The dashed lines show $\psi(\gamma)$ for increments of α of 0.2 where increasing α causes increasing deviations from the diagonal. The heavy line shows the $\psi(\gamma)$ curve for the α and γ_c (in degrees) values given in parentheses at the top of each box. Also shown are data from forearcs as labeled. The first three plots show that the Aleutian data can be fit with a range of α and γ_c values. The remaining plots show data from selected forearcs that display a variety of behaviors.

and

$$\psi = \gamma \quad \text{for } \gamma \leq \gamma_c. \tag{4b}$$

This relationship is shown in Figure 6 for different values of γ_c and α, along with observations from some forearcs. Equation (4) predicts the slip vector obliquity as a function of convergence obliquity for different forearc rheologies, depending on the values of α and γ_c as follows (see also Fig. 7a):

Elastic: $\alpha \approx 0$ and γ_c is any value, or $\gamma_c \approx \gamma_{max}$ (where γ_{max} is the maximum obliquity at a trench) and α is any value. In this case no significant anelastic strain occurs in the forearc and slip vectors are everywhere parallel to the plate vector. The Mid America forearc is an example of this case (Figure 6).

Perfectly plastic: $\gamma_c \approx 0$ and $\alpha \approx 1.0$. The forearc deforms at small convergence obliquity and moves along the arc at near the rate of the arc-parallel component of relative plate motion. Slip vectors are everywhere perpendicular to the trench. The Izu-Mariana forearc shows this behavior (Figure 6).

Elastic-perfectly plastic: $\gamma_c > 0$ and $\alpha \approx 1.0$. The forearc has some strength in that it can withstand some shear elastically and the slip vector is parallel to the plate vector at obliquity less than γ_c. At obliquities beyond γ_c the forearc deforms at constant stress and ψ equals γ_c. Java-Sumatra is an example of this case (Figure 6).

Elastic-plastic with slip hardening: $\gamma_c > 0$ and $0 < \alpha < 1$. The forearc has some elastic strength and beyond the critical angle γ_c the strength of the forearc increases with increasing slip, causing an increase in ψ with γ. The Aleutian forearc is an example of this case (Figure 6).

Elastic-plastic with slip softening: $\gamma_c > 0$ and $\alpha > 1.0$. The forearc has some elastic strength and beyond the critical angle γ_c the strength of the forearc decreases with increasing slip. The pattern in the slip vectors expected for this behavior can be seen in forearcs at which the slip vectors follow the plate vector out to some critical obliquity γ_c and then rotate back toward the trench normal as obliquity exceeds γ_c (i.e., $\psi < \gamma_c$ when $\gamma > \gamma_c$; examples in Figure 4 are Solomon near $x = 2000$, New Hebrides near $x = 800$, and Sumatra near $x = 2500$). This pattern suggests that the peak (elastic) yield stress in the forearc just prior to the onset of failure is greater than the stress during plastic deformation at obliquities greater than the critical value. If the onset of plastic deformation in the forearc is caused by the development of faults then these faults apparently lower the effective strength of the forearc, as one would expect (e.g., SCHOLZ, 1990). This may be an important mechanism in weakening forearcs that are actively deforming, which may also lower the maximum stress on the thrust fault.

The case of plastic rheology with slip hardening is a special case of elastic-plastic deformation with slip hardening where $\gamma_c = 0$. The case of plastic with slip softening rheology ($\gamma_c \approx 0$ and $\alpha > 1.0$) will give slip vectors that will also fall in the perfectly-plastic field due to the requirement that ψ must have the same sign as γ or be zero. Plastic with slip softening rheology results in slip vectors pointing on the opposite side of the trench-normal than the plate vector because the forearc would be moving along the arc faster than the arc-parallel component of relative plate motion; this is not impossible but is not consistent with the model of forearc deformation driven by oblique convergence.

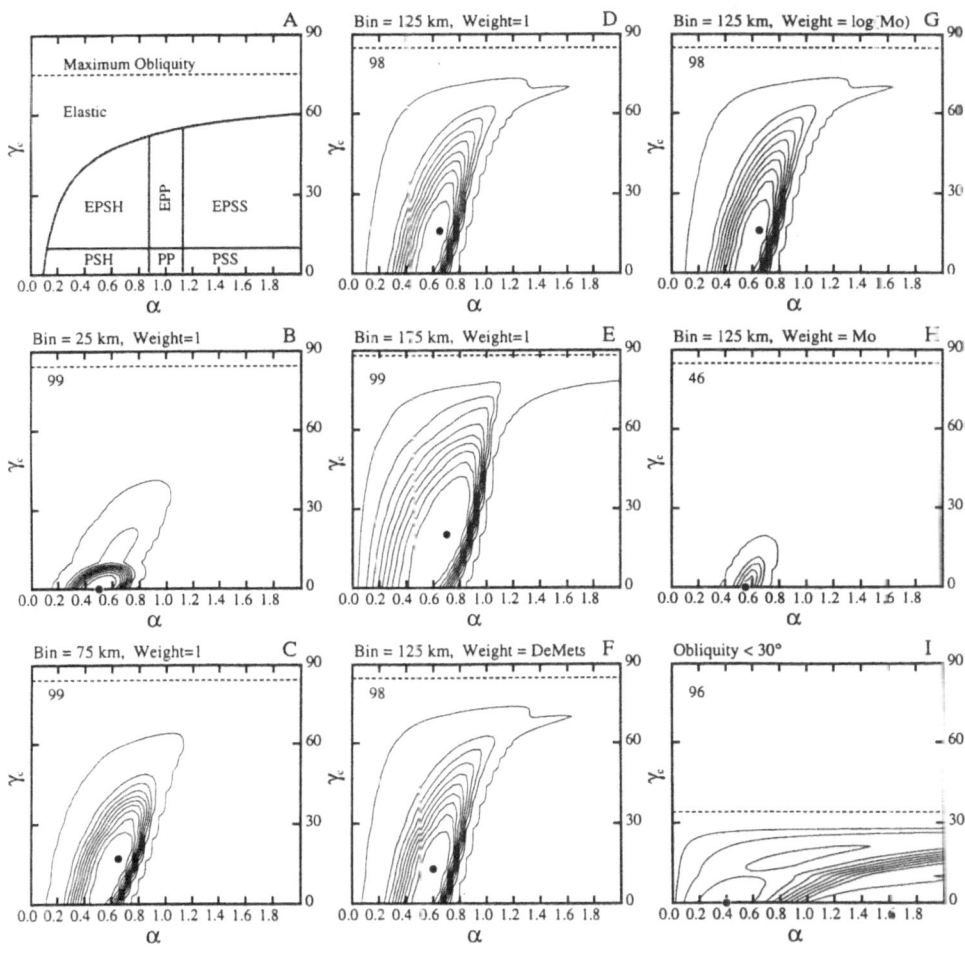

Figure 7

Contours of probability in α-γ_c space corresponding to the seven characterizations of the Aleutian data shown in Figure 2. Number at the upper left corner of the box gives the maximum probability (minimum χ^2), whose position is shown by the dot. The contour interval is 10% in increments of 10% (except the contour farthest from the dot which is the 0.1% contour) decreasing from the maximum. The dashed horizontal line shows the maximum obliquity for the trench. The last box shows the results of using data for obliquities less than 30 only. Plot A shows approximate rheological descriptions for areas of α-γ_c space. EPSH = elastic-plastic with slip hardening; EPP = elastic-perfectly plastic; EPSS = Elastic-plastic with slip softening; PSH = plastic with slip hardening; PSS = plastic with slip softening; PP = perfectly plastic.

Statistical Approach to Constraining Forearc Rheology

A range of rheologies can be represented by equation (4) and the parameters α and γ_c. (Figure 7a shows a schematic interpretation of the relation of rheology to α and γ_c.) Due to the complex tradeoffs between α and γ_c, I estimate them for each forearc by a grid search. For each trench I calculate probabilities for a range of α and γ_c values based on the fit of binned ψ to the predicted $\psi(\gamma, \alpha, \gamma_c)$ (Equation (4)). χ^2 is calculated throughout the range $0 \le \alpha \le 2$ and $0 \le \gamma_c \le \gamma_{max}$, where γ_{max} is the maximum obliquity at the trench. Because both ψ and γ are random variables, each residual of ψ is weighted by the inverse of the sum of the variance of γ and ψ.

$$\chi^2 = \sum_{i=1,N} f[\psi_i - \psi(\gamma_i, \alpha, \gamma_c)]^2/[f(\sigma_{\psi i})^2 + f(\sigma_{\gamma i})^2]$$

where N is the number of bins at the trench, ψ_i is the mean observed ψ in bin i, γ_i is the mean observed γ in bin i, $\sigma_{\psi i}$ is the standard deviation of ψ in bin i, $\sigma_{\gamma i}$ is the standard deviation of γ in bin i, and $f(\xi) = 2 \sin \xi/2$ defines the residual for a difference in angles (see DeMets *et al.*, 1990). The probability $P_\chi(\chi^2, n)$ where n is the number of degrees of freedom ($n = N - 2$), is that any random set of N observations would yield a value of χ^2 as large or larger than the measured χ^2 (Bevington, 1969). A low maximum probability for a trench indicates either that the standard deviations of the observations are underestimated or that the physical model is inappropriate.

The effects of binning and weighting on the computed probability contours for the seven sets of Aleutian data (Figure 2) are shown in Figure 7. Changing bin widths changes the means and standard deviations for γ and ψ and this modifies the probability contours. The sharpest peak in P_χ is for a bin width of 125 km (Figure 7d) and all bin widths show similar results except for 25 km (Figure 7b). Weighting has little effect on the probability contours except when weight is equal to seismic moment which produces a low maximum probability because of the low variance of the observations in the bins (Figure 7h) and a more restricted solution in α-γ_c space. Earlier I argued against use of the seismic moment-weighting because it is more likely to produce a biased mean and to underestimate the variance.

Figure 8 shows the contoured probabilities P_χ in α-γ_c space for most of the world's trenches and Figure 9 shows similar P_χ contours for some trenches using recalculated poles of rotation. Two observations are clear from these plots: (1) the responses of the forearcs to oblique convergence, evidenced by the most probable values of α and γ_c (compare to Figure 7a), are quite variable, and (2) the calculated responses are in some cases sensitive to the choice of the pole of rotation.

Looking at the second point first, small maximum probabilities at some forearcs are caused by slip vectors that are inconsistent with the forearc deformation and plate motion models; maximum P_χ for the Antilles, Izu-Mariana, New Hebrides, and Ryukyu trenches are all less than 80% (Figure 8). The low probability could be caused by underestimating the variance of the slip vectors but I suggest this is unlikely because variance is calculated the same way for all trenches and it is typically larger than the slip vector residual variance in NUVEL-1. Instead, some trenches show inconsistent $\psi(\gamma)$ data in that the slip vectors are either more oblique than the plate vector or point on the other side on the trench-normal from the plate vector (i.e., ψ and γ are of opposite sign). In Figure 4 such inconsistencies appear as slip vector obliquity values (dots) that are not between the shaded (or dashed) line (plate obliquity) and the horizontal line at $0°$ (the trench-normal). For example, at the New Hebrides trench there are no inconsistent points relative to the shaded obliquity curve (from the published pole) but several relative to the new pole (dashed curve) which results in a smaller maximum P_χ for the new pole. At the Izu-Mariana trench the opposite case is seen; the new pole has fewer inconsistencies and a larger maximum P_χ. Other trenches show inconsistent data but they are of low importance and do not influence the solution.

The new poles of rotation in most cases improve the model fit but in others (Java-Sumatra, New Hebrides, and S. Tonga) they degrade it (compare Figure 9 to

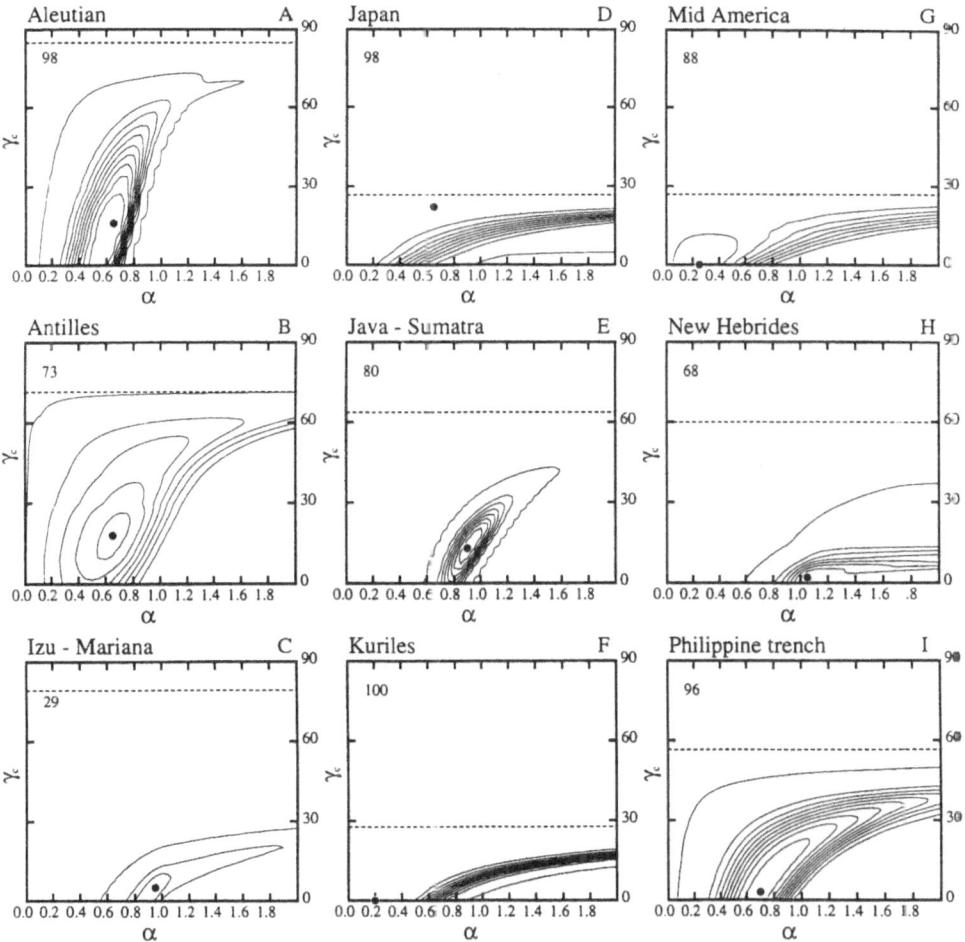

Figure 8

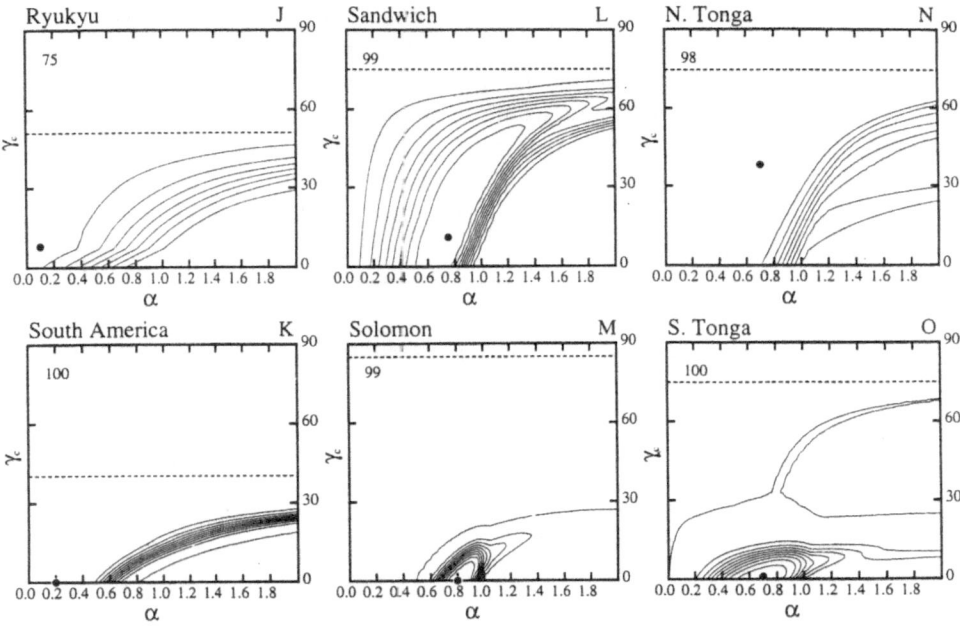

Figure 8
Contours of probability in α-γ_c space for the world's trenches. See Figure 7 for explanation.

8) although they provide a better fit to the slip vector azimuths themselves (Table 3). This apparent contradiction arises because the least-squares solution for the new pole of rotation matches the slip vector azimuths with the plate convergence direction while the partitioning model merely requires that the plate convergence direction be more oblique than the slip vector. Moreover, the pole of rotation estimated by least-squares fit to the slip vectors can increase the inconsistencies between the obliquities and the slip vectors. For example, consider the case, such as Java-Sumatra, where slip vectors along the trench align with the plate convergence direction at low obliquities (in the range $0 \leq x \leq x_c$ where x_c is the position along the trench where $\gamma = \gamma_c$) and at greater obliquities (at $x > x_c$) they are rotated away from the plate vector toward the trench-normal. A least-squares solution for the pole of rotation will try to match the slip vectors in the range $x > x_c$ at the expense of fitting those in the range $x \leq x_c$, which in fact show the true convergence direction. The new pole of rotation will predict that the convergence direction in the range $x \leq x_c$ (where x_c retains its original definition) will be more normal to the trench than the true convergence direction and therefore the slip vector will be more oblique than the new calculated plate convergence direction. Such slip vector and obliquity data are inconsistent with the deformation model and arise from using the wrong pole of rotation, even though this pole minimizes misfit to the slip vector azimuths.

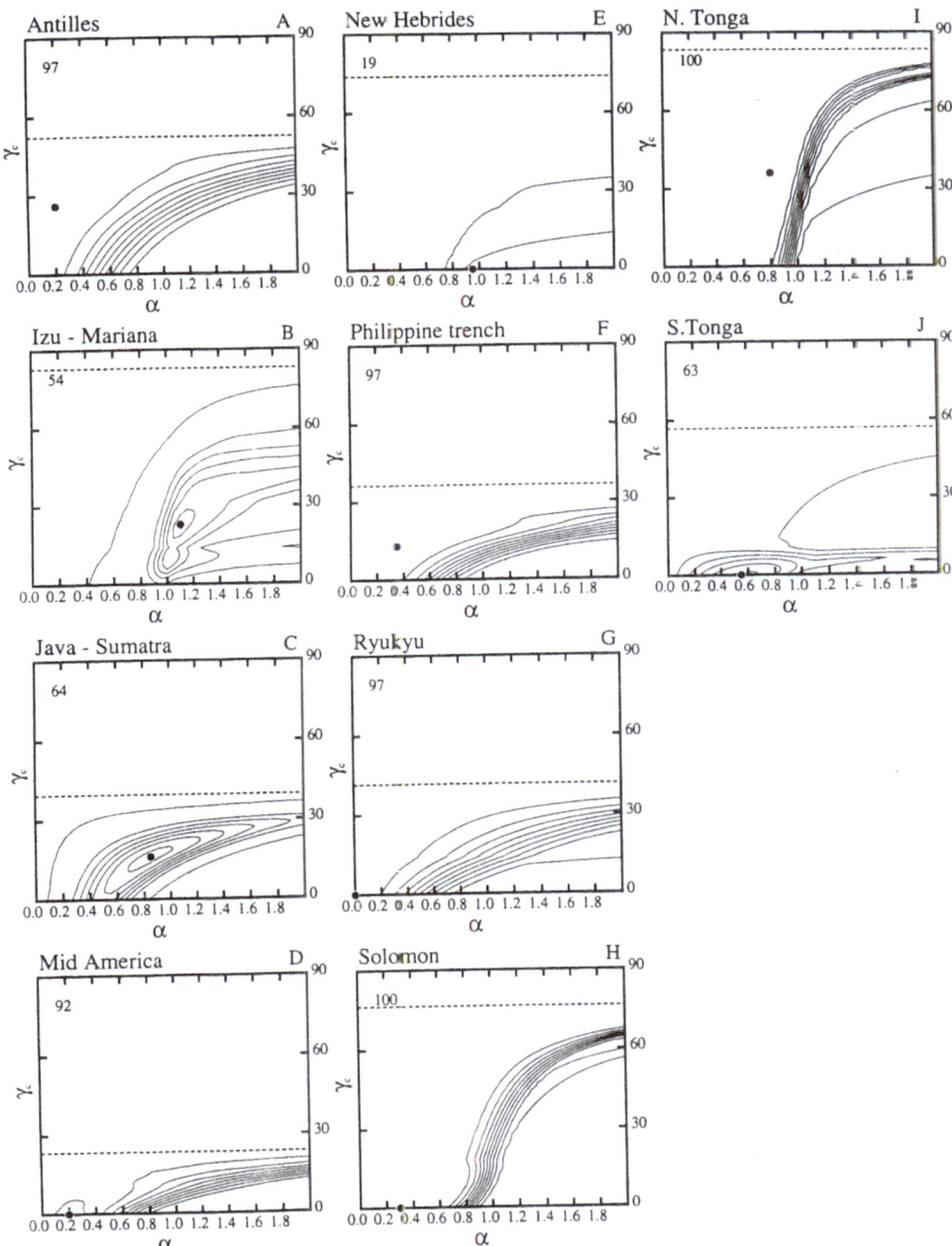

Figure 9
Contours of probability for some of the world's trenches using recomputed poles of rotations. Format
is same as Figure 7.

Discussion

A variety of behaviors of forearcs to oblique convergence is evident in Figure 4 and shown statistically in Figure 8. My best estimate of rheology for each forearc is listed in Table 4 (by comparing the most probable region of $\alpha - \gamma_c$ space in Figure 8 to Figure 7a). Elastic behavior is exhibited by Japan, Kuriles, Mid America, Ryukyu, N. Tonga and South America. In these cases, the slip vectors follow the plate convergence vector through all convergence obliquities. The clearest example of elastic behavior at high obliquity is N. Tonga where obliquity reaches 74° and the slip vectors remain within a few degrees of the Pacific-Australia plate vector (Figure 4; Table 5), although the S. Tonga trench shows different

Table 4

Rheology estimation for forearcs

Forearc	Bins	γ_{max}	α	γ_c	χ^2	P_{max}	Rheology
			Using published poles of rotation				
Aleutian	27	85	0.65	16	0.49	98	EPSH
Antilles	6	72	0.65	18	0.51	73	EPSH
Izu-Mariana	19	79	0.95	5	1.16	29	PP
Japan	9	27	0.65	22	0.19	98	E
Java-Sumatra	20	64	0.90	13	0.72	80	EPP
Kuriles	22	28	0.20	0	0.28	100	E
Mid America	25	27	0.25	0	0.66	88	E
New Hebrides	7	60	1.05	2	0.57	68	PP
Philippine	10	57	0.70	3	0.32	96	PSH
Ryukyu	10	51	0.10	8	0.64	75	E
S. America	30	40	0.20	0	0.29	100	E
Sandwich	11	75	0.75	11	0.20	99	EPSH
Solomon	13	85	0.80	0	0.24	99	PSH
N. Tonga	8	75	0.70	38	0.09	100	E
S. Tonga	24	75	0.70	1	0.50	98	PSH
			Using New poles of rotation				
Antilles	6	52	0.20	26	0.13	97	E
Izu-Mariana	19	83	1.10	23	0.93	54	EPP
Java-Sumatra	21	40	0.85	16	0.85	64	EPSH
Mid America	25	24	0.20	0	0.61	92	E
New Hebrides	10	74	0.95	0	1.40	19	PP
Philippine	10	36	0.35	13	0.28	97	E
Ryukyu	10	42	0.00	0	0.29	97	E
Solomon	14	76	0.30	0	0.17	100	E
N. Tonga	8	83	0.80	36	0.10	100	E
S. Tonga	24	57	0.55	0	0.87	63	PSH

Bins is the number of bins along the trench. γ_{max} is the maximum obliquity at the trench for which slip vectors are available. χ^2 is the weighted variance of the residuals in ψ. P_{max} is the probability for the α and γ_c values listed and is the maximum. Rheology: E = elastic; EPSH = elastic - plastic with slip hardening; EPP = elastic - perfectly plastic; EPSS = Elastic - plastic with slip softening; PSH = plastic with slip hardening; PSS = plastic with slip softening; PP = perfectly plastic.

Table 5.

Forearcs sorted by rheology (decreasing amount of elastic component)

Trench	SVR N	SVR Average	Arc Age	Rheo- logy		Arc Type	Slab Age	Conv. rate	Max. Obl.	Ave. Dip	W (km)	R (°)	Mw≥8	Strain Class
Japan	120	0±11	115	E		C	90-120	86	27	28	109	9	Y	6
N. Tonga	59	-2±29	24	E	(E)	O	120	87	74	21	68	3	Y	1
Mid America	161	2±12	95	E	(E)	C	50	60-80	26	22	90	10-36	Y	3
Kuriles	233	3±10	153	E		C	130	82	28	25	80	23	Y	5
S. America	149	4±14	226	E		C	5-45	80	41	24	107	14-70	Y	6
Ryukyu	24	5 ± 9	55	E	(E)	O	60	45-60	50	22		7-10	Y	2
Aleutian	240	8±16	56	EPSH		O	55	70	80	23	53-113	15	Y	4b
Sandwich	53	1±21	30	EPSH		O	45	18	75	28	53	3	N	1
Antilles	10	-21±11	48	EPSH	(E)	O	100	12	71	25		6	N	4a
Java-Sumatra	80	24±20	27	EPP	(EPSH)	C	70	60	77	20	140	23	N	5
Philippine	74	-25±17	6	PSH	(E)	T	55	85	56	26	-	57	?	
S. Tonga	306	13±11	30	PSH	(PSH)	O	110	59	26	25	97	66	N	1
Solomon	87	-22±19	8	PSH	(E)	O	55	96	89	27	107	18-29	Y	4a
Izu-Mariana	73	-16±26	45	PP	(EPP)	O	150	40-60	68	26	74	7	N	1-2
N. Hebrides	81	-33±31	8	PP	(PP)	O	40	79	88	28	66	5	N	1

SVR gives the statistics of slip vector residuals (in degrees) along the trench. N is the number of slip vectors. All ages in Ma. Arc ages and types from Jarrard (1986). Rheology: E=elastic, EPP=elastic-perfectly plastic, PP=perfectly plastic, EPSH=elastic-plastic with slip hardening, PSH=plastic with slip hardening, EPSS=elastic-plastic with slip softening. Rheologies are sorted first by putting all elastic forearcs at the top (in order of decreasing SVR), then in order of increasing deformation (increasing α and decreasing γ_c using minimum χ^2 values). Rheologies in parentheses are based on new poles of rotation. Arc type: C=continental, O=oceanic, T=transitional. Convergence rates in mm/yr. Maximum obliquity and Dip angle of slab in degrees. The dip angle is the average dip of all thrust fault planes beneath the forearc to a depth of 60 km. W is the down-dip length of the thrust fault estimated by Pacheco et al. (1993). R is the radius of curvature of the arc in degrees. Mw≥8 indicates whether or not the arc has great subduction earthquakes. Strain class describes deformation of the overriding plate (Jarrard, 1986; Table 3); smaller numbers indicate tension (1 for active back arc spreading), 4 is neutral, and larger numbers indicate compression.

behavior (Figure 8). In addition, using new poles of rotation, the Antilles, Philippine, and Solomon forearcs also show elastic behavior (Figure 9).

The other commonly observed rheology is plastic with slip hardening (PSH), sometimes with an elastic component (EPSH). PSH is seen at the Philippine, Solomon, and S. Tonga forearcs and possibly at the Aleutian and Sandwich forearcs. These forearcs deform at small obliquities but are not completely partitioned; i.e., slip vector obliquity increases with increasing plate convergence obliquity. It is shown below that this behavior could also be caused by a viscous forearc rheology. EPSH is seen at the Aleutian, Antilles, and Sandwich forearcs. These forearcs respond elastically at low convergence obliquities and show plastic with slip hardening behavior at obliquities higher than about 30° (Figure 8).

The Java-Sumatra forearc displays elastic-perfectly plastic behavior (EPP). It is elastic up to obliquity of about 20° and at higher obliquities it deforms at a rapid enough rate to keep the slip vectors at a nearly constant angle relative to the trench-normal (Figure 6). Earlier (McCAFFREY, 1992), I used the Java-Sumatra case to support the use of EPP rheology in the presence of an arc-parallel strike-slip fault that takes up the motion of the forearc relative to the upper plate. Java-Sumatra is apparently the only one of the world's forearcs that displays this behavior (see also YU et al., 1993).

The Izu-Mariana and New Hebrides forearcs fit the category of perfectly plastic ($\alpha \approx 1, \gamma_c \approx 0$) although Izu-Mariana displays some elastic component when the new pole is used (Figure 9). These forearcs withstand relatively little arc-parallel shear stress and slip vectors are largely perpendicular to the trench. This behavior is referred to as complete partitioning.

The values of $0 \ll \alpha \ll 1$ and $\gamma_c \approx 0$ (PSH), may represent, at a macroscopic scale, a viscous response of the forearc to the stress of oblique convergence. This range of α indicates either that the forearc moves as a coherent block relative to the upper plate at a rate lower than V_r, or that there is an arc-perpendicular gradient in the arc-parallel velocity V_s. The latter could be due to viscous deformation. For example, consider an oblique convergent margin with a linear viscous upper forearc (without an arc-parallel strike-slip fault). For simplicity assume that the plate boundary is vertical so that the results of ENGLAND et al. (1985) provide the arc-parallel velocity gradients within the interior of the upper plate. Using the x direction parallel to the trench, y perpendicular to the trench ($y = 0$ at the trench and increases arcward), and again as a reference frame the upper plate far from the boundary, within the forearc the velocity parallel to the trench is $v_x(y) = V \sin \gamma \exp(-y/\iota)$ which decays exponentially (with a length scale of ι) away from the trench and toward the interior of the upper plate (this includes a no-slip condition at the trench). V_s is the average velocity across the width of the forearc (w) over which the slip vectors are averaged:

$$V_s = 1/w \int_0^w v_x \, dy = \iota/w(1 - \exp(-w/\iota))V \sin \gamma.$$

Comparing this result to (3), for $\gamma_c = 0$, we see that $\alpha = \iota/w[1 - \exp(-w/\iota)]$. In the limit as ι/w gets large, α approaches 1, approximating a rigid forearc sliding at the full arc-parallel plate rate ($V \sin \gamma$). ENGLAND and WELLS (1991) estimate values for ι of 106 and 153 km for the United States Pacific Northwest on the basis of paleomagnetic rotations. Forearcs are typically 2 to 3 times this wide so that a reasonable range for α might be 0.3 ($w = 3\iota$) to 0.6 ($w = \iota$). Alternatively, if v_x varies linearly across the forearc (i.e., $v_x(y) = V \sin \gamma(w - y)/w$), by carrying out a similar averaging, $\alpha = 0.5$. Because a viscous forearc will have a gradient in v_x in the y direction, α cannot equal 0 or 1 (unless the viscosity is very low, $\alpha = 0$, or very high, $\alpha = 1$). Thus the Aleutian, Philippine, Sandwich, Solomon, and S. Tonga forearcs possibly show viscous behavior.

The global variability in the kinematic behavior of forearcs is clear. This behavior depends on the magnitude of the stress being applied to the forearc, which we do not know, as well as the rheology, so that comparisons among the arcs and any global explanations must be taken with caution. Because the rheology is difficult to quantify, I compare the forearcs qualitatively to other subduction zone parameters in Table 5, in which forearcs are sorted from least deformable to most deformable, from top to bottom.

First, note that half of the elastic forearcs (Kurile, Japan, and Mid America) are characterized by maximum obliquities of less than 30°. Figure 7i demonstrates that the plastic component of the Aleutian forearc rheology is revealed using only plate obliquities less than 30°, suggesting that a lack of convergence obliquities larger than 30° does not alone result in apparent elastic behavior. It is possible that γ_c values for the forearcs with small obliquities are greater than 30°. Nevertheless, because few if any of the forearcs with larger maximum plate obliquities could have γ_c as high as 30° (Figure 8), the Kurile, Japan, and Mid America forearcs must be among the more elastic of the forearcs.

The strength and rheology of the lithosphere varies greatly with depth; the strength of the upper brittle part increases with depth but once ductile behavior predominates, strength decreases rapidly with depth due to increasing temperature. It the thrust earthquakes used here do not occur in the same depth ranges for all trenches, then they may be showing deformation at different levels of this strength-depth relationship. This possibility cannot be addressed confidently with these data, but can be tested by improving the depth estimates of the earthquakes by detailed study. HCMT depths are too uncertain to address this question, considering that the global range in maximum depths of thrust earthquakes beneath forearcs varies by only a few tens of kilometers (TICHELAAR and RUFF, 1993).

Table 5 shows the average dip angle of the thrust plane for each forearc, taken from earthquake mechanisms, and the down-dip width of the seismically active part of the thrust plane, estimated by PACHECO et al. (1993). Neither of these measures, that should influence the force on the thrust plane, appears to correlate with the rheology. TICHELAAR and RUFF (1993), by waveform modeling, show that the maximum depths of thrust earthquakes beneath sections of the Japan, Kurile, Chile, and Aleutian forearcs are in the 35 to 50 km range while for the Mexico forearc it is 20 to 30 km. Nevertheless, Mexico slip vectors show similar behavior to the others (Figure 4), although the maximum obliquity there is small.

The age of the subducting plate or convergence rate or both may be expected to influence forearc behavior because these parameters have been causally linked to other island arc characteristics, such as the occurrence of great earthquakes (RUFF and KANAMORI, 1980), seismic moment release rate (PETERSON and SENO, 1984), and the length of the seismic zone (MOLNAR et al., 1979). However, a few comparisons demonstrate that the link of slab age or convergence rate to forearc rheology is poor (Table 5). The Mariana and New Hebrides arcs display similar rheology and have nearly the same convergence velocity (≈ 80 mm/yr, including 40 mm/yr spreading in the Mariana backarc), yet some of the oldest oceanic lithosphere in the world is subducted beneath the Marianas while 40 Ma lithosphere subducts beneath New Hebrides. The northern Tonga arc also subducts old (≈ 120 Ma) lithosphere at 87 mm/yr and northern South America subducts young (≈ 40 Ma) lithosphere at 80 mm/yr through its big bend (10°S to 30°S), yet these forearcs show similar behavior. However, slip vector residuals are systematically smaller for faster rates of subduction normal to the trench (MCCAFFREY, 1993) which suggests that the effects of convergence rate are more localized than on the scale of entire forearcs. MCCAFFREY (1993) interprets the trench-normal convergence rate effect as being caused by cooling of the forearc by faster subduction and from this infers that the shear stress on subduction thrust faults is lower than 40 MPa.

The upper third or so of the continental lithosphere is thought to be weaker than oceanic lithosphere, owing to the presence of a thick, weak crust (CHEN and MOLNAR, 1983; VINK et al., 1984), so that forearcs built on continental lithosphere may be expected to deform more easily than those on oceanic lithosphere. Based on geologic evidence, JARRARD (1986) showed that continental forearcs were considerably more likely than oceanic forearcs to have forearc faults. This tendency is not evident in the behavior of the arcs inferred from slip vectors, and if any inference of this type can be drawn it is that continental arcs deform less (Table 5). This apparent contradiction can be explained by the generally greater ages of the continental forearcs; slowly slipping forearc faults may be geologically obvious in continents but will not have much effect on the slip vectors. For example, JARRARD (1986) cites the Atacama and Guayanquil faults in South America as continental forearc faults. Slip vector deflections west of the Atacama fault in Chile (bins CS11 to CS15) indicate a right-lateral slip rate of less than 10 mm/yr and those west of the Guayanquil fault in Ecuador (bins NS10 to NS12) indicate 5 mm/yr left-lateral to 6 mm/yr right-lateral slip. These slip rates are very low compared to the tens of mm/yr needed in the New Hebrides and Mariana forearcs to deflect the slip vectors (Table 2). No forearc faults are apparent geologically in the Marianas and New Hebrides (JARRARD, 1986).

The difference between continental and oceanic arcs can be seen locally at the Alaska-Aleutian trench. At the eastern end of the trench, where the arc is continental, the obliquity is $-40°$ and the slip vectors are parallel to the plate vector (Figure 4). In the middle, near Rat Island where the arc is oceanic, the obliquity is $40°$ but the slip vectors are clearly rotated toward the trench normal, suggesting that under similar conditions the oceanic arc deforms while the continental arc does not.

The age of the volcanic arc, for the current subduction episode (JARRARD, 1986), appears to correlate with rheology; older arcs are generally more in the elastic field (E and EPSH) than younger arcs that appear to be PSH and PP (Table 5). This correlation may be secondary, however, as there is a strong correlation between arc age and arc type (Table 5) and the oldest arcs appear to be built on continents. Forearc rheology also does not appear to correlate with the presence of active backarc spreading (see also YU et al., 1993) nor with the upper plate 'strain classes' in general defined by JARRARD (1986) (Table 5).

If the horizontal shear stress acting on the forearc increases with obliquity then forearcs with smaller radii of curvature are expected to deform more because gradients in plate obliquity and stress along these will be larger. Radii of curvature estimated from the trench outlines do not reveal such a correlation (Table 5).

As mentioned earlier, it is difficult to separate the effects of forearc rheology from stress levels on the thrust fault by examination of forearc deformation alone. One could reasonably argue that the global variability in the thrust zone–forearc system is caused by variations in the magnitudes of the stress on the thrust faults rather than by the rheologies of forearcs. In this case, forearcs that deform most should be regions of highest stress and those that deform least have lowest stress. An observation that pertains to this question is that all major thrust earthquakes ($M_w \geq 8.0$) in this century nucleated at sections of forearcs with small slip vector residuals (i.e., little forearc deformation) compared to global averages (MCCAFFREY, 1993). A similar pattern is seen in Table 5 in that the more elastic (less deforming) forearcs are more likely to produce great earthquakes than deforming forearcs. While Table 5 suggests that great earthquakes occur in forearcs that display anelastic deformation, these earthquakes generally nucleate at sections of the forearcs that show largely elastic behavior (MCCAFFREY, 1993).

At supposedly 'decoupled' subduction zones (those without great earthquakes), such as the Mariana and New Hebrides, forearcs appear to be weak relative to the stress on the thrust fault in that they deform readily. First, assuming that forearcs are all identical, the deformation of the Mariana-type forearc suggests that the thrust fault at the Mariana trench sustains a larger than average stress, but this is inconsistent with the notion that a lack of great earthquakes indicates low stress (RUFF and KANAMORI, 1980). By the same reasoning, forearcs where there is little anelastic deformation, such as South America and Japan, must have a lower than average stress on the thrust fault, yet these arcs are the sites of great earthquakes. Thus the ideas that forearc deformation and the occurrence of great thrust earthquakes are both indicators of high stress on the thrust fault are inconsistent.

If, instead, all subduction thrust faults are assumed to be similar, then deforming Mariana-type forearcs are apparently weaker than average and South America-type forearcs (i.e., elastic) are stronger than average. This view is consistent with the distribution of great earthquakes, in that they occur at forearcs with greater elastic components and less permanent deformation (Table 5). Therefore, great earthquakes may not be as much indicators of high stress as they are indicators of stiff, elastic forearcs. Great earthquakes may occur preferentially beneath rigid forearcs either because such forearcs can store elastic strain energy more effectively or because the thrust fault may be physically smoother, and hence more frictionally unstable (see SCHOLZ, 1990), beneath rigid forearcs (MCCAFFREY, 1993). In either case, unless great subduction earthquakes occur at thrust faults with lower than average stress, then forearc deformation is globally controlled by forearc rheology rather than stress on the thrust fault.

Conclusions

By examining the kinematic response of forearcs to along-strike variations in oblique convergence through the study of earthquake slip vectors, it is seen that forearcs display a great range of apparent rheologies. The apparent rheology of the forearc as a whole does not appear to depend on convergence rate, subducting plate age, downdip width of the thrust fault, arc curvature, or the presence of backarc spreading. Forearc rheology appears to correlate with arc type (whether the upper plate is oceanic or continental) and arc age. It is unclear at this time what determines the response of a particular forearc to oblique convergence, but elastic forearcs appear to correlate with the occurrence of great earthquakes, suggesting that rheology rather than stress controls forearc deformation and where great subduction zone thrust earthquakes occur.

Acknowledgements

Thanks to Chuck DeMets and Peter Molnar for many helpful discussions, to Renata Dmowska and Göran Ekström for reviews, and to DeMets and Don Argus for slip vectors. This work would not have been possible without the dependable HCMT data unselfishly supplied by Adam Dziewonski, Göran Ekström, and others at Harvard. Supported by NSF Grants EAR–8903762, EAR–8908759, and EAR–9105050.

REFERENCES

BECK, M. (1991), *Coastwise Transport Reconsidered: Lateral Displacements in Oblique Subduction Zones, and Tectonic Consequences*, Phys. Earth Planet. Int. *68*, 1–8.

BEVINGTON, P. R., *Data Reduction and Error Analysis for the Physical Sciences* (McGraw-Hill, New York 1969) 336 pp.

CHEN, W.-P., and MOLNAR, P. (1983), *Focal Depths of Intracontinental and Interplate Earthquakes and their Implications for the Thermal and Mechanical Properties of the Lithosphere*, J. Geophys Res. *88*, 4183–4214.

DEMETS, C., GORDON, R. G., ARGUS, D. F., and STEIN, S. (1990), *Current Plate Motions*, Geophys. J. Int. *101*, 425–478.

DEMETS, C. (1992), *Oblique Convergence and Deformation along the Kuril and Japan Trenches* J. Geophys. Res. *97*, 17,615–17,625.

DEMETS, C. (1993), *Earthquake Slip Vectors and Estimates of Present-day Plate Motions*, J. Geophys. Res. *98*, 6703–6714.

DZIEWONSKI, A. M., CHOU, T.-A., and WOODHOUSE, J. H. (1981), *Determination of Earthquake Source Parameters from Waveform Data for Studies of Global and Regional Seismicity*, J. Geophys. Res. *86*, 2825–2852.

EKSTRÖM, G., and ENGDAHL, E. R. (1989), *Earthquake Source Parameters and Stress Distribution in the Adak Island Region of the Central Aleutian Islands, Alaska*, J. Geophys. Res. *94*, 15,499–15,519.

ENGLAND, P., HOUSEMAN, G., and SONDER, L. (1985), *Length Scales for Continental Deformation in Convergent, Divergent, and Strike-slip Environments: Analytical and Approximate Solutions for a Thin Viscous Sheet Model*, J. Geophys. Res. *90*, 3551–3557.

ENGLAND, P., and WELLS, R. E. (1991), *Neogene Rotations and Quasicontinuous Deformation of the Pacific Northwest Continental Margin*, Geology *19*, 978–981.

FITCH, T. J. (1972), *Plate Convergence, Transcurrent Faults and Internal Deformation Adjacent to Southeast Asia and the Western Pacific*, J. Geophys. Res. *77*, 4432–4460.

GEIST, E. L., CHILDS, J. R., and SCHOLL, D. W. (1988), *The Origin of Summit Basins of the Aleutian Ridge: Implications for Block Rotation of an Arc Massif*, Tectonics *7*, 327–341.

GEIST, E. L., and SCHOLL, D. W. (1992), *Application of Continuum Models to Deformation of the Aleutian Island Arc*, J. Geophys. Res. *97*, 4953–4967.

JARRARD, R. D. (1986), *Relations among Subduction Parameters*, Rev. Geophys. *24*, 217–284

MCCAFFREY, R. (1988), *Active Tectonics of the Eastern Sunda and Banda Arcs*, J. Geophys. Res. *93*, 15,163–15,182.

MCCAFFREY, R. (1991), *Slip Vectors and Stretching of the Sumatran Forearc*, Geology *19*, 881–884.

MCCAFFREY, R. (1992), *Oblique Plate Convergence, Slip Vectors, and Forearc Deformation*, J. Geophys. Res. *97*, 8905–8915.

MCCAFFREY, R. (1993), *On the Role of the Upper Plate in Great Subduction Zone Earthquakes*, J. Geophys. Res. *98*, 11,953–11,966.

MOLNAR, P., FREEDMAN, D., and SHIH, J S. F. (1979), *Lengths of Intermediate and Deep Seismic Zones and Temperatures in Downgoing Slabs of Lithosphere*, Geophys. J. Roy. Astr. Soc. *56*, 41–54.

NATIONAL GEOPHYSICAL DATA CENTER, *ETOPO-5 Bathymetry/topography Data*, Data Announc. 88-MGG-02, Natl. Oceanic and Atmos. Admin., U.S. Dep. Commer., Boulder, CO, 1988.

PACHECO, J. F., SYKES, L. R., and SCHOLZ, C. H. (1993), *Nature of Seismic Coupling along Simple Plate Boundaries of the Subduction Type*, J. Geophys. Res., *98*, 14,133–14,159.

PELAYO, A. M., and WIENS, D. A. (1989), *Seismotectonics and Relative Plate Motions of the Scotia Sea Region*, J. Geophys. Res. *94*, 7293–7320.

PETERSON, E. T. and SENO, T. (1984), *Factors Affecting Seismic Moment Release Rates in Subduction Zones*, J. Geophys. Res. *89*, 10,233–10,248.

RANKEN, B., CARDWELL, R. K., and KARIG, D. E. (1984), *Kinematics of the Philippine Sea Plate*, Tectonics *3*, 555–575.

RUFF, L., and KANAMORI, H. (1980), *Seismicity and the Subduction Process*, Phys. Earth. Planet. Int. *23*, 240–252.

SCHOLZ, C. H., *The Mechanics of Earthquakes and Faulting* (Cambridge University Press, New York 1990) 439 pp.

SENO, T., MORIYAMA, T., STEIN, S., WOODS, D. F., DEMETS, C., ARGUS, D., and GORDON, R. (1987), *Redetermination of the Philippine Sea Plate Motion*, EOS Trans. AGU *68*, 1474.

SMITH, W. H. F. (1993), *On the Accuracy of Digital Bathymetric Data*, J. Geophys. Res. *98*, 9591–9603.

TICHELAAR, B. W., and RUFF, L. J. (1993), *Depth of Seismic Coupling along Subduction Zones*, J. Geophys. Res. *98*, 2017–2037.

VINK, G. E., MORGAN, W. J., and ZHAO, W.-L. (1984), *Preferential Rifting of Continents: A Source of Displaced Terranes*, J. Geophys. Res. *89*, 10,072–10,076.

WDOWINSKI, S., O'CONNELL, R. J., and ENGLAND, P. (1989), *A Continuum Model of Continental Deformation above Subduction Zones: Application to the Andes and the Aegean*, J. Geophys. Res. *94*, 10,331–10,346.

YU, G., WESNOUSKY, S., and EKSTRÖM, G. (1993), *Slip Partitioning along Major Convergent Plate Boundaries*, Pure and Appl. Geophys. *140* (2), 183–210.

(Received March 16, 1993, revised/accepted July 26, 1993)

PAGEOPH, Vol. 142, No. 1 (1994)

0033-4553/94/010225-13$1.50 + 0.20/0

Large Thrust Earthquakes and Volcanic Eruptions

SERGIO E. BARRIENTOS[1]

Abstract —Forty-eight hours after the occurrence of the May 22, 1960 ($M_W = 9.5$) Chile earthquake, Puyehue volcano initiated its eruptive activity. The closeness in space and time of both phenomena provides us with a unique opportunity to examine the possible causal relationship between the sudden strain change and the mechanism of the eruption. From the slip distribution of the 1960 event (BARRIENTOS and WARD, 1990) and a static propagator technique, which allows for variable slip faults in vertically heterogeneous media, I calculate the strain field and its depth dependence in the region beneath the volcano. The presented semi-analytical formalism can be applied to any two-dimensional dipping fault. Calculations show extension at the surface of the order of 40 μ strain, in agreement with what was observed in triangulation networks in the central valley about 50 km oceanward from the line of volcanoes. The amplitude of the strain field beneath the volcano is uniform up to a depth of 20 km and decreases downward. The sudden extension of the region is thought to be the main factor in facilitating the eruption of the volcano. It is postulated that strain beneath the volcano triggered the eruption of the Puyehue–Cordón Caulle volcanic system because it was in a mature stage of its eruptive cycle and there was lack of eruptive activity in other volcanoes located along the 1960 rupture region in the immediate period following the earthquake.

Key words: Chile, volcanism, underthrusting earthquakes.

Introduction

A causal relationship between earthquakes and volcanoes has been largely sought in different tectonic environments (CARR, 1977; ACHARYA, 1987, 1982; BJÖRNSSON *et al.*, 1977; GUDMUNDSSON and SAEMUNDSSON, 1980; OURA *et al.*, 1992). In the case of eruptions prior to large earthquakes, it has been thought that the slowly accumulating stresses affect the magma chambers increasing the volcanic activity. ACHARYA (1982) proposed that the time interval between the beginning of the enhanced volcanic activity and a large earthquake depends on the magnitude of the subsequent earthquake. RIKITAKE and SATO (1988) studied the volume change of a spherical cavity under uniform compressive stress to relate the activity of the Izu-Oshima volcano and the occurrence of great interplate earthquakes in the Sagami trough, off the Pacific coast of central Japan. Modulation of eruptive

---

[1] Department of Geophysics, University of Chile, Casilla 2777, Santiago, Chile.

activity by ocean load and tectonic stress has been proposed by McNUTT and BEAVAN (1987) in relation to the behavior of the Pavlof volcano during the period 1973–1984.

The thesis proposed in this article is that the extensional deformation associated with the occurrence of a large earthquake alters the stress field, allowing magma to migrate from the deep chambers toward the surface. A somewhat similar

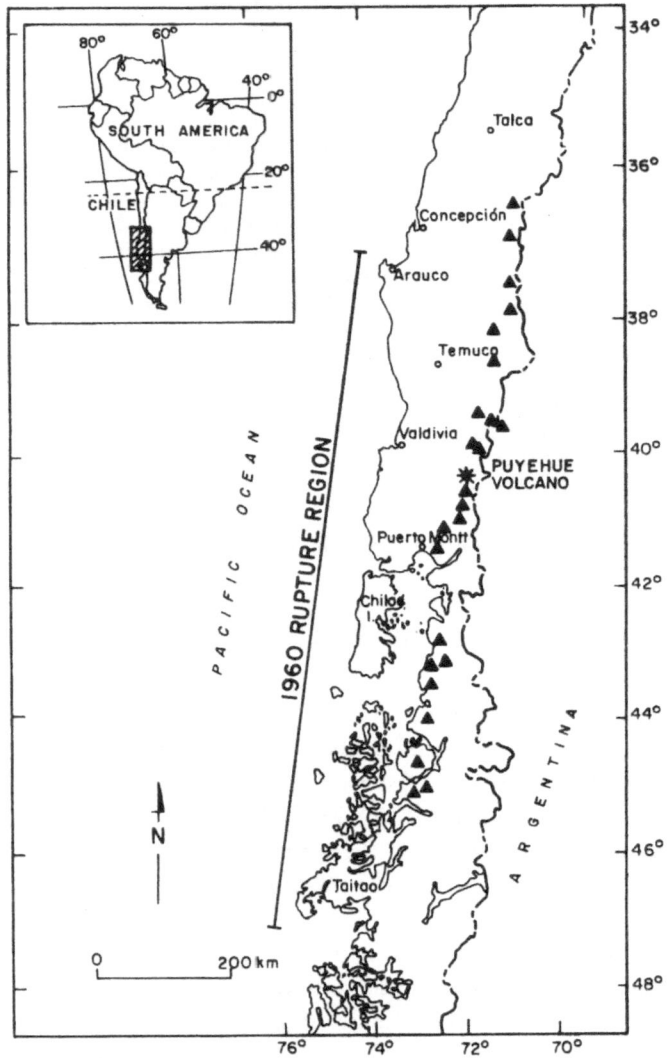

Figure 1

Map view of south-central Chile, site of the great 1960 earthquake. Only volcanoes located within the extent of the rupture zone (Arauco Peninsula in the north to Taitao Peninsula in the south) are shown by solid triangles. Puyehue volcano lies close to the center of the ∼1000 km long coseismic rupture.

approach was adopted by OURA et al. (1992) to relate the shallow $M_J = 5.5$ Ito-Oki, July 9, 1989, earthquake and the volcanic eruption on the Teishi Knoll four days later.

The May 22, 1960 ($M_W = 9.5$) Chile earthquake was followed, two days later, by the eruption of the Puyehue volcano (Figure 1). The closeness in space and time of both phenomena provides us with a unique opportunity to examine the possible causal relationship between the sudden strain change and the mechanism of the eruption. This earthquake, the result of the convergence between the Nazca and South American plates at a rate of 8.4 cm/yr (DEMETS et al., 1990) has been the largest seismic event recorded this century (KANAMORI, 1977). Remarkable changes in land levels were observed in a region 1000 km long by 200 km wide. Extreme coseismic sea level changes ranged from 5.7 m of uplift in Guamblin Island to 2.7 m of subsidence in the city of Valdivia (PLAFKER and SAVAGE, 1970).

The Puyehue volcanic system consists of two main features: The Puyehue composite stratovolcano itself and the Cordón Caulle fissural range (Figure 2). Basement rocks in the region are undifferentiated igneous and metamorphic complexes. The main edifice of the volcano (2236 m above sea level) is flat topped and has a 2.5 km diameter summit caldera. Most of the edifice was constructed during the late glacial time, mostly by basaltic eruptions with less frequent flows and dikes of basaltic andesite and dacite (GERLACH et al., 1988; MORENO-ROA, 1977). The 1960 eruption produced lava flows (0.2 km^3) at the southeast end of the Cordón Caulle (Figure 2) at about 5 km from the Puyehue caldera. These flows are similar to the late-stage rhyodacites and rhyolites from Puyehue volcano (GERLACH et al., 1988).

The population in the nearby farming regions was not only alarmed by the landslides, a consequence of the earthquake, on May 22, but also by the violent explosion that took place near the crest of Puyehue volcano at 3:15 p.m. (local time) on May 24. A brief eruption occurred in March of the following year (WRIGHT and MELLA, 1963).

Method

Because Puyehue volcano lies roughly in the center of the almost 1000 km-long rupture zone associated with the 1960 earthquake (Figure 1), it is possible to describe the strain field as independent of the third coordinate (i.e., approximately north-south). Thus, there is explicit dependence of displacements and stresses only on depth and distance from the trench (which corresponds to the intersection of the fault and the earth's surface).

When considering the static case in an elastic, isotropic and transversely homogeneous medium, the equation of motion

$$\sigma_{ji,j} + f_i = \rho \frac{\partial^2 u_i}{\partial t^2} \tag{1}$$

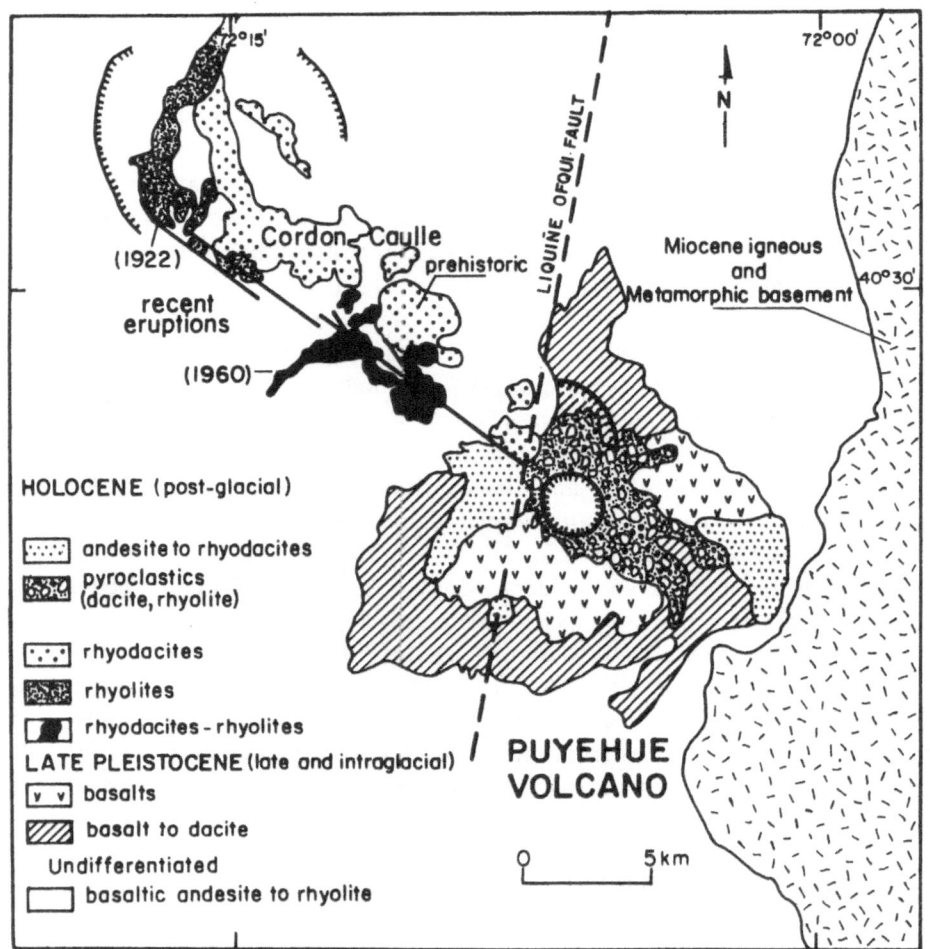

Figure 2
Generalized geologic map of the Puyehue–Cordón Caulle area after MORENO-ROA (1977). The darkest shading indicates lava flows erupted from the 1960 fissural eruption.

reduces to

$$\sigma_{ji,j} + f_i = 0.$$

In a two-dimensional representation we neglect derivatives with respect to y, therefore there is only explicit dependence on the horizontal coordinate (x) and depth (z). Transforming the horizontal coordinate into wavenumber domain by means of

$$\mathscr{F}\{f(x,z)\} = \tilde{f}(k_x, z) = \int_{-\infty}^{\infty} f(x,z)\, e^{-ik_x x}\, dx$$

the three second-order equations in displacement (1) can be reduced to six first-order equations in transformed (tilded) displacements and stresses:

$$\frac{\partial}{\partial z}\begin{bmatrix} i\tilde{u}_x \\ \tilde{u}_z \\ i\tilde{\sigma}_{xz} \\ \tilde{\sigma}_{zz} \\ \tilde{u}_y \\ \tilde{\sigma}_{yz} \end{bmatrix} = \begin{bmatrix} 0 & k & 1/\mu & 0 & 0 & 0 \\ -k_x\lambda/\gamma & 0 & 0 & 1/\gamma & 0 & 0 \\ A & 0 & 0 & k_x\lambda/\gamma & 0 & 0 \\ 0 & 0 & -k_x & 0 & 0 & 0 \\ 0 & 0 & 0 & 0 & 0 & 1/\mu \\ 0 & 0 & 0 & 0 & \mu k_x^2 & 0 \end{bmatrix}\begin{bmatrix} i\tilde{u}_x \\ \tilde{u}_z \\ i\tilde{\sigma}_{xz} \\ \tilde{\sigma}_{zz} \\ \tilde{u}_y \\ \tilde{\sigma}_{yz} \end{bmatrix} - \begin{bmatrix} 0 \\ 0 \\ i\tilde{f}_x \\ \tilde{f}_z \\ 0 \\ \tilde{f}_y \end{bmatrix} \tag{2}$$

where $A = 4\mu k^2(\lambda + \mu)/\gamma$ and $\gamma = \lambda + 2\mu$, with λ and μ being the elastic constants of the medium.

In a compact form, the system of differential equations can be written as

$$\frac{\partial}{\partial z}\tilde{v}(k_x, z) = A(k_x, z)\tilde{v}(k_x, z) - \tilde{f}(k_x, z). \tag{3}$$

The solution of the system of equations is expressed in terms of the propagator matrix $P(z, z_0)$ which allows to describe the solution at any depth.

$$\tilde{v}(k_x, z) = P(k_x, z, z_0)\tilde{v}(k_x, z_0) - \int_{z_0}^{z} P(k_x, z, \hat{z})\tilde{f}(k_x, \hat{z}) d\hat{z} \tag{4}$$

with $P(z, z_0)$, the propagator matrix, for the equivalent P-SV (4 × 4) case given by (WARD, 1984)

$$P(k_x, z, z_0) = \frac{1}{\gamma}\begin{bmatrix} \gamma C + \zeta_s & \mu S + \zeta_c & ((\lambda + 3\mu)S + \zeta_c)/\xi & \zeta_s/\xi \\ \mu S + \zeta_c & \gamma C - \zeta_s & -\zeta_s/\xi & ((\lambda + 3\mu)S - \zeta_c)/\xi \\ \xi(\lambda + \mu)(S + k_x\Delta C) & \xi\zeta_s & \gamma C + \zeta_s & -\mu S - \zeta_c \\ -\xi\zeta_s & \xi(\lambda + \mu)(S - k_x\Delta C) & -\mu S - \zeta_c & \gamma C - \zeta_s \end{bmatrix}$$

with

$$\xi = 2\mu k_x,$$

$$\zeta_s = (\lambda + \mu)k_x\Delta S,$$

$$\zeta_c = (\lambda + \mu)k_x\Delta C,$$

$$S = \sinh\{k_x\Delta\},$$

$$C = \cosh\{k_x\Delta\},$$

and

$$\Delta = (z - z_0).$$

In equation (4), the first term on the right-hand side propagates the known solution from depth z_0 to depth z and the second term contains the information about the source, which also has to be propagated. This system is subjected to the boundary conditions of free stresses at the surface $\sigma_{xz} = \sigma_{yz} = \sigma_{zz} = 0$, and decaying solutions with depth (no displacement at infinity) $u_x, u_y, u_z \to 0$ as $z \to \infty$.

If the source term $\tilde{f}(k_x, \hat{z})$ represents a line source at a depth z_s, given by $\tilde{f}(k_x)\delta(z_s - \hat{z})$, then the integral term in eq. (4) reduces to $\mathbf{P}(k_x, z, z_s)[\mathbf{v}_s]^+_-$ in which the discontinuity in displacements and stresses produced by the source, in terms of the elements of the moment tensor M_{ij}, is described as:

$$[\mathbf{v}_s]^+_- = \begin{pmatrix} iM_{xz}/\mu \\ M_{zz}/\gamma \\ -k_x(M_{xx} - \lambda M_{zz}/\gamma) \\ -k_x(M_{xz} - M_{zx}) \end{pmatrix}.$$

Two depth-decaying solutions to the homogeneous (no body forces) version of equation (3) are: $\mathbf{e}^{(1)} = [1, 1, -2\mu k_x, -2\mu k_x]^T$ and $\mathbf{e}^{(2)} = [-\gamma, \mu, 2\mu k_x(\lambda + \mu), 0]^T$. The stress-displacement vector can be written then as

$$\mathbf{v}(k_x, z) = a_1 \mathbf{e}^{(1)} + a_2 \mathbf{e}^{(2)}, \quad z > z_s.$$

We can propagate these solutions to the surface (crossing the source depth z_s) by

$$\mathbf{v}(k_x, 0) = \mathbf{P}(k_x, 0, z_0)\mathbf{v}(k_x, z_0) - \mathbf{P}(k_x, 0, z_s)[\mathbf{v}_s]^+_-$$

or

$$\begin{bmatrix} i\tilde{u}_x(k_x, 0) \\ \tilde{u}_z(k_x, 0) \\ 0 \\ 0 \end{bmatrix} = a_1 \mathbf{P}(k_x, 0, z_s) \begin{bmatrix} 1 \\ 1 \\ -2\mu k_x \\ -2\mu k_x \end{bmatrix} + a_2 \mathbf{P}(k_x, 0, z_s) \begin{bmatrix} -\gamma \\ \mu \\ 2\mu k_x(\lambda + \mu) \\ 0 \end{bmatrix}$$

$$- \mathbf{P}(k_x, 0, z_s) \begin{bmatrix} iM_{xz}/\mu \\ M_{zz}/\gamma \\ -k_x(M_{xx} - \lambda M_{zz}/\gamma) \\ -k_x(M_{xz} - M_{zx}) \end{bmatrix}$$

if we select z_0 as z_s^+, i.e., a depth right below the source depth. This is a system of four equations and four unknowns, therefore it is fairly simple to invert for $a_1, a_2, i\tilde{u}_x, \tilde{u}_z$.

Once the stress-displacement vector $\mathbf{v}(k_x, 0)$ is known, it can be propagated to any desired depth z by means of $\mathbf{V}(z) = \mathbf{P}(z, z_n)\mathbf{P}(z_n, z_{n-1}) \cdots \mathbf{P}(z_1, 0)\mathbf{V}(0)$ in which $z_1, \ldots, z_n$ are the bottom depths of each layer.

What makes this approach unique is that displacements and strains as a function of depth are possible to evaluate for variable slip faults even in vertically heterogeneous media. To allow for variable slip as a function of position on the fault, we first consider the effect of one line source (two-dimensional description) then we integrate the weighted individual contributions according to

$$u(x_i, z_i) = \sum_{j=1}^{M} \mathcal{K}_{ij} s_j \tag{5}$$

where u is the displacement (or stresses) at (x_i, z_i), $\mathcal{K}_{ij}$ is the displacement (or stresses) produced at the ith observation by the jth line source, which has an associated slip s_j. The kernel $\mathcal{K}_{ij}$ contains information on both vertical and horizontal displacements. Since we are interested in the strains produced by the dislocation, we calculate the numerical derivative of the horizontal displacement ($e_{xx} = u_{x,x}$) in the horizontal direction in the transformed domain or just a premultiplication by ik_x in the wavenumber space.

Application

To illustrate the possibilities of this method, Figure 3 plots the displacement field associated with a uniform slip dislocation (constant $s_j = 1$ m, $25°$ dip reverse fault, width = 100 km) embedded in an elastic half-space. As expected, greater displacements are located in the neighborhood of the fault and close to the free surface. At one fault-width away from the fault, the displacement field has substantially decreased and presents a smoothly-varying horizontal character. In the region located under the bottom end of the fault, material is vertically displaced

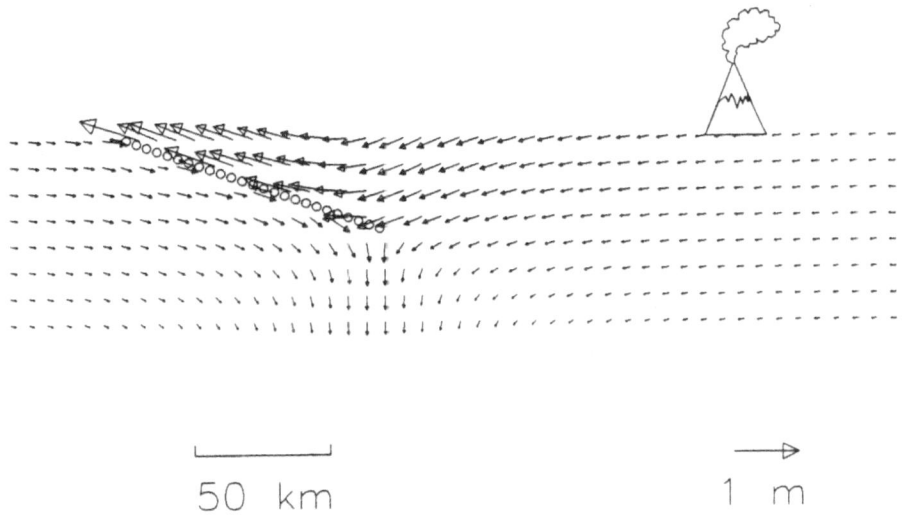

50 km   1 m

Figure 3
Cross section of the expected displacement field for a 25° dipping fault (1 m of reverse slip) embedded in a half-space.

downward. The difference of the displacement field at the two ends of the fault reflects the influence of the free surface.

The results of this formulation can be easily compared with standard output at the surface for 2-D uniform slip models embedded in elastic half-spaces (e.g., FREUND and BARNETT, 1976). Figure 4 shows the results obtained with the standard method and those emerging from the propagator technique for a 25° dipping reverse fault. Elevation change in Figure 4 is well reproduced by the propagator technique except in the region close to the fault trace where it presents an oscillating character due to numerical truncation. For a better fit in this region, a smaller spatial interval would be required. The vertical displacement is well reproduced in its entirety half fault length away of the fault trace.

The formalism allows for the existence of different layers. In this case, the model consists of an elastic layer overlying an elastic half-space with different rigidities, therefore only four parameters are needed: both rigidities, layer thickness and slip distribution. We assume standard values for rigidities ($\mu_L = 1/2\mu_H = 3 \times 10^{10}$ Pa,

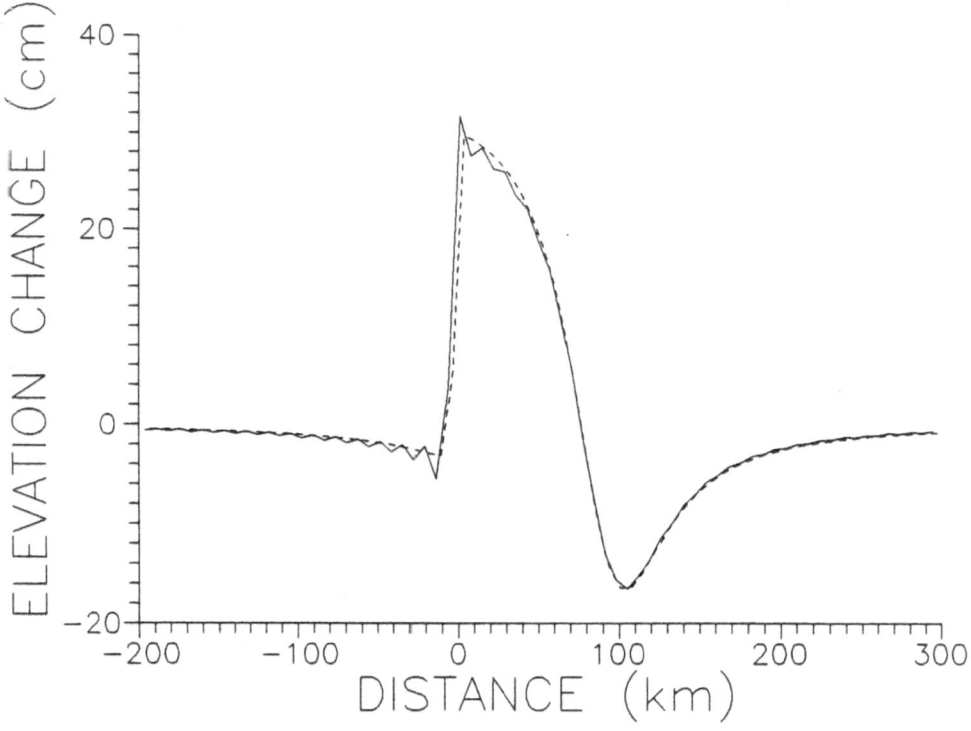

Figure 4
Surface elevation change due to a 25° dipping reverse fault embedded in a half-space. Results of the static propagator technique (solid line) are compared with results of standard techniques (e.g., FREUND and BARNETT, 1976) in dashed line.

where μ_L corresponds to the layer, and μ_H to the underlying elastic half-space) and layer thickness (50 km). The slip distribution along the fault is that obtained by BARRIENTOS and WARD (1990). Their inversion shows that the slip is mainly concentrated on a 900 km long, 150 km wide band parallel to the coast with maximum amplitude of 40 m. For our 2-D purposes, we averaged the slip over the length of the fault, allowing variability only along the down-dip direction (Figure 5). Because the seismic coupling extends in that region to a depth of ~50 km (TICHELAAR and RUFF, 1991) and the slip inversion was performed over data including 8 years of post-seismic readjustment, the down-dip patches of slip (see Figure 6 of BARRIENTOS and WARD, 1990) are thought to be the result of slow post-seismic creep. BARRIENTOS et al. (1992) based on tide gage records at Puerto Montt, located on the center of the nearly 1000 km long rupture, offer further support for large post-seismic adjustments.

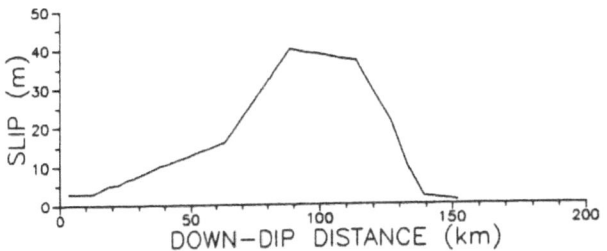

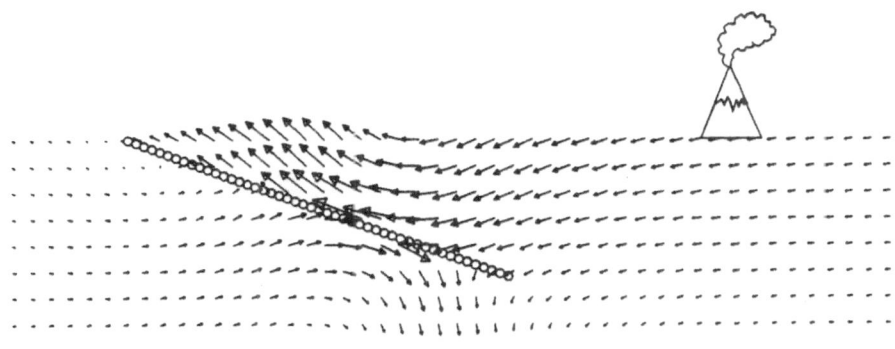

50 km   20 m

Figure 5

Displacement field associated with the 1960 faulting (bottom). Two-dimensional averaged cross-sectional slip distribution as a function of down-dip distance along the fault (adapted from BARRIENTOS and WARD, 1990) is shown on the top.

The contribution of 48 line sources are used to model the dislocation. Each line source slips the amount shown in Figure 5 (top) and the displacement field is calculated using eq. (5). Because the displacement field under the volcano is mainly horizontal, in our approach we will be considering just the e_{xx} component of strain which is shown in Figure 6.

The variable slip models produce high, almost uniform extension to 20 km depth in the region beneath the volcano. At the surface, the amount of extension (40 μ strain) agrees with that observed in triangulation networks about 50 km oceanward from the line of volcanoes (50 μ strain; PLAFKER and SAVAGE, 1970). Extension decreases with depth. The strains calculated with this model, for the region under the volcano, do not differ significantly from those resulting from uniform slip calculations. The differences between uniform and variable slip models become apparent at close distances to the fault, depending on the scale of variability of slip, i.e., size of patches of high and low slip.

Historical data prior to 1960 in the Puyehue–Cordón Caulle system indicate eruptive activity in the years 1893, 1905, 1921–1922, 1922, 1929, 1934, and 1960 (MORENO-ROA, 1977). Inter-occurrence intervals (12, 16, 1, 7, 5, and 26 yr) are highly irregular and the pre-1960 eruption interval has been the longest dormant period, probably indicating a mature stage at the time of the great 1960 earthquake.

The sudden extension of the region down to a 40 km depth is thought to be the main factor in facilitating the eruption of the volcano. But the lack of eruptive activity on other volcanoes located along the 1960 rupture region in the immediate period following the earthquake and the eruptive history of this volcanic system

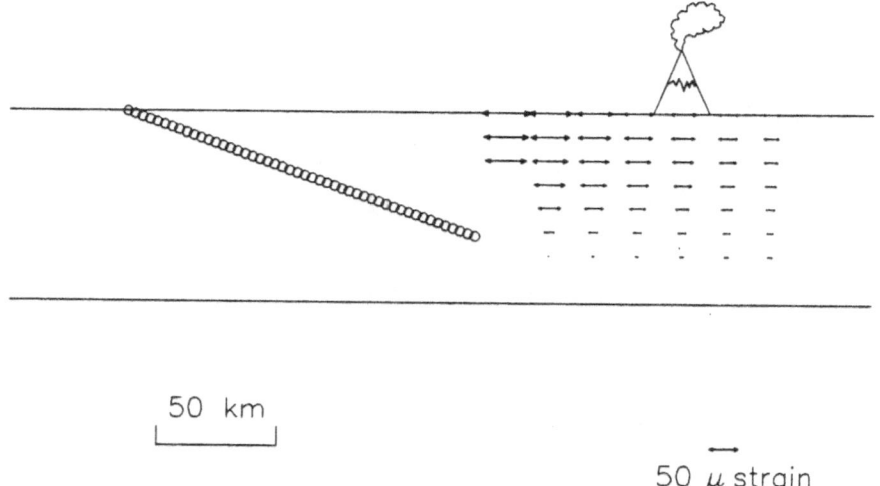

50 km

50 μ strain

Figure 6
Strain field (e_{xx}) produced by a dislocation with slip distribution of Figure 5. Extension dominates the area surrounding the volcano, the highest values extending to around 20 km depth.

suggest that this is a triggering effect. It is postulated that the strain produced by the 1960 rupture triggered the eruption of the Puyehue–Cordón Caulle volcanic system because it was in a mature stage of its cycle. Therefore, the strain produced by the earthquake is not sufficient, by itself, to initiate an eruption; favorable conditions in the volcanic system must exist prior to the occurrence of the earthquake.

The most widely adopted model to relate strain changes and eruptive activity is the toothpaste tube analogy in which compressive stress around magma reservoirs squeezes up the magma to the surface (NAKAMURA, 1971, 1975; McNUTT and BEAVAN, 1987). Tensile strains two orders of magnitude smaller than the ones estimated here were proposed by YAMASHINA and NAKAMURA (1978) to act as a source of excitation for the activity on the Izu-Oshima volcano (Japan). Under critical conditions (late in its eruptive cycle) tensile strains act as a trigger of eruptions when considering the short-term (less than a few weeks) response of the volcano. This idea was further supported by OURA et al. (1992) when analyzing the tensile strain produced by the 1989 Ito-Oki earthquake beneath the Teishi Knoll (Japan). These two cases, in addition to the one presented in this paper, imply that it is also possible that the tensile strain changes can act as a triggering mechanism of eruptive activity.

Conclusions

The static propagator matrix approach has been successfully used to model the displacement field associated with faulting. Results obtained with this semi-analytical technique reproduce the expected shape and amount of subsidence and uplift at the surface due to a two-dimensional dislocation. Furthermore, displacement and strain associated with faulting can be easily computed in vertically heterogeneous media for variable slip faults.

The slip distribution associated with the 1960 earthquake is incorporated to the presented formalism and the strain field at depth under the Puyehue volcano is calculated. Surface extension (40 μ strain, uniform to 20 km depth) agrees with that observed in triangulation networks about 50 km oceanward from the line of volcanoes (50 μ strain; PLAFKER and SAVAGE, 1970). The amount of extension decreases with depth. In summary, the sudden extension of the region down to a 40 km depth is thought to be the main factor in facilitating the eruption, mainly as a triggering effect because the volcanic system was in a mature stage of its eruptive cycle.

Acknowledgments

I would like to thank Steven Ward who introduced me to the fundamentals of propagator matrix theory. Valuable suggestions of an anonymous reviewer, E.

Kausel, T. Monfret and L. Ponce are also acknowledged. This work has been partially supported by grants from the Fondo de Desarrollo Científico y Tecnológico (FONDECYT) and Fundación ANDES.

REFERENCES

ACHARYA, H. (1982), *Volcanic Activity and Large Earthquakes*, J. Volcanology and Geothermal Res. *86*, 335–344.

ACHARYA, H. (1987), *Spatial Changes in Volcanic and Seismic Activity Prior to Great Earthquakes*, Pure Appl. Geophys. *125*, 1097–1118.

BARAZANGI, M. and ISACKS, B. L. (1976), *Spatial Distribution of Earthquakes and Subduction of the Nazca Plate Beneath South America*, Geology *4*, 686–692.

BARRIENTOS, S. E., PLAFKER, G., and LORCA, E. (1992), *Postseismic Coastal Uplift in Southern Chile*, Geophys. Res. Lett. *19*, 701–704.

BARRIENTOS, S. E., and WARD, S. N. (1990), *The 1960 Chile Earthquake: Coseismic Slip from Surface Deformation*, Geophys. J. Int. *103*, 589–598.

BJÖRNSSON, A., SAEMUNDSSON, K., EINARSSON, P., TRYGGVASON, P., and GRÖNVOLD, K. (1977), *Current Rifting Episode in North Iceland*, Nature *266*, 318–323.

CARR, M. J. (1977), *Volcanic Activity and Great Earthquakes at Convergent Plate Margins*, Science *197*, 655–657.

DEMETS, C., GORDON, R. G., ARGUS, D. F., and STEIN, S. (1990), *Current Plate Motions*, Geophys. J. Int. *101*, 425–478.

FREUND, L. B., and BARNETT, D. M. (1976), *A Two-dimensional Analysis of Surface Deformation due to Dip-slip Faulting*, Bull. Seismol. Soc. Am. *66*, 667–675.

GERLACH, D., FREY, F. F., MORENO-ROA, H., and LÓPEZ-ESCOBAR, L. (1988), *Recent Volcanism in the Puyehue–Cordón Caulle Region, Southern Andes, Chile (40.5°): Petrogenesis of Evolved Lavas*, J. of Petrology *29*, part 2, 333–382.

GUDMUNDSSON, G., and SAEMUNDSSON, K. (1980), *Statistical Analysis of Damaging Earthquakes and Volcanic Eruptions in Iceland from 1550–1978*, J. Geophys. *47*, 99–109.

KADINSKY-CADE, K. (1985), *Seismotectonics of the Chile Margin*, Ph.D. Thesis. 253 pp. Cornell University, Ithaca, New York.

KANAMORI, H. (1977), *The Energy Release in Great Earthquakes*, J. Geophys. Res. *82*, 2981–2987.

KANAMORI, H., and CIPAR, J. (1974), *Focal Process of the Great Chilean Earthquake May 22, 1960*, Phys. Earth Planet. Interiors *9*, 128–136.

LINDE, A. T., and SILVER, P. G. (1989), *Elevation Changes and the Great 1960 Earthquake: Support for Aseismic Slip*, Geophys. Res. Lett. *16*, 1305–1308.

MCNUTT, S. R., and BEAVAN, R. J. (1987), *Eruptions of Pavlof Volcano and their Possible Modulation by Ocean Load and Tectonic Stresses*, J. Geophys. Res. *92*, 11509–11523.

MORENO-ROA, H. (1977), *Geología del área volcánica Puyehue Carrán en Los Andes del sur de Chile*, Thesis, Univ. of Chile, Santiago, 170 pp.

NAKAMURA, K. (1971), *Volcano as a Possible Indicator of Crustal Strain*, Bull. Volcanol. Soc. Japan *16*, 63–71 (in Japanese with English abstract).

NAKAMURA, K. (1975), *Volcano Structure and Possible Mechanical Correlation between Volcanic Eruptions and Earthquakes*, Bull. Volcanol. Soc. Japan *20*, 229–240 (in Japanese with English abstract).

OURA, A., YOSHIDA, S., and KUDO, K. (1992), *Rupture Process of the Ito-Oki, Japan, Earthquake of 1989 July 9 and Interpretation as a Trigger of Volcanic Eruption*, Geophys. J. Int. *109*, 241–248.

PLAFKER, G., and SAVAGE, J. C. (1970), *Mechanism of the Chilean Earthquakes of May 21 and 22, 1960*, Geol. Soc. Am. Bull. *81*, 1001–1030.

PLAFKER, G. (1972), *Alaskan Earthquake of 1964 and Chilean Earthquake of 1960: Implications for Arc Tectonics*, J. Geophys. Res. *77*, 901–925.

RIKITAKE, T., and SATO, R. (1989), *Up-squeezing of Magma under Tectonic Stress*, J. Phys. Earth *37*, 303–311.

TICHELAAR, B. W., and RUFF, L. (1991), *Variability in the Depth of Seismic Coupling along the Chilean Subduction Zone*, J. Geophys. Res. *96*, 11997–12022.

WARD, S. N. (1984), *A Note on Lithospheric Bending Calculations*, Geophys. J. R. Astr. Soc. *78*, 241–253.

WRIGHT, C., and MELLA, A. (1963), *Modifications to the Soil Pattern of South-central Chile Resulting from Seismic and Associated Phenomena during the Period May to August, 1960*, Bull. Seismol. Soc. Am. *53*, 1367–1402.

YAMASHINA, K., and NAKAMURA, K. (1978), *Correlations between Tectonic Earthquakes and Volcanic Activity of Izu-Oshima Volcano, Japan*, J. Vocanol. Geotherm. Res. *4*, 233–250.

(Received February 16, 1993, revised July 2, 1993, accepted July 11, 1993)

PAGEOPH

Reprints from Pure and Applied Geophysics

Campbell, W.H. (Ed.)
**Deep Earth
Electrical Conductivity**
1990. 96 pages. Softcover
ISBN 3-7643-2564-X

Campbell, W.H. (Ed.)
**Quiet Daily
Geomagnetic Fields**
1989. 244 pages. Softcover
ISBN 3-7643-2338-8

Dmowska, R. /
Ekström, G. (Eds)
**Shallow
Subduction Zones**
Part I:
1993. 220 pages. Softcover
ISBN 3-7643-2962-9

Part II:
1994. 220 pages. Softcover
ISBN 3-7643-2963-7

Gibowicz, S.J. (Ed.)
Seismicity in Mines
1989. 404 pages. Softcover
ISBN 3-7643-2273-X

Hughes, R.L. (Ed.)
Oceanic Hydraulics
1990. 184 pages. Softcover
ISBN 3-7643-2498-8

Liebermann, R.C. /
Sondergeld, C.H. (Eds)
**Experimental
Techniques in Mineral
and Rock Physics**
1994. Approx. 430 pages.
Softcover
ISBN 3-7643-5028-8

McGarr, A. (Ed)
Induced Seismicity
1992. 460 pages. Softcover
ISBN 3-7643-2918-1

Okal, E.A. (Ed.)
**Aspects of
Pacific Seismicity**
1991. 200 pages. Softcover
ISBN 3-7643-2589-5

Ord, A. / Hobbs, B.E. /
Mühlhaus, H.-B. (Eds)
**Localization of
Deformation in Rocks
and Metals**
1992. 158 pages. Softcover
ISBN 3-7643-2772-3

Plumb, A.R. /
Vincent, R.A. (Eds)
Middle Atmosphere
1989. 472 pages. Softcover
ISBN 3-7643-2290-X

Ruff, L.J. / Kanamori, H. (Eds)
Subduction Zones
Part I:
1988. 352 pages. Softcover
ISBN 3-7643-1928-3

Part II:
1989. 282 pages. Softcover
ISBN 3-7643-2272-1

Sammis, C.G. / Saito, M. /
King, G.C.P. (Eds)
**Fractals and Chaos
in the Earth Sciences**
1993. 188 pages. Softcover
ISBN 3-7643-2878-9

Scholz, C.H. /
Mandelbrot, B.B. (Eds)
Fractals in Geophysics
1989. 313 pages. Softcover
ISBN 3-7643-2206-3

Stuart, W.D. / Aki, K. (Eds)
**Intermediate-Term
Earthquake Prediction**
1988. 550 pages. Softcover
ISBN 3-7643-1978-X

Udias, A. / Buform, E. (Eds)
**Source Mechanism and
Seismotectonics**
1991. 214 pages. Softcover
ISBN 3-7643-2709-X

Wu, R.-S. / Aki, K. (Eds)
**Scattering and
Attenuation of
Seismic Waves**
Part I:
1988. 448 pages. Softcover
ISBN 3-7643-2254-3

Part II:
1989. 198 pages. Softcover
ISBN 3-7643-2341-8

Part III:
1990. 438 pages. Softcover
ISBN 3-7643-2342-6

Birkhäuser

Birkhäuser Verlag AG
Basel · Boston · Berlin